THE CAR CARE BOOK

Third Edition

Ron Haefner

Columbus High School
Columbus, NE

THOMSON

DELMAR LEARNING

Australia Canada Mexico Singapore Spain United Kingdom United States

THOMSON

DELMAR LEARNING

The Car Care Book, 3E

Ron Haefner

Vice President, Technology and Trades SBU:

Alar Elken

Editorial Director:

Sandy Clark

Acquisitions Editor:

Dave Boelio

Development Editor:

Matthew Thouin

Marketing Director:

Cyndi Eichelman

Channel Manager:

William Lawrensen

Marketing Coordinator:

Mark Pierro

Production Director:

Mary Ellen Black

Production Manager:

Larry Main

Production Editor:

Thomas Stover

Art & Design Specialist:

Francis Hogan

Editorial Assistant:

Kevin Rivenburg

Cover & Interior Design:

Alex Vasilakos

Library of Congress Cataloging-in-Publication Data:

ISBN: 1-4018-3553-8

NOTICE TO THE READER

CONTENTS

ABOUT THE BOOK

Owners of automobiles can be categorized in many different ways. For some, their car is a personal extension of themselves, for others it may be simply a matter of transportation, and some may view their automobile as a hobby or a toy.

No matter which of these categories a car owner may fall into, and regardless of the motivation for owning an automobile, the car will need maintenance. *The Car Care Book*, *3E*, forms a strong basis for starting a maintenance program, particularly for the first time car owner or for one who has owned several cars and wishes to learn more about the automobile.

This book addresses the three primary areas of concern for the average car owner. The first is the basics of how the various systems on the automobile work. The second is the maintenance required for the automobile, and the third is the financial concerns of owning the automobile. This last area also touches on some of the buying, leasing, and insurance considerations for ownership, or potential ownership.

I wrote *The Car Care Book*, *3E* for a class that I developed called Consumer Auto's 101. Many of my students were either in the process of buying a car or had just bought their first car. I have also had adults interested in the class as they, too, did not have a comfortable understanding of what their cars needed, even though they had owned many cars over the years.

My goal was to write a book that a reader could pick up, read, and immediately apply the information, perhaps with a little guided practice. It still amazes me when someone does their first engine oil change and is apprehensive, but by the end of the class they become comfortable and even develop a routine.

NEW TO THIS EDITION

This updated edition introduces readers to the three areas of concern that car consumers typically face: learning about the major systems of a car; the basics of inspection and maintenance; and the financial aspects of owning and operating an automobile. Each section of the car is broken down into systems and subsystems to help students digest important concepts.

The majority of the content has been updated to reflect the technological advancements of the modern automobile. Specific vehicle types and brands have been avoided to allow the text to best fit the majority of the market and the individual, and tailor the delivery of the information to meet specific geographic areas and needs. Also new to this edition is the inclusion of chapter objectives and review questions at the end of the chapters. Readers familiar with the second edition may also note that the chapter on wheels, brakes, suspension and steering has been divided into two chapters (8 and 9) for easier reading. A chapter has been added for leasing, and the chapter covering buying tips for parts now includes buying tips for insurance.

ORGANIZATION

A key feature of the book is a set of objectives at the beginning of the chapter and review questions at the end of the chapter. Upon beginning a chapter, the reader will notice a list of things he or she should accomplish after reading the chapter. The reader can answer the questions at the end of the chapter to test his or her comprehension of the content.

This book begins with a general overview of the systems of the car. The overview is general in nature and tries not to scare you off by being too technical. Next is a discussion of the basic tools that one may need for maintenance of the car.

The following chapters, 3 through 10, deal with the engine, fuel, electrical, lubrication and cooling systems. The powertrain, tires and brakes, suspension and steering, air conditioning, and optional equipment round out the balance of the vehicle systems. These chapters discuss the respective system and discuss some common problems that a car owner may encounter, as well as the general maintenance that the particular system might require.

The maintenance chapter, Chapter 11, helps the reader build a preventive maintenance program that can be tailored to car owner's needs, driving styles, and climate. It covers the common scheduled maintenance intervals as well as spark plug inspection and replacement.

Chapter 12 deals with recognizing the repair needs of the car and covers some of the more common repairs a car owner may find including changing a tire, alternator, valve cover gaskets, and the thermostat.

Deciding to keep a car and repair it versus getting a different one is the focus of Chapter 13. It looks at motivations in owning a car, estimating the value of a vehicle, setting a budget, and some of the financial impacts of those decisions.

Finally, coverage is devoted to the financial aspects of the car, including buying and selling, insurance considerations, and the benefits of modern day leasing.

SUPPLEMENTS

An Instructor's CD-ROM is available for the book, which includes new chapter questions, as well as the existing chapter-end questions found in the book, in Examview, which instructors can use to generate customized tests or quizzes for students. A set of lab sheets is also available and covers the most common maintenance tasks. These offer flexibility to the instructor, allowing customization of maintenance procedures to fit a given geographical area and facility.

ABOUT THE AUTHOR

Ron Haefner has a Bachelor of Science Degree in both Industrial Technology and Industrial Education from Fort Hays State University. Having graduated Magna Cum Laude, Haefner went on to receive a Master of Science in Industrial Education. He is currently teaching at Columbus High School in Career and Technical Education. He serves as a Secondary Education Specialist as well as an Automotive Technology Instructor. Ron Haefner is also on the advisory board for Central Community College in the discipline of Automotive Technology, and also serves on the college's Alternate Fuels Advisory Board. Ron is a member of ACTE (Association of Career and Technical Education), and has won numerous awards, including State Instructor of the Year for the Chrysler/AAA Troubleshooting in 1993, Instructor of the Ford/AAA Student Auto Skills State winner 1995, 1997, 1999, 2000, 2001, 2002, 2003, and last, National Finalist Instructor of the Ford/AAA Student Auto Skills 5th Place National Finalist.

ACKNOWLEDGEMENTS

Thanks go to the following people:

Jane Haefner for helping to type the majority of the rough draft of my manuscript. Her experience of manuscript preparation with other authors has been a great asset.

Ramona Kluth, for her experience, assistance, and advice in proofing the chapters.

Gordon Steinbrook, whose perseverance and dream of completing his book provided inspiration.

Tom Donahy, from whom I learned automotive repair in my youth, and whose motto is, "Anything is possible with enough time and effort."

Alex Haefner and Dustin Woodside for their assistance in the photo shoots for the preliminary photos used in the rough draft of the book.

REVIEWERS

I would like to acknowledge and thank the following educators for their comments, criticism, and suggestions during the review process:

Roberta Allen
American School
Lansing, IL

Debra Anderson
Montgomery College
Rockville, MD

Stephen Fowler
Mendocino Community College
Ukia, CA

Ken Mays
Central Oregon Community College
Bend, OR

Jerry Mumms
BYU Idaho
Rexford, ID

Tom Reynolds
Lee's Summit North High School
Lee's Summit, MO

WHAT MAKES THE WHEELS GO?

CHAPTER 1

After reading this chapter, the reader should be able to:

- List the major systems and components of the automobile.
- Describe how energy, work, force, and power relate to one another.
- Explain the conversion of energy into motion.

HOW IT ALL FITS TOGETHER

The automobile has been around for more than a hundred years. It has evolved from gas buggies built in the late 1800s by Daimler and Benz in Germany and by Henry Ford, Ransom Olds, and others in the United States to our present day automobiles composed of more than 15,000 individual parts. The early era cars were, at their time, complex machines that could only be enjoyed by the wealthy and repaired by dedicated and trained personnel known as *mechanics*. The traditional mechanic no longer exists today and has been replaced by highly trained personnel known as *technicians*. Technicians often specialize in a specific area of the automobile such as drivability/diagnostics, drivetrain, transmission, suspension/steering/brakes, and body systems. This specialization has been made necessary by the many complex electronic systems found in today's cars. The relatively recent advent of electronic systems on the modern car has transformed it into a self-diagnosing and, to an extent, self-adjusting machine.

When you look under the hood of today's automobile, the maze of tubes, wires, hoses, belts, and other parts may seem intimidating. Don't let this first view alarm you. You can become more comfortable with what you're looking at by becoming familiar with your vehicle. Let's start by dividing the car into its major sections and their subsystems (see Figure 1–1):

- The **chassis** is the underlying structure (comprised of the passenger compartment, engine compartment, doors, fenders) on which all other parts are mounted, including the steering, suspension, and brake systems.
- The **engine** is an internal combustion device that intakes air and fuel and converts the expansive force of the burning air and fuel into rotary motion used to supply power to the automobile.
- The **drivetrain** consists of the transmission and other parts that transfer the power produced by the engine to propel the car. *Note:* The engine and drivetrain together comprise the **powertrain.**

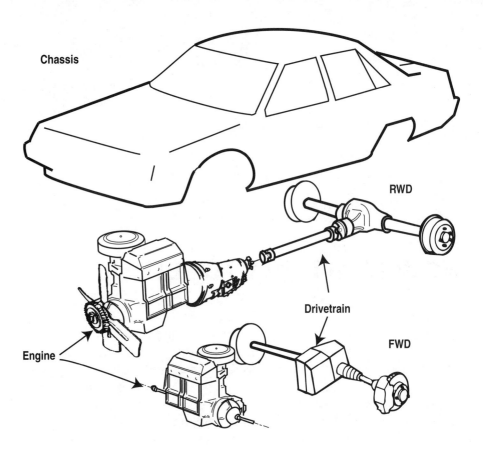

Figure 1–1: A chassis (or unibody) and engine placement and drivetrain for both rear wheel drive (RWD) and front wheel drive (FWD).

Chassis

RWD

Drivetrain

FWD

Engine

Sometimes these terms are used for one another, but they mean different things.

In this book, you'll learn about many of the parts and systems that make up chassis, engine, powertrain, and body components. You can use the basic understanding to perform preventive maintenance, diagnose common problems, and do some simple repairs.

ENERGY, FORCE, WORK, AND POWER

Although you may not realize it, you already understand the basic physical processes that make the wheels go. For example, consider what happens if your car stalls in traffic. If you are strong enough by yourself or get enough people to help you (it's said that the average human can produce about $\frac{1}{3}$ of a horsepower), you can push the car off to the side of the road. The movement of a car results from the use of **energy**.

Energy can take six forms: chemical, mechanical, heat, light, electrical, and radiation. A tightly wound spring in an alarm clock stores mechanical energy. Human beings and animals store energy chemically in their muscles. A battery stores chemical energy and coverts it to electrical energy on demand. When the flame on a gas stove is lit, heat energy is present for cooking. When a light is turned on electrical current passes through a filament that heats up and gives off light energy. An example of radiation energy would be X-rays.

When you push the car, you are using energy to apply **force** to it. Force applied to an object causes motion. If you have to push the car very far, you will have done a good bit of **work.** Basically, work is the result of applying force to an object and causing it to move. The formula to calculate work (W) is to multiply the distance traveled (D) by the force (F) applied:

$$W = D \times F$$

Power is the rate at which work is done, for example, how fast you are able to do the work of pushing a car. Power (P) is calculated by using the value you get for work (W) and dividing it by the time (T) it took:

$$P = \frac{W}{T}$$

Often, the value we get for power is not immediately meaningful to us so we convert power into **horsepower** (Hp). One horsepower is 33,000 foot-pounds per minute. So we divide power by 33,000 to get Hp:

$$Hp = \frac{P}{33,000}$$

Now the Hp rating for an engine has more meaning when we talk about work and power.

If you have more energy, you can apply more force and do more work. The faster the work is done, the more power you are using. These concepts are summarized in Figure 1–2.

A car engine uses the energy present in its fuel to apply force and move the vehicle. The engine produces **torque,** a twisting force that turns the driving wheels. This torque acts on the crankshaft that will convert the reciprocating or up and down motion of the piston into rotary motion for use in the car. As the engine applies torque and moves the car, work is performed.

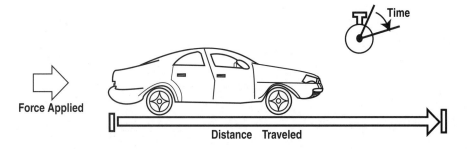

Force Applied

Distance Traveled

Figure 1–2: Work is the force applied to a vehicle that causes it to move a given distance. The rate at which the distance can be covered is an indication of power; the less time, the more power.

CONVERTING HEAT ENERGY TO MOTION

A car's engine is a machine, or mechanical device, that changes chemical energy into heat energy and the resulting heat energy into mechanical energy. The burning, or **combustion,** of chemical-energy-rich gasoline or other fuel produces heat energy. This heat energy is converted, or changed, into mechanical energy that turns the wheels and moves the car.

External Combustion Engine

Two types of engines commonly have been used to convert heat energy to motion. In one type, fuel is burned outside of the engine. These engines are called *external combustion* engines. External combustion means "outside burning."

An old-fashioned steam locomotive is an example of an external combustion engine. The burning of fuel in a firebox turns water into steam. The steam is routed to a power **cylinder** mounted on the side of the locomotive. The steam pushes against a tight-fitting plug, called a **piston,** inside the cylinder (see Figure 1–3).

The hot steam creates **pressure** against the top, or head, of the piston. The pressure forces the piston from one end of the cylinder to the other. A metal rod connects the piston to the drive wheel of the locomotive, causing it to turn.

The firebox and the boiler tubes take up the main bulk of a locomotive. The actual steam engine, the cylinders, is usually located just ahead of the drive wheels of the train. Thus external combustion engines are generally too big, too noisy, and too messy to be used for personal transportation. But people tried anyway. Steam-driven automobiles of the early 1900s used

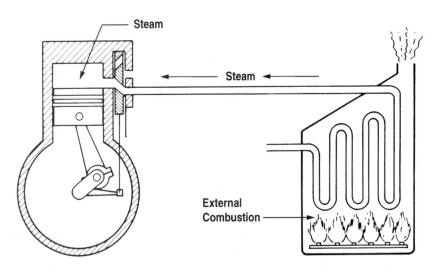

Figure 1–3: An external combustion engine uses pressure from steam created by boiling water. The steam pressure pushes the piston in the engine.

a liquid fuel instead of coal or wood to heat the water. Still, steam engines were too cumbersome and took too long to produce steam to be practical for everyday use.

Internal Combustion Engine

The personal vehicle that generated its own horsepower became feasible with the introduction of the **internal-combustion engine**. Internal combustion means "inside burning." The heat is generated by burning an air/fuel mixture inside a cylinder within the engine, above the piston. The burning produces heat that, in turn, produces pressure. The pressure forces the piston down the cylinder.

The piston is connected by a rod to a **crankshaft**. If you have ridden a bicycle, you are familiar with the action of a crankshaft. Your legs move up and down to pump the pedals mounted on a bicycle crank, or crankshaft. The crank turns in a rotating motion to drive a chain that powers the bicycle. The crankshaft and pedals rotate within **bearings** that allow them to turn freely.

In an automobile engine, several pistons are connected to a common crankshaft to push downward and provide turning force. Today's automobile engines typically have four, six, or eight cylinders.

A simplified drawing in Figure 1–4 illustrates what happens in the cylinder of an internal combustion engine. It shows heat being generated at the top of a cylinder as a mixture of fuel and air is burned. The burning, or

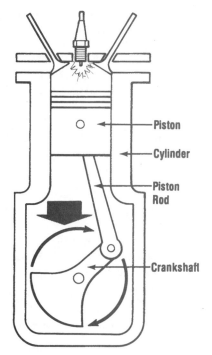

— Piston

— Cylinder

— Piston Rod

— Crankshaft

Figure 1–4: In an internal combustion engine, fuel is burned inside the mechanism that produces force.

combustion, produces energy that causes the piston to move to the bottom of the cylinder. A **connecting rod** attached to the piston is also connected to the crankshaft. The turning of the crankshaft provides the force that drives the wheels.

THE ENGINE AS A GROUP OF SYSTEMS

Controlling the burning of the air and fuel is simple in theory but gets a little complicated in reality. A number of systems are necessary to make an internal combustion engine run and allow it to handle the various functions of converting the chemical power of the fuel into usable torque (mechanical turning force) and power.

The Electrical System

Every automobile has an electrical system consisting of a series of complex parts. Chemicals and metals within the battery store chemical energy and create electric power on demand. This power is used to operate a **starter,** or electric motor, that turns, or cranks, the crankshaft and starts the engine. This electrical energy must also power the electric fuel pump to deliver the fuel to the engine. In addition, the computers and ignition system deliver the high voltage spark used to ignite the air and fuel mixture inside the cylinders.

Once the engine is running, a belt attached to the front of the rotating crankshaft turns an **alternator,** or electric **generator,** that produces electricity. This electricity is used for several purposes:

- To operate the vehicle's electrical systems necessary to run the engine
- To maintain or recharge the battery as a source of starting power
- To operate the lights, blower motors, radio, and other accessories

Figure 1–5 highlights typical parts of an electrical system.

Modern automotive electrical systems typically include one or more **electronic control units,** or computers, that monitor and operate many other parts and systems. Automotive computers are discussed in Chapter 5, "The Electrical System."

The Fuel System

To make an engine run, there must be a system that handles fuel. A **fuel system** includes a tank to hold the fuel, an electric fuel pump that delivers the fuel out of the tank, a filter to trap particles that could damage the inside of an engine, fuel lines to carry the fuel from the tank to the engine, and a **fuel injection system**. Figure 1–6 shows the parts of a fuel injection system. Older cars made use of a mechanical device called a **carburetor** to

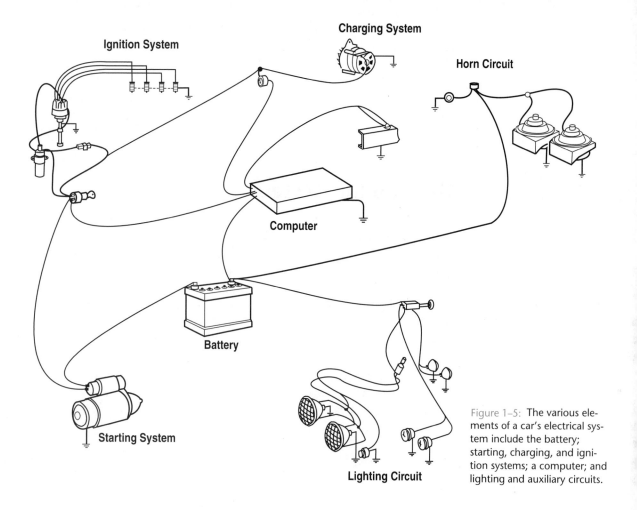

Ignition System

Charging System

Horn Circuit

Computer

Battery

Starting System

Lighting Circuit

Figure 1–5: The various elements of a car's electrical system include the battery; starting, charging, and ignition systems; a computer; and lighting and auxiliary circuits.

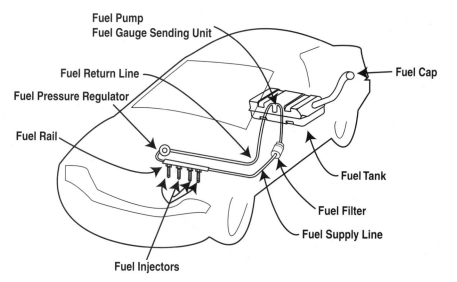

Fuel Pump
Fuel Gauge Sending Unit

Fuel Return Line

Fuel Pressure Regulator

Fuel Rail

Fuel Cap

Fuel Tank

Fuel Filter

Fuel Supply Line

Fuel Injectors

Figure 1–6: A typical fuel injection system for a late model vehicle.

make the final delivery of the fuel to the engine. Carburetors were phased out during the 1980s and replaced by the more reliable and efficient fuel injection systems.

Carburetors and fuel injection systems both do the same job: They provide a fine mist of fuel and air in measured amounts to be burned in the cylinders. On today's cars, the operation of all fuel injection systems is monitored and controlled by computers, or electronic control units. Fuel systems are discussed in Chapter 4, "The Fuel System."

The Cooling System

The burning of the fuel/air mixture produces heat, sometimes to several thousand degrees Fahrenheit. The heat generated is so great that, without special controls, engine parts could warp or melt. To dissipate the excess heat, your car has a **cooling system.** An engine-driven coolant pump provides a continuous flow of liquid, or **coolant,** that circulates through hollow spaces in the nonmoving parts of the engine called **water jackets.**

In liquid-cooled engines, the heat generated within the engine is transferred to the circulating coolant, which is a mixture of water plus **antifreeze.** The coolant is circulated through hollow tubes within a **radiator.** The radiator transfers heat to the air that flows around the tubes. Air is moved through the radiator by an electric fan or, on some cars, by a fan driven by a belt turned by the rotating crankshaft. At highway speeds, most of the cooling air is simply forced through the radiator by the movement of the vehicle.

To summarize, heat generated within the engine is transferred to the circulating coolant and then to the outside air. This transfer of heat keeps internal engine parts below the temperature at which they would be damaged and the coolant below its boiling point.

Figure 1–7 shows, in simplified form, the parts of an automotive cooling system and the flow patterns of air and coolant.

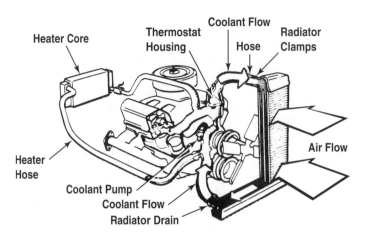

Figure 1–7: A liquid cooling system circulates hot coolant from the engine to the radiator, where heat is transferred to the outside air.

The Lubrication System

Moving parts rub against each other and also rub against the nonmoving housings in which they are contained. Every moving part within the engine eventually wears out through **friction,** or the resistance of materials that rub against one another.

To minimize friction wear, every engine has a **lubrication system** (Figure 1–8). A reservoir of oil is contained in an **oil pan** at the bottom of the engine. An **oil pump** circulates the oil through an **oil filter** and then to moving parts via **galleries.** All moving parts are bathed continuously in filtered oil to reduce friction and the resulting heat.

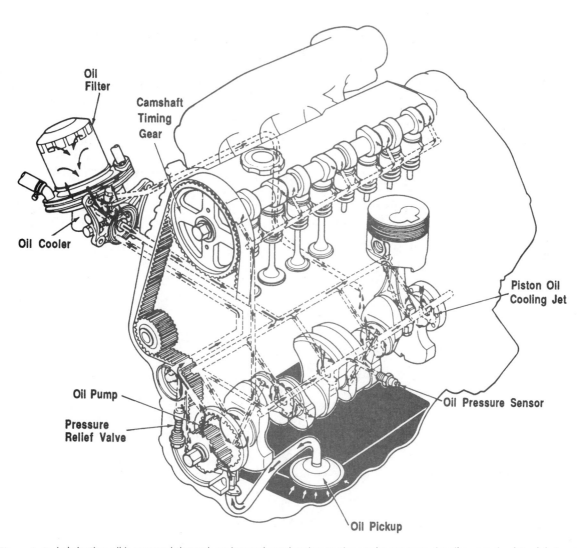

Figure 1–8: Lubricating oil is pumped through an internal combustion engine under pressure. An oil pump circulates lubricant from the pan at the bottom of the engine through a filter to moving parts within an engine.

In removing the heat from many of the moving parts, the oil may reach 250 degrees Fahrenheit, (120 degrees Centigrade). Air flowing over the oil pan, or reservoir, at the bottom of the engine helps to cool the oil. Some cars and trucks may be equipped with a special oil cooler to reduce heat further and keep the oil from breaking down at high temperatures.

TRANSFERRING TORQUE

The power from the rotating engine crankshaft represents torque, or turning force. From the crankshaft, the torque is transferred through a drivetrain, or group of parts and systems, that eventually provides torque to the driving wheels. The main drivetrain component is the **transmission,** which transmits engine torque to other parts connected to the wheels.

Front-Wheel Drive

In the typical passenger car, only two wheels are turned by torque created by the engine. These **driving wheels** can be either front or rear wheels. Today, the majority of new cars have **front-wheel drive** (FWD). Power is transferred from the crankshaft of an engine (mounted transversely, or sideways, on most front-wheel drive cars) into a **transaxle** (Figure 1–9). The transaxle contains many of the parts and functions of the drivetrain in a single housing.

A transaxle contains a transmission to provide multiple forward speeds and reverse. In addition, the transaxle contains a special gear axle mecha-

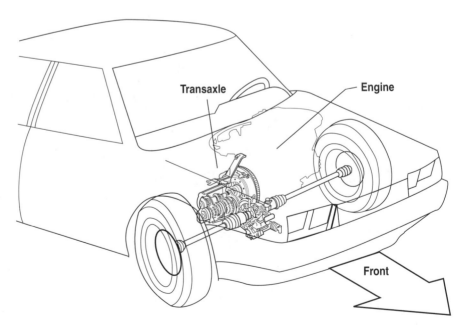

Figure 1–9: Most new cars have front-wheel drive (FWD). Engines are commonly mounted transversely (sideways).

nism known as a **differential.** The name "differential" comes from the fact that this mechanism makes it possible for your driving wheels to turn at different speeds around a corner while continuing to supply the wheels with torque. This function is described in greater detail in Chapter 7, "The Drivetrain."

As you shift gears manually or as your automatic transmission operates, torque is transmitted, or transferred, from the engine to the drivetrain parts. Torque from the engine passes into the transaxle, to the differential, to two metal **drive shafts** connected to the front wheels. The torque turns the front wheels and your car is pulled along the road.

Rear-Wheel Drive

Until the mid-1970s, most passenger cars were designed with **rear-wheel drive** (RWD). Trucks, buses, utility vehicles, and some cars continue to use this design. If your car has rear-wheel drive, the crankshaft is linked via a **clutch** mechanism or a **torque converter** to a transmission and then into a single long drive shaft that carries the power to a rear-mounted differential. Metal axle shafts are connected between the differential and the two rear wheels. See Figure 1–10.

Four-Wheel/All-Wheel Drive

Front-wheel drive and rear-wheel drive can be combined on a single vehicle to power all four wheels. This configuration of **four-wheel/all-wheel drive** (4WD/AWD) is useful for off-road driving or driving through heavy snow or mud. More information on 4WD/AWD is provided in Chapter 7, "The Drivetrain."

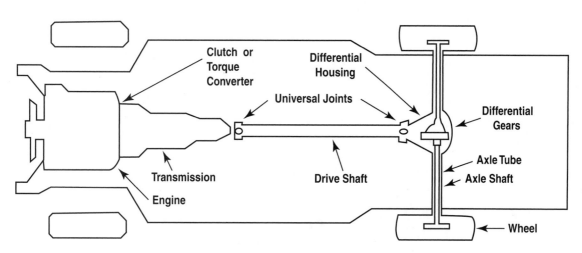

Figure 1–10: In a rear-wheel-drive (RWD) powertrain, torque is transmitted from the engine through the clutch, transmission, drive shaft, and differential to the rear wheels.

Now you know, essentially, what makes the wheels turn. However, your preliminary understanding of the operation of a car should also include the parts that permit you to operate your car safely and comfortably, so read on.

BRAKES

Making a car move is one thing. But to operate a vehicle safely, you also have to be able to make it slow down and stop, which is the job of your car's **brake system.** The brake system is an excellent example of energy conversion. Brakes operate by converting the motion of the car, mechanical energy, into heat energy. As this conversion occurs, the car is slowed down. That's why your brakes get hot with repeated usage.

Two general types of brakes are used on automobiles: **disc brakes** and **drum brakes.** Many new cars have disc brakes on the front wheels and either disc or drum brakes on the rear wheels. A majority of new cars, pickups, and SUVs have disc brakes on the rear.

The **service brakes** refer to the normal braking system excluding the **emergency** or **parking brake** and are **hydraulic.** A fluid under pressure operates hydraulic devices. When you press the brake pedal with your foot, you operate a simple pump that forces brake fluid through metal tubing and rubber hoses to operate the disc and drum brakes. When you lift your foot from the brake pedal, hydraulic pressure decreases and the brakes no longer are applied. Figure 1–11 shows the hydraulic connections that operate the brakes.

Each of the four wheels on your car has its own individual brake unit to slow and stop the wheel from turning. All of the brake units operate together to stop your car. Many brake systems are power assisted. The power assist is usually accomplished by vacuum supplied by the engine to a special chamber called a *vacuum booster*. Disc brakes require more hydraulic

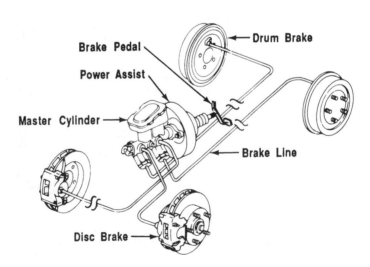

Figure 1–11: The hydraulic brake system.

pressure to operate than drum brakes, which is why most cars today have power-assisted brakes.

Disc Brakes

If you have ever ridden a bicycle with handbrakes, you are already familiar with the principles that make disc brakes stop your car. As on a bicycle, friction parts called **pads** are held in an assembly called a **caliper**. When bicycle brakes are operated, the caliper forces the pads to rub against the spinning wheel rim. The friction of the pads rubbing against the moving rim slows the wheel.

On a bicycle, the pads rub directly against the metal wheel rim on which the tire is mounted. On your car, however, a separate flat part, called a **brake rotor,** is attached to and rotates with the wheel. When you step on the brake pedal, forces are applied to the hydraulic caliper and clamp the friction pads against the spinning disc to slow it and the attached wheel. As the disc stops rotating, so does the wheel to which it is attached, and the vehicle stops. Figure 1–12 shows the components of a disc brake assembly.

Disc brakes work better under heavy usage and wet conditions than drum brakes of similar size. Thus, most cars have disc brakes on the front wheels where most braking force is needed. Because drum brakes can provide a better *parking* or *emergency* brake to lock the rear wheels, most cars have drum brakes on the rear wheels. Some vehicles with rear disc brakes will use a separate set of smaller drum brakes inside the disc brake rotor for a parking brake.

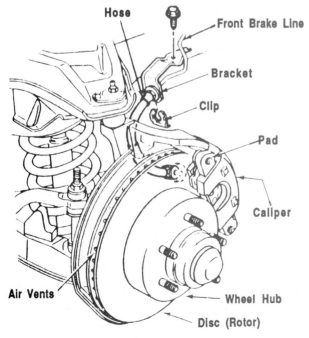

Hose
Front Brake Line
Bracket
Clip
Pad
Caliper
Air Vents
Wheel Hub
Disc (Rotor)

Figure 1–12: Disc brakes operate on the same principles as bicycle hand brakes. To slow and stop the vehicle, a caliper assembly forces two friction pads against a disc that rotates with the wheel.

Drum Brakes

A drum brake uses a hollow cast-iron drum that encloses an expanding mechanism (Figure 1–13). Inside the drum, which is shaped like a shallow cooking pot, there is a set of semicircular frictional **brake shoes**. The brake shoes are curved to fit the inside of the drum. When you step on the brake pedal, the brake shoes are forced outward to rub against the drum. The friction between the shoes and the drum slows or stops the rear wheels. When you remove your foot from the brake pedal, springs retract the brake shoes from their expanded positions.

More information on brake systems is provided in Chapter 8, "Tires and Brakes."

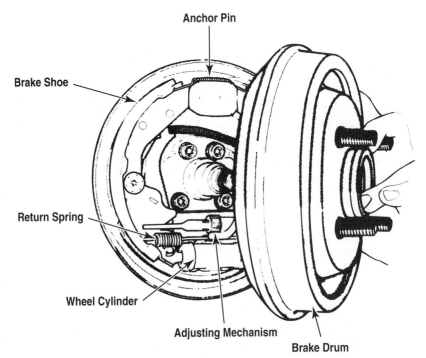

Anchor Pin

Brake Shoe

Return Spring

Wheel Cylinder

Adjusting Mechanism

Brake Drum

Figure 1–13: In a drum brake, a mechanism expands against the inner side of a brake drum that rotates with a wheel. To slow and stop the vehicle, friction materials are pressed against the inside of the drum.

STEERING

Generating and utilizing the power to move and having a braking system to stop moving are two primary necessities for every automobile. A third critical ability is controlling the direction of movement.

The early horseless carriages had very simple steering systems that relied on the driver's strength to turn the wheels. As cars grew larger, heavier, and more complicated, better steering systems were developed that utilized mechanical leverage to assist the driver. Technological advances brought power-assisted steering or so-called power steering to the marketplace. Today's cars also may include electronically controlled power steering mechanisms and variable power steering systems. These power steering systems utilize an engine-driven pump to supply hydraulic pressure to the steering mechanism to assist the driver and reduce steering-wheel effort. A new development is an all-electric power steering system that promises to further decrease driver effort.

Better cars and better highways permit greater speeds. As speeds increased, it became necessary to improve steering systems to provide fast, accurate response to driver input. Today's steering systems provide this kind of response from only minimum driver effort.

A type of steering system, rack and pinion, common on modern cars is shown in Figure 1–14. More information is provided in Chapter 9, "Suspension and Steering."

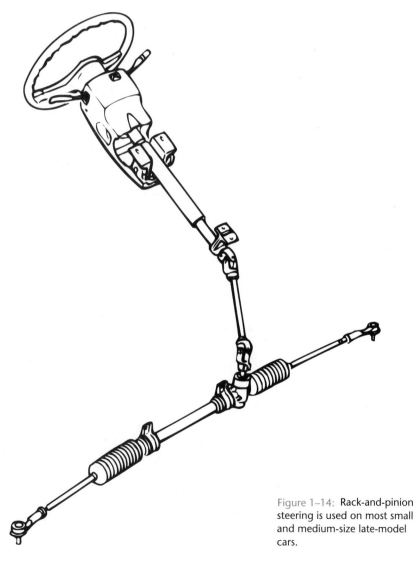

Figure 1–14: **Rack-and-pinion steering is used on most small and medium-size late-model cars.**

SUSPENSION

A car's wheels are attached to its chassis, or underlying structure, by the **suspension system.** Your car's safety, stability, and ride depend on the proper functioning of that system. Most of today's cars use an independent suspension system, which allows the wheels to travel up and down independently of the rest of the car and independently of other wheels. This system also makes it possible for the wheels to follow curves and ride the bumps in the pavement while the body of the car maintains a smooth, stable ride. Engineers design suspension systems for specific vehicles. An SUV or pickup may have increased wheel travel compared to a passenger car, a sports car may have a stiffer suspension and antiroll bars, while a large luxury car may have softer springs for a smoother ride.

Different suspension and shock-absorbing devices are used at the front and rear of the car. The front end, which supports the engine, has a lot more weight to handle. The front suspension typically includes a combination of devices, including coil springs, shock absorbers, and a stabilizer or antiroll bar that keeps the car from rolling or leaning too far during a turn on curves (Figure 1–15). Most newer passenger vehicles also use coil springs to suspend the rear wheels. A coil-spring-type rear suspension for front-wheel drive vehicles is shown in Figure 1–16.

Many trucks and some cars use **leaf springs** in their rear suspensions (Figure 1–17). Leaf springs are built up from a series of metal strips made from spring steel. These are bolted and banded together. Leaf springs typically are used with rear-wheel-drive vehicles to help transfer torque to the

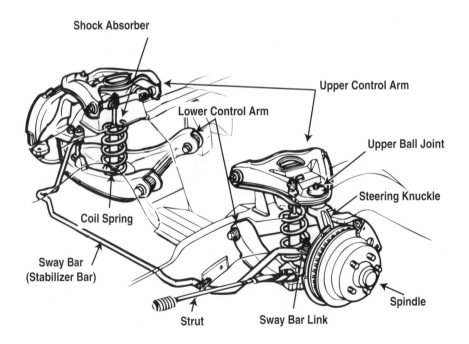

Figure 1–15: A typical long arm/short arm suspension that includes a stabilizer bar.

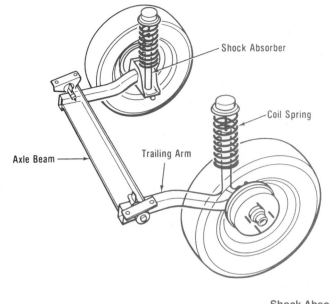

Figure 1–16: Coil springs are used for rear suspension on many late model cars. A connecting beam axle helps stabilize the car around curves and over bumps.

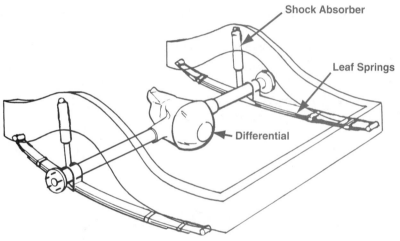

Figure 1–17: Leaf spring suspensions can be used to help transfer driving torque. This type of suspension is used chiefly on RWD and 4WD vehicles.

driving wheels. In truck-type 4WD vehicles, leaf springs may be used in both the front and rear suspensions.

In addition to the springs, there are shock absorbers at each wheel. These dampen the tendency of the springs to continue bouncing after the car hits a bump and help to stabilize your car's ride.

See Chapter 9, "Suspension and Steering," for more details.

TIRES

The tires are the single most important component of the car. Your car's safety and comfort depend on the traction and flexibility of the tires. The steering and traction control of your vehicle depend on the area of each tire, about the size of your extended hand and palm, that rests on the ground. Everything needed to operate the car—to accelerate and steer, the suspension and braking—ultimately rests on the fate of the tires.

Tires are made from a sandwich of layers of fabric and rubber (Figure 1–18). The tires are held in place on metal wheel rims by air pressure. The tire and rim together make up an assembly called a *wheel*. The kind of tires you use, the size you select, and the pressure to which you inflate them are vitally important to the safety, dependability, and economy of your car. Possibly the greatest waste of money experienced by motorists is having to replace the tires because of uneven and unnecessary wear on tires. You can learn to monitor tire pressures and perform simple checks to determine if they are wearing evenly. This type of maintenance will increase your tires' safety, performance, and overall wear.

Specific information on types of tires is provided in Chapter 8, "Tires and Brakes." Chapter 11 describes how to inflate your own tires. And in Chapter 12, you can learn how to change tires.

You step on the gas; the tires turn. Between the gas pedal and the tires, many parts must function together properly to provide safe, dependable, and economical car operation.

This chapter presents a simplified look at the systems that make up your car. The next chapter discusses what tools are needed for maintaining and repairing cars. From there, you can build your understanding of what you can do to check your own car for problems, fix the car yourself, or become a wiser shopper when you have others fix it.

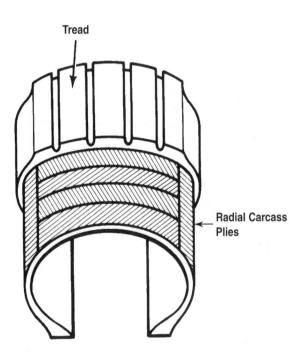

Tread

Radial Carcass
Plies

Figure 1–18: A typical cross-section of a passenger car and light truck tire.

1. What is the term used to refer to the trained professional that repairs automobiles today?

2. What are the three major sections of the automobile?

3. In terms of a calculation, how is work determined?

4. How is power determined?

5. How much is one horsepower?

6. Name the six forms of energy.

7. Gasoline and diesel fuel are examples of which of the six forms of energy?

8. Describe external combustion.

9. Describe internal combustion.

10. How is the fuel delivered to the engine?

11. What produces power in the internal combustion engine?

12. What converts recriprocating motion into rotary motion in an engine?

13. Name the systems an engine needs to operate.

14. Which system reduces wear while removing heat?

15. Name the major components of the FWD drivetrain system.

16. Name the major components of the RWD drivetrain system.

17. What are the two type of brakes used to stop a vehicle? How do they operate?

18. What is the purpose of the steering system?

19. What is the purpose of the suspension system?

20. What types of springs are used in suspension systems?

21. Why are tires an important component of the car?

22. What materials are used to make a tire?

REVIEW QUESTIONS

TOOLING UP

CHAPTER 2

After reading this chapter, the reader should be able to:

- Explain the importance of awareness when approaching the automobile.
- List the levels of commitment for maintenance of the automobile.
- List the safety precautions for car care.
- List the different tasks and tools necessary for the different levels of commitment.

AWARENESS—A STARTING POINT

If you want to save money and increase the reliability of your car, you should try to take an active role in its upkeep and maintenance. If you don't, you will eventually be forced to seek help for a major problem or breakdown. There are some simple things you can do to greatly reduce the chances of a major problem or breakdown. This book assumes that you are starting with little or no knowledge and that you want to learn what maintenance should be done for maximum dependability and economy. You also want to know what tools you will need.

Whatever your reasons for learning about your car, the first tool you need is an increased awareness of your car's maintenance and repair needs. Even if you don't want to become a full-fledged technician, you should try to learn what maintenance should be done regularly. Take a few minutes to read through your owner's manual and note the recommended maintenance schedules.

Nothing thrives on neglect. Preventive maintenance can stop problems from occurring. With a small investment of time and interest, you can become aware of some of your car's needs and help to avoid expensive breakdowns. For example, there are two simple checks you can make as you walk toward your parked car:

1. *Look at the tires.* Soft tires spell trouble. It pays to be sure that your car's tires are inflated properly, according to the car manufacturer's recommendations. After your tires are inflated, look at them carefully. It's more difficult to judge when radial tires are properly inflated because their sidewalls always have a characteristic bulge. Even so, each time you approach the car, look at each of the tires and compare what you see with your mental image of properly filled tires. If you're in doubt, check them with a good-quality tire-pressure gauge. At the very least, you risk damage to expensive tires by driving on them while they are underinflated. At worst, you are risking your life.

2. *Look for wet spots or puddles under or around the car.* Puddles can spell trouble. They can be:

- Engine oil (brownish, oily)
- Automatic transmission fluid (reddish, oily)
- Power steering fluid (reddish or light brownish, oily)
- Engine coolant (green, red, blue, or other watery, sticky liquid)
- Brake fluid (clear to brownish, with the smell of ether)
- Gasoline (clear, strong smelling, also a fire hazard)
- Water (condensation from air conditioning system, no problem)

Under a front-wheel drive transaxle, or under a rear-wheel drive manual transmission or differential, there may be grease spots. Inside the wheels, you might notice leaking brake fluid. Any of these signs could be warning you of trouble. But don't jump to conclusions; spots or fluids on the ground could have been left by another vehicle. If you are in doubt, check further wherever you normally park your car. Place a large piece of cardboard or several sheets of newspaper under the area of the car where you found the spots. Leave it there overnight. If something is leaking, you will find a fresh puddle on top of the cardboard or paper the next morning. Also try to note the location of the leak relative to the car. It may help you determine where the leak is coming from. But remember that as you drive, air flowing past the car can push these fluids toward the rear of the vehicle before it drips off. Just to be sure, you should check the fluid levels as described in Chapters 3 through 10.

If you find brake fluid leaking onto the ground, you should have the car's brakes inspected by a licensed professional technician as soon as possible. Brakes are your car's most important safety system.

Fuel leaks are extremely dangerous as gasoline is very flammable. *Don't start your car if you find a puddle of gasoline.* Have the leak repaired immediately.

In the chapters that follow, you will find more information about what these leaks may mean and what you can do about them. Just by checking regularly for tire condition and for leaks, you have developed and used the first, most essential tool for effective car care—your awareness.

LEVEL OF COMMITMENT

It is up to you how involved you get in maintaining your own car. Think about how much time you can invest. Some repairs can be time-consuming, but even a little time could save you a lot of money. If you find it interesting to work on mechanical devices, or if you can gain satisfaction from fixing things yourself, you will probably enjoy a commitment to car maintenance and repair.

Think about these things *before* you start buying tools. A commitment to get involved in the repair of your own car can lead to an investment of from less than a hundred dollars to several thousand dollars. You may already have some of the tools needed for basic maintenance of your car, or you may want to add to your tool set. You really can't do much maintenance on your car unless you are willing to invest at least $100 on tools. For the same $100, you can hire people to change the oil and lubricate your car for as long as a year, but this approach only takes care of the short-term engine maintenance.

If you are uncertain, move ahead in stages: Get acquainted with simple car checks and maintenance before you plunge ahead to advanced work. This chapter is organized to help you to understand the tools you need in relation to the level of commitment you want to make.

Your commitment to car care can involve a number of different levels. You can start at the basic level, supported by a few common, basic tools. After that, you add tools at each level. The decision levels open to you are:

- Inspection and basic maintenance—under the hood
- Inspection and basic maintenance—under the car
- Maintenance and minor repairs—under the hood
- Maintenance and minor repairs—under the car
- Tune-up and basic electrical work
- Diagnostics, advanced electrical, and fuel injection system work
- Major mechanical repairs
- Engine and transmission overhaul jobs.

The last three items on this list—diagnostics and advanced electrical, major mechanical repairs, and overhaul jobs—are not for novices. You may want to get into this kind of work only if you find that cars hold a real fascination for you. But before you do any kind of work on your car, you should become familiar with the safety precautions that follow.

SAFETY PRECAUTIONS FOR CAR CARE

You should realize that there are potential dangers to be avoided when you check or work on cars. For example, an engine that has been running has parts that are hot enough to cause serious burns. In addition you can be injured seriously if a raised car, which is supported only by a jack, falls on you.

A little common sense will go a long way toward avoiding problems. *Don't take chances.* Avoid unsafe situations and follow all safety recommendations. To learn about specific safety hazards to be avoided and specific safety precautions to be taken on a particular job, ask your instructor or refer to the manufacturer's service manual *before* you begin work.

As a general guide for your continuing safety and health, read about and follow the safety practices listed here. Learn and remember these rules before you go on to the review of maintenance levels that follow.

- Wear eye protection whenever you work on cars. Safety glasses, goggles, or a full-face shield can prevent eye injuries, infections, and possible blindness (Figure 2–1).
- Remove watches, rings, neck or ankle chains, and all other jewelry from your body before you start work. These items can be caught by moving parts and cause serious injury.
- Do not wear loose clothing that can be caught in moving parts. Tuck in shirttails and roll up long sleeves or wear short-sleeved shirts or blouses when you work on a running engine.
- If you have long hair, tie it into a tight ponytail. Then pin the ponytail to the top of your head or stuff it down the back of your shirt or blouse.
- Work in well-ventilated areas. Never run the engine in a closed garage or other enclosed area. Carbon monoxide gas from the exhaust is poisonous and cannot be seen or detected by smell.
- Work on cars only when you're feeling fit and well. Never work on cars if you are tired, ill, or have taken drugs that can cause drowsiness or that can impair mental or physical functioning.
- Whenever possible, work with friends or partners. Difficulties and safety problems may arise with which you may need assistance.
- Perform all repairs and maintenance in accordance with the car manufacturer's recommendations, as printed in the manufacturer's

Figure 2–1: Special glasses or face protectors are among the types of eye protection available for work on a car. Always wear eye protection when you do any work on or around a car.

service manual. *The Car Care Book* includes generalized procedures for common maintenance jobs, but it cannot anticipate all the variations that may be found on different makes and models of cars.

- Make sure you have the correct tools, parts, supplies, and materials before you begin work. Do not substitute or try to improvise with makeshift tools, parts, or supplies.
- Have a source of fresh water nearby to flush out eyes and minor wounds. It is also a good idea to have a first-aid kit handy whenever you work on cars.
- Whenever possible, avoid working on a running engine or one that is still hot. Serious burns can result from touching a hot exhaust system, cooling system, or other engine parts. If possible, allow the engine to cool down for several hours before you begin work.
- If you must work on a running engine, stay clear of all moving parts, including belts, pulleys, and fan blades. A spinning fan may be impossible to see. Serious injury or loss of limbs can result if you are caught in moving machinery.
- Work in well-lighted areas and use a safety droplight with a cage around the bulb. If a bare bulb breaks, the hot filament can ignite fuel or other flammable vapors and cause a fire.
- Position the car on a flat and level surface before you begin work. Never work on a car that is on an incline.
- Put the transmission in PARK (automatic transmission) or in a gear (manual transmissions) and set the parking brake firmly.
- Always place large wheel chocks or blocks both in front and in back of the wheels that remain on the ground to prevent the car from rolling, especially if you are going to raise the car.
- Raise and support the car properly and safely, according to the recommendations in the manufacturer's service manual.
- Never put any part of your body under a car that is supported only by a jack. A jack is only used to raise or lower the car. Always use safety stands, or **jack stands,** to safely support a car.
- Always disconnect the ground cable, typically the negative (–) battery terminal, when you work on the electrical system. This precaution will prevent sparks, fires, and damage to electrical parts.
- Have a fire extinguisher that will extinguish Classes A, B, and C fires (Figure 2–2).
- Do not use a heater with an open flame to heat the work area. Use an electric heater placed as far away as possible from the car and from any gasoline or other flammable liquids. Heat and other ignition sources can ignite fuel vapors, flammable liquids, and hydrogen gas produced by the battery.
- Collect oil, fuel, brake fluid, and other liquids only in approved metal or heavy plastic containers that can hold more liquid than you expect to drain (Figure 2–3). Do not use food or beverage containers that

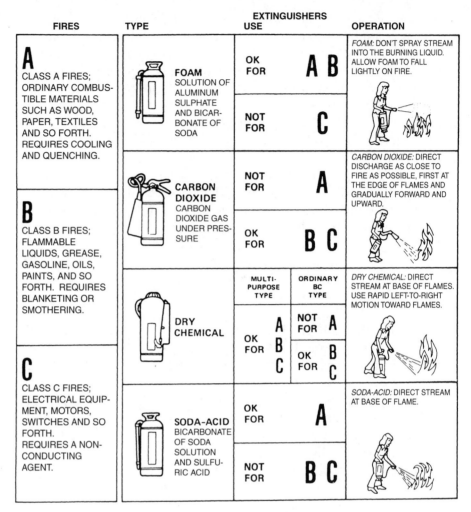

Figure 2–2: To extinguish different types of fires, fire extinguishers with Class A, B, and C ratings are required. This chart identifies the different types of extinguishers and identifies situations in which they should be used.

might be reused. Dispose of liquids in accordance with all local and federal environmental and safety standards.

- Avoid spilling or storing fuel or other flammable liquids near any source of open flame or spark, including gas water heaters and electrical switches. Wipe up all spills immediately. Store soaked rags and flammable liquids in approved metal safety containers with tight-fitting lids to prevent fires (Figure 2–3).

If you have any doubt about your own ability to perform repairs or checks correctly and safely on your own car, especially on safety systems such as brakes, leave such jobs to a licensed professional technician.

The following sections provide lists of maintenance procedures according to the number of tools and the amount of expertise needed to perform them. Details on the procedures are reserved for following chapters.

DO NOT USE NONAPPROVED CONTAINERS

Figure 2–3: Gasoline should only be stored in approved containers. Never store flammable liquids in glass or other nonapproved containers. Rags soaked with gasoline, oil, or other flammable liquids should be stored in safety containers. Never store rags in an open container.

INSPECTION AND BASIC MAINTENANCE—UNDER THE HOOD

At this level, tools are virtually no problem. You should have a flashlight, a sturdy screwdriver, and a pair of pliers. Beyond that, you simply need to learn what to look for.

Basic under-the-hood maintenance includes:

- Checking/refilling the engine oil in your crankcase
- Checking/refilling the coolant in your cooling system
- Checking/refilling brake fluid level in the reservoir
- Checking/refilling battery electrolyte (if possible)
- Checking battery cables for corrosion and tightness
- Checking/refilling fluid levels in the power steering system, if your car has one
- Checking/replacing the air and breather filters
- Looking for wear or cracks on belts
- Looking for wear or leaks on hoses and clamps
- Looking for oil or coolant leaks
- Checking/refilling the fluid in your automatic transmission (if equipped with a dipstick).

On most of these jobs, the only thing you need is a rag or paper towel to wipe off the **dipstick** as you check oil levels. As you will see later in the book, you may need a screwdriver or pair of pliers to check the brake fluid or the air filter.

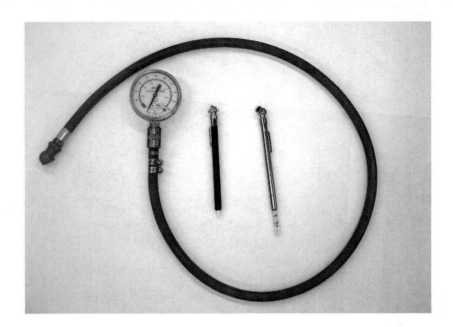

Figure 2–4: A tire pressure gauge with a range of 0–60 psi should be used to check automobile tire pressures.

There is one very important, basic maintenance function that doesn't take place under the hood: checking tire pressure. A good-quality tire-pressure gauge will pay for itself many times over if you take a few minutes to use it once a week, or any time when your tires look low (Figure 2–4). Over the life of an average car, tires are the most expensive parts you will replace on a regular basis.

INSPECTION AND BASIC MAINTENANCE—UNDER THE CAR

Many of your car's parts, especially those of the suspension, steering, and exhaust systems, can be seen and reached only from underneath. To assure your safety, it is necessary, periodically, to check these systems and other parts on the underside of the car.

If you decide to do the basic maintenance checks under the car, you will need some items to raise the car safely. You need a set of blocks to hold the wheels. You can use a car jack and some sturdy jack stands (Figure 2–5). Be sure to buy the correct capacity jacks and jack stands to support the weight of your car. A small portable hydraulic floor jack, which is easier to use than the standard equipment jack that comes with the car, is shown in Figure 2–6. Such jacks come in several sizes and weight-lifting capacities.

To get under the car comfortably, you may want to buy a creeper, which is a board that rests on four or six casters (Figure 2–7). You lie on the

Figure 2–5: For under-the-car work, a set of jack stands should be used to support the car safely. Never place any part of your body under a car supported only by a jack.

Figure 2–6: A 2¼ ton floor jack may be used to jack up a vehicle. A small portable floor jack may also be used in a garage or carried in the vehicle.

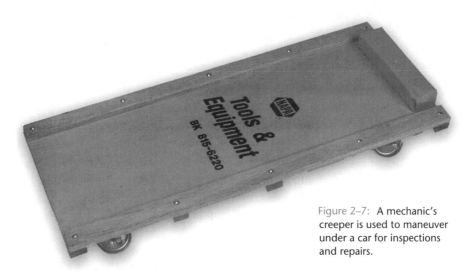

Figure 2–7: A mechanic's creeper is used to maneuver under a car for inspections and repairs.

creeper, usually face up, and push yourself into position under the elevated car with your feet.

Once you have invested in the equipment you need to get under your car, the basic maintenance steps you perform can add greatly to your safety and reduce repair costs sharply. Maintenance checks include:

- Search for engine oil leaks that involve gaskets, seals, or both.
- Change your oil filter and oil (Figure 2–8). This is the most important maintenance your engine can receive. To do an oil change, you

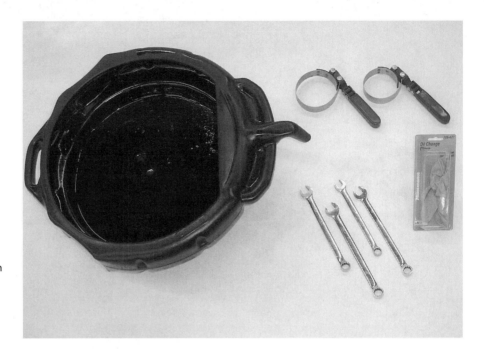

Figure 2–8: Items used for an oil change include a drain pan, oil filter, wrenches, and funnels. Quart containers of oil have a funnel shaped top to aid in pouring.

need an open-end or box wrench to remove and retighten the oil drain plug. You need some type of container to catch the oil as it is drained from the engine. You also need an oil-filter wrench; be sure to get the correct size wrench for your filter. To refill the engine with oil, you may need a funnel or pouring spout. Most plastic containers of engine oil have funnel-shaped tops to aid in pouring.

- Remove the automatic transmission pan to change fluid and replace the filter and pan gasket.
- Do your own lubrication. There are a number of lubrication points on a car that most people overlook and don't service until they start to give you trouble. These include body hinges, door striker plates, door tracks, hood latch, trunk latch, fuel door hinges and latch, linkage pivot points, and weather stripping. Always consult your owner's manual for the correct lubricant and service intervals for these components. Most cars today require no lubrication of the chassis and suspension thanks to permanently lubricated components. Some lubricants are shown in Figure 2–9. These include penetrating spray lubricants, oils, and greases.
- Check the steering linkage for play or looseness. If you can wiggle or rattle the steering mechanism, have it checked.
- Check for automatic transmission leaks. If you find any leaks, take the car to a repair center or a specialty transmission facility.

Figure 2–9: A variety of supplies are available for car maintenance, including lubricants, oils, greases, spray cleaners, hand cleaners, hand wipes, latex gloves, and floor dry. These may be purchased at retail parts stores.

- Check the lubricant level in a manual or automatic transmission, differential, or transaxle.
- Check the external clutch and transmission linkage to be sure that no looseness exists and that a manual transmission's clutch linkage is adjusted properly (some clutch mechanisms are self-adjusting).
- Inspect the drivetrain, including the universal joints or constant velocity joints, for looseness and play. Drive axle boots should not be torn or leaking.
- Inspect the insides of your tires for signs that either grease or brake fluid is leaking. As mentioned earlier, if you find leaking brake fluid inside a wheel, you should have the brake system checked by a licensed professional technician as soon as possible.
- Inspect the emergency brake cable to see if it needs tightening.
- Check your shock absorbers or **MacPherson** struts for leakage to see if they need replacement.
- Inspect the muffler, tailpipe, and exhaust system for smashed pipes, holes, rusted-out parts, and broken straps and hangers.

Some of the special purpose tools mentioned pay for themselves immediately through the money you save. For example, tools to change oil or oil filters will be paid for after one or two changes. It will take a little longer to pay for the jack and jack stands with savings from oil and filter changes. However, because a typical car weighs more than a ton, you can't afford to work under a car without the protection of proper support.

BASIC MECHANICAL TOOLS

If you decide to do some of the mechanical work listed in the "Maintenance and Minor Repairs" sections following, you will need a basic set of tools. The descriptions provided here will help you to become familiar with tools, and some accessories that may come in handy, that you should consider including in that set.

Open-End, Box, and Combination Wrenches

Wrenches, *not* pliers, should be used to loosen and tighten nuts and bolts (Figure 2–10). Combination wrenches have one open end and one box end, both designed to fit the same size nuts and bolts. Wrenches with two different-size ends are also available. There are some narrow spaces on an automobile where you can only slip in an open-end wrench. An open-end wrench is convenient to use, but, unlike a box-end wrench, it can slip or round off the corners of a bolt head or nut. Combination, open-end, and box wrenches are available in complete sets as well as individually. They come in Society of Automatic Engineers (SAE) (inch) and metric (millimeter) measurements. All foreign cars require metric tools. Domestic car manufacturers have been shipping parts to and from Europe and Asia since the 1970s. The number of metric fasteners used in domestic cars has been steadily increasing since the 1970s. Today most fasteners found in a new car will be metric; however you may still find a few SAE fasteners as well.

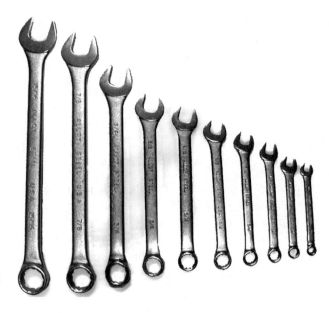

Figure 2–10: Combination wrenches include both an open end and a box end. The wrenches are sized for SAE (American standard measurements) and for metric sizes. At the top of the photo are examples of box-end wrenches (both ends are box ends) and open-end wrenches (both ends are open ends).

Socket and Ratchet Set

Sockets fit on a ratchet handle with a square drive piece. They are very useful when you are working in areas of the vehicle where you don't have room to turn a wrench. In such situations, you can use an extension that fits between the ratchet and the socket to clear the obstacles. Sockets usually are bought in complete sets. As with other wrenches, they are available in either SAE or metric sizes (Figure 2–11). The standardized square drive on the ratchet (¼, ⅜, or ½ inch square) fits many sizes of sockets, as well as extensions of different lengths. A ⅜ drive socket set is probably the most useful for most automotive jobs.

The term **ratchet** means that you can operate the device in short strokes, during which the mechanism applies no torque on the backstroke. You need only to move the handle back and forth and the socket will only turn in one direction. A knob or lever is used to set the ratchet for tightening or for loosening. With a ratchet, you have the advantage of being able to leave the socket mounted on the nut or bolt during the entire tightening or loosening operation. You don't have to remove and reposition the tool to turn the nut or bolt head, as you do if you are using wrenches.

Figure 2–11: A socket set can include extensions, ratchets, and breaker bars (*top*). These are used to drive, or turn, various types of sockets, deep well, shallow, and flex sockets.

Allen and Torx Wrenches

Allen and Torx wrenches are used on bolts or screws that have sockets, or shaped holes, in their heads (Figure 2–12). Allen wrenches are hexagonal (six-sided). Torx wrenches fit special hexagonal, star-shaped holes like their Torx screwdriver counterparts. Larger Torx fasteners can be found in the body, chassis, and many seat-belt and other safety equipment fasteners. Allen and Torx wrenches are usually purchased in sets, some of which open and close like blades on a penknife.

Screwdrivers

To do any type of maintenance work, you should have a good selection of screwdrivers for slotted and Phillips screws. Sometimes a few Torx screwdrivers are useful especially if you need to remove screws from headlamps, body trim, and other areas. Torx fasteners have become popular in recent years. You need a long screwdriver for reaching screws surrounded by other equipment. You also should have a short, or stubby, screwdriver to work on screws in places where you have limited clearance. Usually, a set of screwdrivers has one or two in-between sizes (Figure 2–13).

Pliers

You probably should have three or four pairs of pliers in different styles (Figure 2–14). A good assortment includes a pair of 6- or 8-inch slip-joint

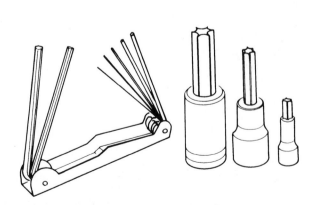

Figure 2–12: Allen wrenches (*left*) and Torx drivers (*right*) are used on bolts or screws with specially shaped holes in their heads.

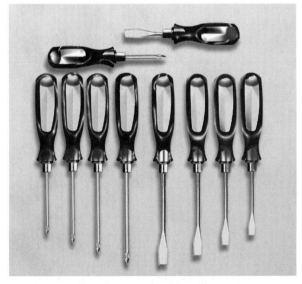

Figure 2–13: Screwdrivers can be purchased in sets. You should have Phillips and slotted screwdrivers, and you may want some Torx screwdrivers as well.

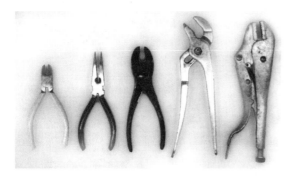

Figure 2–14: Pliers come in a variety of shapes and sizes for a number of different jobs. Shown here, from left, are wire-cutting pliers, long-nose pliers, slip-joint pliers, arc-joint pliers, and locking pliers.

pliers. These have jaws that extend straight from the handles and two positions to adjust the opening of the jaws. In addition, consider buying a pair of 12-inch adjustable arc-joint or channel lock pliers that have a number of adjustment positions. The jaws of these pliers are set at angles to the handles. Another valuable addition to your tool collection might be a pair of 8- or 10-inch locking pliers sometimes commonly called vise grips. These are pliers that can be locked in position on a device you want to hold. A pair of 6- or 8-inch diagonal cutting pliers can be used for pulling metal pins, cutting wire, or getting into places where you must grip an object that is close to a flat surface. Long-nose pliers also can come in handy. These have long, thin jaws that fit into tight places.

Miscellaneous Hand Tools

There are other tools that may come in handy for automotive work (Figure 2–15). You may need a hacksaw for cutting metal. A set of files can be useful to smooth out sharp edges of metal or to remove burrs (jagged pieces of clinging metal). A set of chisels and punches can be used with a hammer to cut metal parts and drive out pins or rivets when the heads are broken off. Always use a ball peen hammer, not a claw hammer, when you work with punches and chisels.

CAUTION

Always wear eye protection (safety glasses, goggles, or a face shield, see Figure 2–1) when you work with a hammer and chisel or punches. Flying metal particles can permanently blind you.

You can buy tools individually, a few at a time as needed, or as a large technician's set. You may be able to save quite a bit of money by buying a complete set of tools. Large department stores, hardware stores, and auto accessory and parts stores often have sales, advertised in newspapers and magazines, on individual tools and on sets of tools.

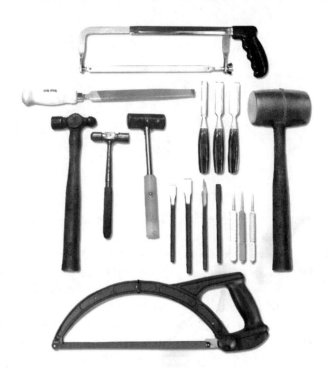

Figure 2–15: Miscellaneous tools that may be needed include a hacksaw, files, punches, chisels, and a ball peen hammer.

You may decide to purchase a toolbox, or a tool chest, or perhaps a tool board to store your tools. Decide which will best suit your needs. Do not consider the preceding sections as advice to go out and buy all of the tools listed. Rather, you should determine your own level of interest and then decide which tools you need.

SUPPLIES AND ACCESSORIES

Besides basic tools, you will need a few additional items before you start maintenance work on your car (Figure 12–16). These items include the following.

Droplight

A droplight is simply a socket on a long electrical cord. There is usually a reflector at the end, as well as a protective cage. Also, there is a hook that can be used to hang the light under the car or in the engine well. Most droplights take standard light bulbs and operate on standard electric current (110 volts AC). However, there are also fluorescent droplights and 12 volts DC units that plug into the cigarette lighter.

Figure 2–16: Additional shop supplies include a reel-type droplight that can be hung in place while you work on your car. Be sure to use light bulbs designed for this purpose. Other tools include pry bars and gasket scrapers.

Gasket Scraper and Sealer

If you are going to replace gaskets or seals, you will need a gasket scraper, a flexible metal tool with a wide edge that looks something like a putty knife. You also will need some gasket sealer, a jellied material that is spread on gaskets to form a reliable seal.

Pry Bar

A pry bar is a heavy metal bar with a flat edge at one end, something like a large screwdriver. Some units have a carved handle at one end, like a cane handle. A pry bar gives you the leverage to move heavy or stuck parts.

Penetrating Oil

Penetrating oil is special oil that is used to help loosen screws or bolts that are stuck in place by rust, or seized. The oil penetrates to help lubricate through the rust.

Floor Cleaner

Not all of the oil or grease will get on your hands and clothes. The floor in your work area can become dirty and slippery. Even if you are working outdoors, you should clean up the unavoidable grease spots. A number of products are available to absorb and sweep away grease spots (see Figure 2–9).

Hand Cleaner

Car maintenance is dirty work. There are a variety of creams, liquids, and borax or pumice soaps that you can use to clean your hands after a work session. Latex gloves specifically designed for use while working on the car are becoming increasingly popular. These gloves can prevent the different lubricants, solvents, and other contaminants from remaining in prolonged contact with your skin. Their cost is nominal and they make cleanup easy. Another effective way to clean your hands after the work is done is the use of a hand cleaner. These are available with a pleasant orange scent and quickly clean. Where water is not available, you may simply wipe your hands off after using them with a towel or rag. Hand towels with cleaner are also available for use (see Figure 2–9).

Solvent

Grease and rust are present in almost any car repair job. They often must be cleared away before you can even find the parts you want to replace. Gasoline gives off explosive vapors and should never be used to clean parts. Instead, use a solvent, which is a less flammable grease-cutting liquid.

⚠ CAUTION

Most solvents are flammable and should be used and stored with care. Both solvents and the rags you use with them should be stored in a metal can or bin with a tight-fitting lid and kept where children can't get to them. Because of their flammable nature, solvents should be kept away from sparks, heat, and open flames. Accidents have resulted from storage of solvents being too close to electrical light switches as well as space and water heaters. Some solvents, such as carburetor cleaners, contain harmful chemicals that can damage clothing, skin, and eyes. When you use such solvents, wear eye protection, gloves, and overalls or a shop coat.

If you have decided to do more than basic maintenance under the hood, you now realize that it will be necessary to spend some money. This decision is a turning point. You may decide to make the investment for economic purposes or because you are fascinated by mechanical things.

Even if you have decided to leave repairs to a professional technician, you still can profit from performing routine checks. You can learn to spot potential problems and have them fixed before they cause a breakdown.

MAINTENANCE AND MINOR REPAIRS—UNDER THE HOOD

Once you have some basic hand tools, you can perform a number of minor, relatively easy maintenance and repair jobs. The skills you develop will save you money and also could be vital in getting you moving again if you break down on the highway.

A few valuable under-the-hood jobs include:

- Replace a leaking valve cover gasket.
- Tighten or replace coolant hoses.
- Flush cooling system.
- Replace windshield-washer tubing and other hoses (other than air conditioning hoses and fuel lines).
- Replace air filter.
- Replace a fuel filter.
- Tighten loose bolts, nuts, and screws.
- Tighten or replace fan and other drive belts.

These jobs, as described in later chapters, can be relatively easy. With the help of a manufacturer's service manual, you also will be able to replace drive belts.

MAINTENANCE AND MINOR REPAIRS—UNDER THE CAR

Minor maintenance and repair jobs that you can do for yourself under the car with a basic set of hand tools and a set of jack stands or ramps and wheel blocks include:

- Rotate tires.
- Inspect disc and drum brakes, hydraulic lines, and fastening parts.
- Inspect exhaust system components.
- Change lubricant in a manual transmission or differential.
- Remove the automatic transmission pan to change fluid and replace the filter and pan gasket.
- Check the clutch, if you have a manual transmission.

With the help of a service manual, you can also do the following:

- Pack and adjust non-driving-wheel bearings.
- Make parking brake adjustments.
- Adjust the clutch.

These jobs relate directly to the reliability, safety, and performance of your car.

TUNE-UP AND BASIC ELECTRICAL WORK

With just a few more tools, you can take the next step in car care: You can do some of your own electrical system, ignition system, and basic fuel system maintenance. You may even be able to improve your car's performance and decrease its exhaust emissions by changing spark plugs and spark plug wires. These maintenance jobs are part of tune ups.

A motivated novice can handle the following, relatively easy, electrical, ignition, and fuel system jobs:

- Clean and tighten battery cables and connections.
- Jump-start the car if your battery is low.
- Check all lights and replace bulbs.
- Remove and replace turn-signal flasher.
- Replace blown fuses.
- Check, remove, and replace alternator.
- Remove and replace spark plugs.
- Perform a compression test on the engine.
- Remove and replace distributor cap and rotor (if equipped).
- Remove and replace ignition or spark plug wires.
- Remove and replace the fuel system filter.

The amount of money to be saved in doing your own electrical work will more than offset the cost of the additional tools needed, if you develop your skills to a level at which you can follow instructions and handle the jobs effectively. Tools you might need (see Figure 2–17 and Figure 2–18) in doing these jobs can include:

- Jumper cables (4- to 6-gauge stranded copper wire)
- Tachometer
- Feeler gauges
- Spark plug wrench or socket
- Spark plug gap gauge
- Compression gauge
- Test light or volt/ohm meter for circuit testing

These tools and their uses are not described in this chapter. Their use represents advanced work. If you are interested in doing your own electrical, ignition, and basic fuel system work, pay close attention to Chapters 4, 5, and 11.

The advice about buying tools is worth repeating: You don't need to buy a lot of tools to get your money's worth out of this book. Don't go out and buy an extensive set of tools unless you are sure that you want to put in the

Figure 2–17: Jumper cables can be used to help start a car with a weak battery.

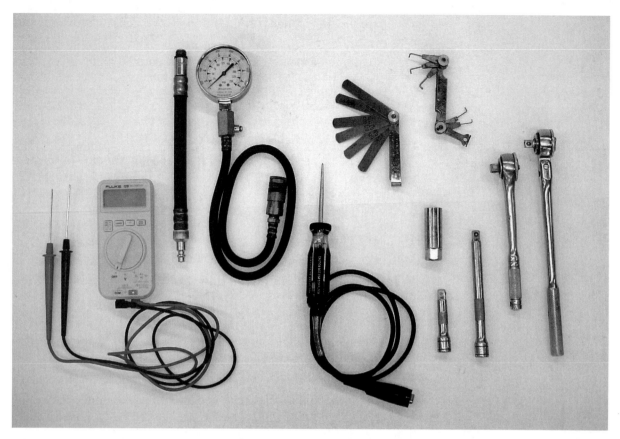

Figure 2–18: A spark plug socket and ratchet are used to remove spark plugs. Feeler gauges are used to gap spark plugs. A test light and voltmeter may be used to test electrical circuits in the car.

time and effort to learn to use them. If you aren't sure, keep reading. You can make up your mind as you learn more about your car. The next chapter will add to your knowledge considerably by describing the operation and maintenance of the engine, as well as simple checks to diagnose engine problems.

1. What are the first things you should look at as you walk toward your car?

2. Describe some of the fluids that a car may contain and how to identify them.

3. What are the different levels of commitment to automobile maintenance?

4. Which levels are suitable for a novice?

5. What should be used to protect your eyes? Why should you protect your eyes?

6. Why should you determine your level of interest before buying tools?

7. What is a sensible way to buy tools?

8. What precautions should you observe while running an engine?

9. What precautions should you follow if you are performing electrical work on a vehicle?

10. What are some precautions for working with or around flammable liquids?

11. List some characteristics of solvents.

12. List some common tools you might need for maintenance of a car.

13. What are some common items to check or inspect under the hood?

14. What are some common items to check or inspect under the car?

15. What are some common items to maintain or repair under the hood?

16. What are some common items to maintain or repair under the car?

17. What are some tasks to perform for a tune up and basic electrical work?

THE ENGINE

After reading this chapter, the reader should be able to:

- Describe the operating principles of the engine.
- Explain the four-stroke cycle and the function of each of the four strokes.
- Describe the coordination and operation of the valvetrain.
- Explain the operation of the camshaft.
- Explain the importance of routine engine maintenance.
- List some common engine problems.

AN OVERVIEW

Your car's engine uses heat to create mechanical energy. In turn, the mechanical energy supplies force to move the car. Modern automotive engines are powered by the internal burning of a fuel, typically gasoline.

The operating principles of your engine can be divided into four basic functions:

1. The engine draws in fuel and air.
2. A fuel and air mixture is compressed, or squeezed, into a small area.
3. The fuel is burned to produce heat and pressure to move the pistons and turn the crankshaft.
4. The burned gases are discharged so the process can be repeated.

Automotive engines consist of multiple power-producing units, called *cylinders*. Most cars built today have four, six, or eight cylinders. Engines can be classified according to the number of cylinders, shape, size, fuel used, and other characteristics.

The discussion of engine operation in this chapter focuses on the four-stroke-cycle gasoline piston engine. This is the engine used in most automobiles, pickup trucks, vans, and utility vehicles. In this engine, an electrical spark ignites a mixture of gasoline and air.

Included in this chapter are descriptions of mechanical parts of the piston engine. Excluded are the fuel, electrical, lubrication, and cooling systems essential to engine operation, covered in Chapters 4, 5, and 6. The chapter begins with a simple explanation of what happens when an engine is operating. It ends with a discussion covering problems that you may be able to spot and presents some troubleshooting advice.

THE BASIC GASOLINE ENGINE

Engines that burn unleaded gasoline today power the great majority of cars on the road. The gasoline is mixed with air and then burned to create heat. The heat expands burned gases to create pressure. This pressure moves several pistons within the engine, each of which transmits a driving force to a crankshaft. The turning power (torque) created by the rotating crankshaft is transmitted to the drivetrain, which turns the driving wheels. The combination of the engine and drivetrain is known as the *powertrain*.

Figure 3–1 is a simplified drawing of how pistons and cylinders are typically are arranged in engines. Together, each piston and cylinder makes up a single power-producing unit. Several pistons and cylinders are used to create enough torque to move your car. Other factors being equal, the larger the pistons and cylinders and the more of them there are, the more powerful the engine will be and the more fuel it will use.

The top drawing in Figure 3–1 shows an **in-line engine** with four cylinders and pistons used in many modern vehicles. The pistons and cylinders are arranged in a single, straight line. Such engines also are called "straight fours." The middle drawing shows a **V-type engine** with six cylinders, or "V6." The bottom drawing shows a flat four-cylinder engine, or a "flat four," as found in older Volkswagens and some Subarus.

From the late 1950s to the 1970s, V8 engines, with eight cylinders arranged in a V-shaped **cylinder block,** were the standard for domestic American cars. When concerns about the price and availability of fuel became important, car manufacturers responded with smaller, lighter vehicles. Four- and six-cylinder engines can power today's automobiles adequately and economically.

Currently, car engines are produced with as few as 3 or as many as 12 cylinders. Engines with an odd number of cylinders, such as 3- and 5-cylinder engines, are not as common as engines with even numbers of cylinders.

No matter how many cylinders an engine has, the basic function of each cylinder remains the same. A cylinder takes in a mixture of atomized fuel and air, compresses these gases, and then burns them. The pressure from this burning air and fuel forces a piston downward to generate the torque on the crankshaft. This action occurs in a four-stroke cycle. (A *cycle* is a series of events that occurs repeatedly.)

The Four-Stroke Cycle

The **four-stroke cycle** gets its name from the fact that the piston goes through four strokes. Each **stroke** is a single downward or upward motion. Thus, in the course of each four-stroke cycle, there are two up-and-down movements. Each up-and-down movement corresponds with one rotation of the power-transmitting crankshaft. The complete four-stroke cycle requires

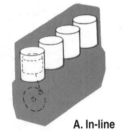

A. In-line

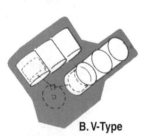

B. V-Type

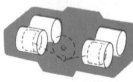

C. Flat

Figure 3–1: The cylinders in an engine can be arranged in a straight line (*top*), in a V-shape (*middle*), or in a flat configuration (*bottom*).

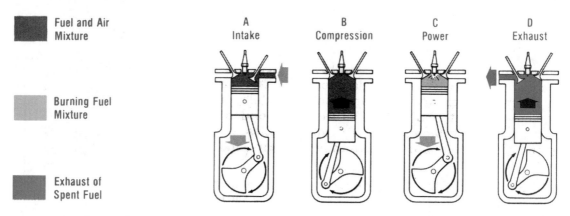

| A Intake | B Compression | C Power | D Exhaust |

Fuel and Air Mixture

Burning Fuel Mixture

Exhaust of Spent Fuel

Figure 3–2: The operating cycle of the engine includes four strokes, or movements, of a piston: intake, compression, power, and exhaust.

two revolutions, or complete 360-degree turns, of the crankshaft. The speed at which the engine operates is measured in **revolutions per minute** (rpm).

A simplified illustration showing the four-stroke cycle and the accompanying two complete turns of the crankshaft is shown in Figure 3–2. Use the diagram as a guide through the discussion that follows. The four strokes in the engine cycle are:

1. Intake stroke
2. Compression stroke
3. Power stroke
4. Exhaust stroke

Intake Stroke   The intake stroke introduces a mixture of atomized gasoline and air into the cylinder (Figure 3–3). The stroke starts when the piston, which is sealed tightly in the cylinder, moves downward from a position near the top of the cylinder. As the piston moves downward, a vacuum, or low-pressure area, is created

The upper part of the engine, called the **cylinder head,** contains two ports, or openings, at the top of each cylinder. During the intake stroke, moving a **valve,** or sealing device, downward into the top of the cylinder opens one of the ports. The other valve remains tightly closed.

At this point, the pressure of the atmosphere is many times greater than that of the low-pressure area created at the top of the cylinder. So, atmospheric pressure pushes fuel and air through the open valve and port into the cylinder. In the process, **atomization** takes place—either a carburetor or a fuel injector delivers a finely atomized spray of gasoline droplets to the incoming air. This mixture of atomized gasoline and air enters and fills the cylinder.

As the air/fuel mixture rushes into the cylinder, it swirls around the open **intake valve.** This action, along with the heat of the engine, helps to mix and vaporize the fuel more thoroughly with the air. The intake valve is opened to permit the passage of atomized gasoline and air into the cylinder.

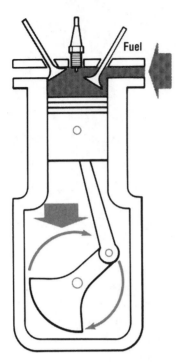

Figure 3–3: **The intake stroke.**

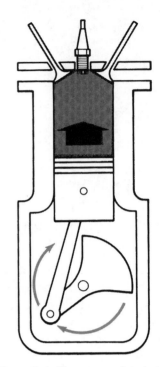

Figure 3–4: **The compression stroke.**

Near the end of the intake stroke, the intake valve closes to trap the air/fuel mixture within the cylinder.

Compression Stroke   As the piston moves upward to compress the fuel mixture trapped in the cylinder, the valves are closed tightly, as shown in Figure 3–4 illustrating the **compression stroke.** This compression, or squeezing, action heats the air/fuel mixture slightly and confines it within a small area formed by the cylinder head and valves, top of the piston, cylinder walls, and gaskets called the **combustion chamber,** the area within which the burning of fuel takes place. Most chemical reactions, such as the burning of fuel, occur more readily when the chemicals are heated. In addition, confining the air/fuel mixture to a small area prepares the way for more concentrated, intense burning that will provide high levels of heat and pressure.

Power Stroke   Just before the piston reaches the top of its compression stroke, an electrical spark is delivered through a spark plug into the cylinder head. (Spark plugs are part of a total **ignition system,** as discussed in Chapter 5.) The spark ignites the compressed and heated mixture of fuel and air in the combustion chamber to cause rapid burning. The burning fuel produces intense heat that causes rapid expansion of the gases compressed within the cylinder. This pressure forces the piston downward. The downward stroke turns the crankshaft with great force.

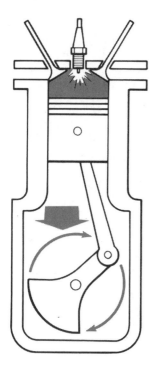

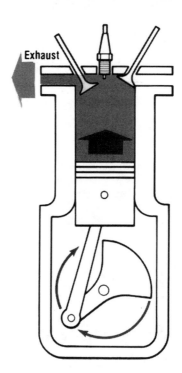

Figure 3–5: The power stroke.

Figure 3–6: The exhaust stroke.

Figure 3–5 illustrates the **power stroke.** Notice that both valves remain closed during the power stroke. Because the valves are closed and the piston is sealed tightly within the cylinder, the expansion of gases drives the piston downward. The power stroke ends just before the piston reaches its lowest point within the cylinder.

Exhaust Stroke   Just before the bottom of the power stroke, the cylinder's other valve, the **exhaust valve,** opens. This action allows the piston, as it moves upward, to push the hot burned gases out through the open exhaust valve, as shown in Figure 3–6 illustrating the **exhaust stroke.**

Then, just before the piston reaches its highest point, the exhaust valve closes and the intake valve opens. As the piston reaches the highest point in the cylinder, known as **top dead center (TDC),** it starts back down again. Thus, one cycle ends and another begins immediately.

Crankshaft Operation

During the four strokes of an individual piston (intake, compression, power, and exhaust—down, up, down, up), the car's crankshaft turns through two complete rotations, or 720 degrees.

In transmitting power during the four-stroke cycle, the upward and downward action of the piston is converted to a rotary, or turning, action

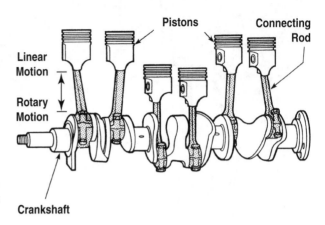

Figure 3–7: Force from the piston's downward movement causes the connecting rod to turn the crankshaft in a circular, rotating motion.

of the crankshaft. Figure 3–7 illustrates how the piston and crankshaft are connected.

Valve Operation

To coordinate the four-stroke cycle, a group of parts called the **valvetrain** opens and closes the valves (moves them down and up, respectively). These valve movements must take place at exactly the right moments. A camshaft controls the opening of each valve.

Overhead Camshaft (OHC) Valvetrain   The **cam** is an egg-shaped piece of metal on a shaft that rotates in coordination with the **crankshaft.** The metal shaft, called the camshaft, typically has individual cams for each valve in the engine. As the camshaft rotates, the lobe, or high spot of the cam, pushes against parts connected to the stem of the valve (Figure 3–8). This action forces the valve to move downward. This action could open an intake valve for an intake stroke or open an exhaust valve for an exhaust stroke. As the camshaft continues to rotate, the high spot moves away from the valve mechanism. As this occurs, **valve springs** pull the valve tightly closed against its opening, called the **valve seat.**

Valves in modern car engines are located in the cylinder head at the top of the engine. This is known as an **overhead valve (OHV)** configuration (Figure 3–9). In addition, when the camshaft is located over the cylinder head, the arrangement is known as an **overhead camshaft (OHC)** design. Some high-performance engines have two separate camshafts, one for each set of intake and exhaust valves. These engines are known as **dual overhead camshaft (DOHC)** engines.

Pushrod Valvetrain   The camshaft can also be located in the lower part of the engine, within the engine block (Figure 3–10). To transfer the motion of the cam upward to the valve, additional parts are needed. In this arrangement, the cam lobes push against round metal cylinders called

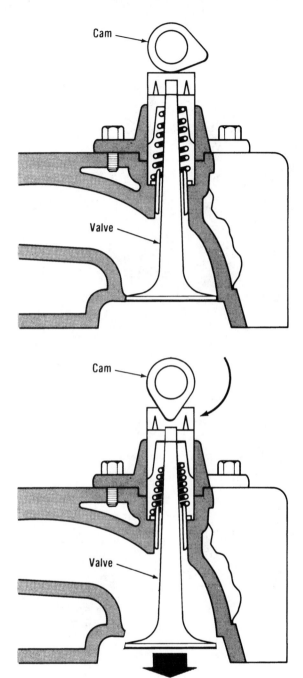

Cam

Valve

Cam

Valve

Figure 3–8: As the cam rotates, the nose of the cam pushes the valve downward to open the port.

valve lifters. As the lobe of the cam comes up under the lifter, it pushes the lifter upward (away from the camshaft). The lifter rides against a **pushrod,** which pushes against a **rocker arm.** The rocker arm pivots on a shaft through its center. As one side of the rocker arm moves up, the other side

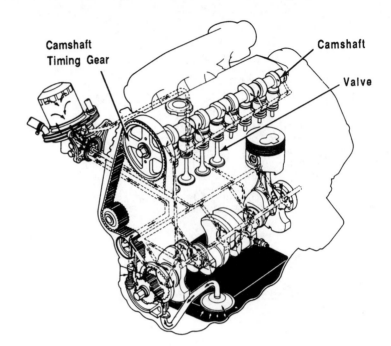

Figure 3–9: On many late-model engines, the camshaft is located over the cylinder head.

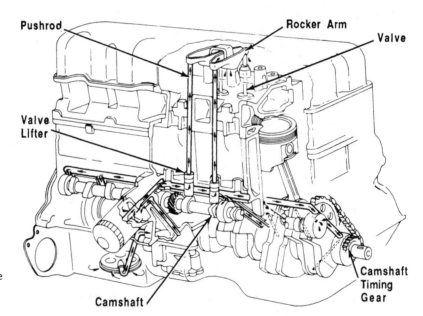

Figure 3–10: Some engines use a pushrod valvetrain with the camshaft located in the block. The motion of the camshaft is transmitted to the valve through intermediate parts.

moves down, just like a seesaw. The downward-moving side of the rocker arm pushes on the valve stem to open the valve.

Because a pushrod valvetrain has additional parts, it is more difficult to run at high speeds. Pushrod engines typically run at slower speeds and, consequently, produce less horsepower than overhead-camshaft designs of equal size. (Remember, power is the rate at which work is done.)

Multiple-Valve Cylinder Heads

As described, most car engines have one intake valve and one exhaust valve per cylinder. However, for more efficient engine operation, more valves can be used in each cylinder. High-performance engines with four valves (two intake, two exhaust) are currently available (Figure 3–11). These engines typically use dual overhead camshaft designs (Figure 3–12). Other engine designs with three valves per cylinder (two intake, one exhaust) also are used to improve engine efficiency and decrease harmful exhaust emissions.

Figure 3–11: In the photo, the underside of this high-performance engine's cylinder head reveals the use of four valves per cylinder, two each for intake and exhaust.

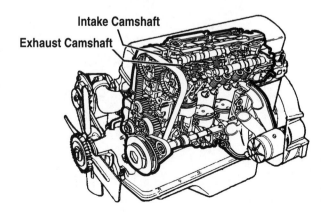

Intake Camshaft

Exhaust Camshaft

Figure 3–12: Dual overhead camshafts (right) are used to activate the two intake valves and two exhaust valves per cylinder in this engine.

Camshaft Drive Mechanism

Each cam must revolve once during the four-stroke cycle to open a valve. A cycle, remember, corresponds with two revolutions of the crankshaft. Therefore, the camshaft must revolve at exactly half the speed of the crankshaft. This ratio is achieved by using a **gear** connected to the camshaft having twice the number of teeth as a gear connected to the crankshaft, or with a 2:1 **gear ratio.** The gears are linked by using one of the three following mechanisms:

1. **Timing belt.** A cog-type belt can be used (Figure 3–13). Such belts are made of synthetic rubber and reinforced with internal steel or fiberglass strands. The belts have teeth, or slotted spaces, to engage and drive matching teeth on gear wheels. A timing belt is typically used on engines with overhead-cam valvetrains.

2. **Timing chain.** On some engines, a metal chain is used to connect the crankshaft and camshaft gears (Figure 3–14). Most pushrod engines and some OHC engines have timing chains.

3. **Timing gear train.** The camshaft and crankshaft gears can be connected directly, or meshed (Figure 3–15). This type of operating linkage is commonly used on older six-cylinder, in-line engines (not V6 units) and on some flat-four engines.

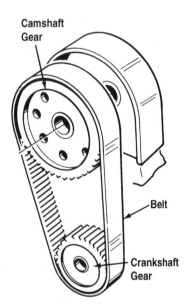

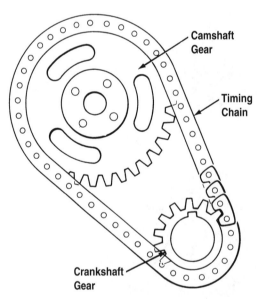

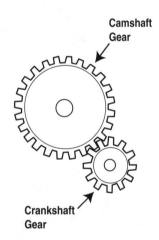

Figure 3–13: Many small OHC engines use a cog-type belt to coordinate the motion of the crankshaft with that of the camshaft.

Figure 3–14: A timing chain can be used on either OHC or pushrod valvetrain engines to link the camshaft to the crankshaft.

Figure 3–15: Gears can provide direct drive linkage between crankshaft and camshaft.

Multiple-Cylinder Coordination

The preceding description of the four-stroke cycle concentrates on the operation of a single cylinder. Just as the movement of valves must be synchronized with the movement of a piston, all of the pistons must be coordinated with each other. Remember that each cylinder operates on a four-stroke, 720-degree cycle of the crankshaft. Smoothness and continuing power come from equal spacing of the timing of the power strokes of different cylinders.

For example, in the design of a 4-cylinder engine, it is possible to divide the 720-degree cycle by four. A power stroke can be applied to the crankshaft each 180 degrees of rotation. In a 6-cylinder engine, it is possible to apply a power stroke each 120 degrees of crankshaft rotation. With 8 cylinders, a power stroke can be applied each 90 degrees of crankshaft rotation. With 12 cylinders, a power stroke can be applied each 60 degrees of crankshaft rotation.

The fewer degrees of rotation between each power stroke, the more smoothly your engine will run. Thus, the greater the number of cylinders in the engine, the smoother and more powerfully it will run.

Except for the variations in smoothness of operation and power delivered, the end result of the operation of all internal combustion engines is similar. Fuel is burned within cylinders to deliver torque through a crankshaft to a drivetrain.

ROUTINE ENGINE MAINTENANCE

To understand how to maintain your engine for maximum life, you have to understand what makes parts wear out. As a car owner and operator, your enemies within the engine are heat, friction, and contaminants (dirt, water, and acid).

Heat comes from the burning of fuel and from the friction caused by the rubbing of parts against one another. You understand friction if you have ever rubbed your hands together to keep them warm. Within an engine, friction increases heat. Heat causes metal parts to expand and rub tighter. Dirt, even dust, is abrasive and contributes to friction. Parts can corrode; sulfur in engine oil and gasoline can combine with water vapor to form sulfuric acid, which eats away at metal parts.

As parts become rough and worn, the precision fit between metal surfaces on the inside of the engine becomes loose and sloppy. Your engine loses power because expanding gases from combustion can leak past the pistons and valves. In addition, the fuel/air mixture doesn't burn as efficiently because pressure is lost during the compression stroke. A worn engine also may consume oil and lose coolant. These oil and coolant losses, in turn, contribute to overheating and to additional wear.

To combat engine wear, you must take steps to avoid the effects of heat, friction, and contamination. One protective measure is to check the engine oil level and coolant level regularly, at least once a week. To check the engine oil level, the car should be on level ground so that you get a true reading. If your engine has been running, shut it off and wait at least five minutes for the oil to drain into the oil pan. Then simply pull out the dipstick, wipe it off, and reinsert it into the dipstick tube. Be sure the dipstick is pushed into the tube as far as it will go. Pull out the dipstick again and take a reading. If the oil level is at or below the ADD line, add oil. Don't fill the crankcase above the FULL line.

Procedures for checking and adding coolant, as well as further information on checking and adding oil, are in Chapter 6, "The Lubrication and Cooling Systems."

ENGINE PROBLEMS

If you understand how engines operate and what causes wear and damage, you will be in a better position to recognize troubles and to deal with them. Understand the difference between use and *abuse*. If checks and maintenance are not performed regularly, you are abusing your car. Short trip driving, hard acceleration, and other severe driving habits also can contribute to the wear of your engine. The following is a review of the internal working parts of an engine that are subject to wear and describes exposures to wear. Bear in mind that problems may show up only after a long period of time or after you have accumulated a substantial number of miles on the odometer.

Rings

Possibly the most important—and perhaps the most expensive—engine wear occurs to parts in the block, the lower part of the engine. Worn **piston rings** and cylinder walls usually indicate that the entire engine has suffered extreme wear.

The exterior surface of a piston has grooves around it. In these grooves, the manufacturers insert metal rings to form a tight seal within the cylinder (Figure 3–16). The rings are made of cast iron or steel and are coated with other metals. In operation, piston rings perform three functions:

1. Compression rings (typically two) form a tight seal between the piston and the cylinder walls and help maintain compression, preventing leakage or blowby, of gases from the combustion chamber into the **crankcase.** The crankcase is at the bottom of the block, above the oil pan, where the crankshaft is supported.
2. An oil control ring, located lower on the piston, prevents oil in the crankcase from being forced upward into the combustion chamber

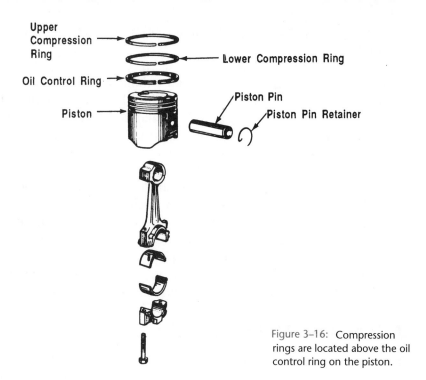

Upper Compression Ring

Lower Compression Ring

Oil Control Ring

Piston

Piston Pin

Piston Pin Retainer

Figure 3–16: Compression rings are located above the oil control ring on the piston.

on the intake stroke. It also wipes excess oil from the cylinder wall, routing it through holes in the ring and piston back down into the crankcase.

3. Piston rings conduct heat from the piston head into the cylinder wall. At the same time, they allow for expansion and contraction of pistons and cylinders caused when the engine heats and cools.

By the time a car has traveled 100,000 miles or more, wear inevitably has taken place. This wear affects all parts of the engine, including the rings and the cylinder wall. With wear, the size of each cylinder increases and the size of each piston and its rings decrease. When this happens, an engine loses compression pressure. Some of the gases that are meant to drive the pistons downward blow by the piston rings, which no longer form a tight seal. In addition, the engine may "pump oil." On each intake stroke, atmospheric pressure forces gases and oil up past worn rings into the combustion chamber. The oil then is burned, typically producing a thick, bluish-white smoke and burnt oil odor.

Typically, the worn cylinder is wider in diameter at the top than it is at the bottom. Combustion takes place at the top of the cylinder, and, therefore, the heat is greatest there. This heat reduces the oil's lubricating ability. As a result, wear is greater at the top than it is at the bottom of the cylinder.

If a compression ring breaks in half, most, if not all, compression pressure is lost. This condition creates a "dead cylinder," one that is no longer producing power. Thus, the engine runs roughly. In such a case, it may be possible to replace the piston rings alone and restore the engine to normal operation.

This wear is not large. The taper of a worn cylinder is likely to be somewhere between 1 thousandth (0.001) and 5 thousandths (0.005) of an inch greater at the top than at the bottom. If the width at the top of the cylinder approaches a point at which it is 10 thousandths (0.010) of an inch wider than at the bottom, ring or piston replacement will not help. It is necessary to rebore the cylinder to create an oversized hole. Then an oversized piston and ring set is inserted.

However, by the time the cylinders are tapered excessively or a ring breaks, the wear on engine crankshaft and connecting rod bearings and other internal engine parts may require a complete rebuild of the engine block. In addition, because the cylinder head must be removed to gain access to the parts within the cylinder block, the head itself should be reconditioned. Thus, excessive wear or damage to pistons and cylinder walls typically requires a complete engine overhaul.

Valves and Valve Guides

An automotive valve can be compared with a tapered cork. When a valve is in good condition and pressed tightly into its seat, the ports, or passages in the cylinder head, are fully sealed. However, if the edge of a valve that fits into the valve seat is roughened, chipped, burned, or broken, its sealing capacity is decreased or lost completely. During the compression stroke, the air and fuel mixture that should be compressed can escape past the valve and result in reduced engine power, as there will be less air and fuel to burn to create power. Also during the power stroke, pressure that should drive the piston downward can flow outward around the valve and be lost through the port. As with a broken ring, damaged valves can create a dead cylinder, which in turn reduces engine power.

Figure 3–17 is a drawing of valves positioned within a cylinder head. The valve guide, valve stem, seat, and other parts are visible. Wear occurs on the valve face—and on the valve seat—as the valve face touches the seat thousands of times per minute. The face and seat of the valve wear from friction and from hot gases and particles of dirt. Severe wear or damage to a valve caused by excessive heat is commonly referred to as "a burnt valve."

Other major wear points are the **valve guide** and the **valve stem.** The valve guide is the narrow cylindrical opening in which the valve stem moves up and down. The valve guide may be part of the cylinder head or may be a removable sleeve. As wear occurs, the gap between the valve stem and the guide increases. Oil can then be drawn down into the combustion chamber, where it is burned.

To repair valve wear, two approaches are possible: repair or replacement. If wear is minimal, it is possible to regrind valve faces and valve seats. If wear is excessive, entire valve assemblies—including valves, valve guides,

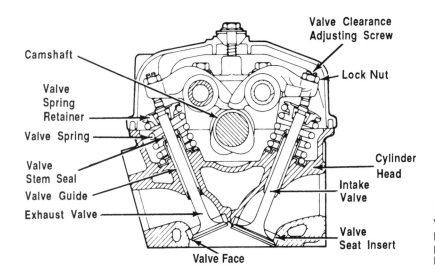

Figure 3–17: This cutaway view illustrates how valves are positioned within guides pressed into the cylinder head.

and valve seats—must be replaced. On a worn engine, if compression is being lost mainly through the valves, it is possible to either recondition the cylinder head alone without overhauling the parts within the cylinder block or to replace the heads with remanufactured ones.

Valve Lifters

You already know that the relationship between the operation of valves and the four-stroke cycle is controlled by a link between the crankshaft and camshaft. In addition, an intermediate part is needed between the camshaft and the valve itself. This part is the valve lifter, or tappet.

There are two main types of valve lifters. One called a *solid lifter*. It consists of a machined piece of metal that simply follows the camshaft. A valve spring keeps the valve in a closed position. The camshaft lobe presses against the solid lifter that, in turn, pushes against the valve and opens it. Valvetrains with solid lifters require periodic checking and adjustment, typically once a year or every 12,000 miles, whichever comes first. This service maintains engine performance and prevents wear and damage to the engine valves and related parts. Solid lifters are usually found on high-performance engines because they have no internal moving parts to break under stress.

The other type of lifter is the *hydraulic lifter*. This system can be used either on overhead-camshaft mechanisms or on pushrod valvetrains. The hydraulic lifter is a hollow cylindrical mechanism that fills with oil when the engine runs. The oil is trapped inside the lifter, which then transfers force from the moving cam to open the valve. The oil cylinder acts as a fluid adjuster taking up any slack in the valvetrain. Hydraulic lifters eliminate the periodic valve adjustments necessary with solid lifters. A pushrod hydraulic-lifter valve assembly is illustrated in Figure 3–18.

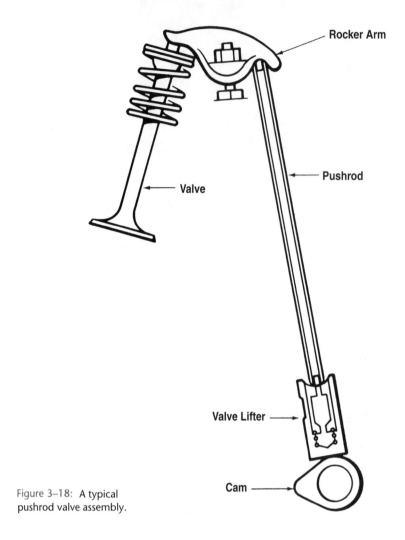

Rocker Arm

Pushrod

Valve

Valve Lifter

Cam

Figure 3–18: A typical pushrod valve assembly.

A lifter, solid or hydraulic, will have either a solid convex bottom or a small metal roller. A lifter equipped with the metal roller on the bottom is referred to as a "roller lifter." It, too, can be found on many engines and is popular for high-performance engines. The roller offers some advantages over the convex bottom. The main purpose is to reduce friction in the valvetrain, and the reduced friction can have some performance advantages. Both types of lifters undergo wear in the course of normal use of the car. The wear is accelerated if dirt or acid contamination is present in the oil. In addition, hydraulic lifters can become clogged with **sludge,** or tarlike deposits, if the engine oil is not changed periodically.

On both types of systems, wear occurs as a result of friction between the camshaft and the lifter mechanisms. When this wear reaches a critical point, the valves do not open fully. At this point, there is a restriction on the passage of gases in or out of the cylinder, and the engine can no longer function with full efficiency. When this occurs, all lifters are usually replaced at once and it is necessary to replace the camshaft. A worn camshaft will wear out the new lifters prematurely. It may also be necessary to re-grind or replace the valve faces and seats in the cylinder head. In addition, the valve guides may need to be replaced.

Bearings

A bearing within the engine does exactly that: it bears a load and absorbs wear. Bearings support rotating metal shafts and allow them to turn freely within their support housings. Bearings also serve to hold parts in position while allowing them to turn freely. Shafts and housings must be made of strong materials, such as steel and cast iron, and will create excessive friction if they rub directly against each other.

For example, the crankshaft and camshaft each revolve within bearings. The bearings, known as *precision insert bearings*, can be replaced when they wear out. It is cheaper and more efficient to permit these bearings to wear than to allow the crankshaft or camshaft to become badly worn.

Figure 3–19 shows the bearings used to support a four-cylinder engine's crankshaft. The bearings that support the crankshaft are known as **main**

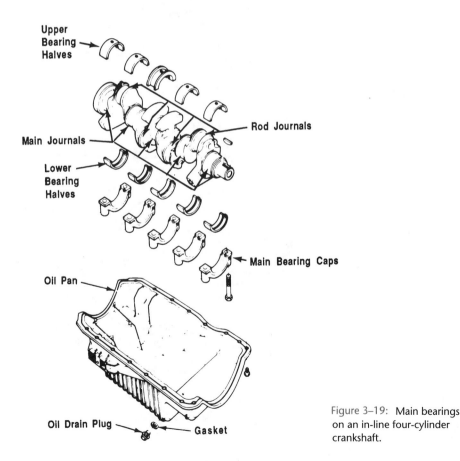

Figure 3–19: Main bearings on an in-line four-cylinder crankshaft.

bearings. These bearings allow the crankshaft to rotate within the engine while preventing the crankshaft from moving excessively along its axis. Other components such as pulleys, flywheels, or torque converters are fitted to the crankshaft. It is necessary to keep these components from moving forward and back too much along the axis of the crankshaft.

Insert bearings also are fitted into connecting rods. These bearings, known as *rod bearings*, allow connecting rods to rotate on the crankshaft (Figure 3–20). The offset portion of the crankshaft, known as the *rod journal*, rotates within the rod bearing. On the compression and exhaust strokes, the crankshaft pushes the piston upward. On the intake stroke, the piston is pulled downward by the crankshaft. On the power stroke, the piston pushes downward with great force against the crankshaft journal. The rod bearings absorb shock created by these up-and-down movements.

Figure 3–21 shows the bearing configuration for a camshaft in a six-cylinder engine. On this camshaft, there are four bearing surfaces in which

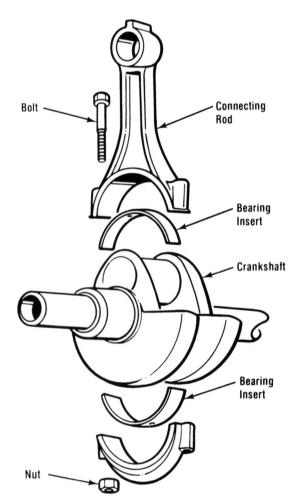

Bolt

Connecting Rod

Bearing Insert

Crankshaft

Bearing Insert

Nut

Figure 3–20: Rod bearing inserts are positioned between the connecting rod halves and the crankshaft rod journal.

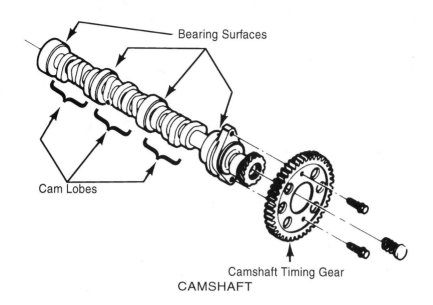

Bearing Surfaces

Cam Lobes

Camshaft Timing Gear

CAMSHAFT

Figure 3–21: **Four bearings support this camshaft from a six-cylinder engine**

the camshaft rides. These bearings allow the camshaft to rotate and allow the lobes to operate the valve lifters.

When the engine is running, oil is supplied to bearings through narrow passages with the block and head called *oil galleries*. The oil flushes away dirt and metal particles and reduces wear on the bearings. However, bearings are designed to wear out; their job is to protect more expensive parts. This wear usually keeps pace with the life of rings, pistons, and cylinder walls. As wear occurs, the gap increases between a bearing's inner surface and the part that rotates within it. In an engine, when bearing clearances, or gaps, increase, oil pressure goes down. As the oil pressure decreases, moving parts within the engine receive less oil. Thus, bearing and other frictional wear increase even faster. In any case, once the engine has been disassembled sufficiently to remove pistons, it is an accepted practice to replace the main and rod bearings.

Bearings will wear excessively if an engine is abused through overloading or if there is not enough oil pressure. The worst abuse a driver can give an engine is to "race" it (accelerate and decelerate rapidly out of gear) when it is cold. Most engine wear occurs during startup. The parts need to warm up and expand slightly to obtain their best and proper fit. Cold parts do not fit well and therefore are subject to more wear. The proper way to warm up an engine is to let it idle just 30 seconds or so after startup, then drive it away *gently*.

"Drag racing" other cars is another type of abuse of the engine and its bearings. Also, if a vehicle is used to pull a heavy trailer up steep grades, the bearings will wear at a faster rate. When main or connecting rod bearings are worn, the engine may make loud thumping or banging noises when it is started, accelerated, or driven up a hill.

Aside from avoiding abusive driving habits, you can best safeguard your engine by never letting it run low on oil.

This chapter has introduced you to the mechanical operation of your car's engine. You need to know still more about the engine and its systems before you can supervise or perform required maintenance and repair. In Chapter 4 you can learn about the essential functions of the fuel system.

1. Name the four strokes of the four-cycle engine.

2. Describe the function of each of the four strokes.

3. What are the common cylinder arrangements we find on today's cars?

4. What are the types of valvetrains used in engines?

5. What opens the valve? What closes the valve?

6. What are the ways in which the crankshaft can drive the camshaft?

7. What is the ratio between the crankshaft and camshaft? Why is this necessary?

8. What produces heat in the engine?

9. What is the most critical fluid for the engine? Why is it so important?

10. What supplies lubricating oil to the engine?

11. Why does an engine "wear"?

12. Describe the piston and the rings used in the engine.

13. What are the types of lifters?

14. What is the purpose of the bearings?

15. What is the purpose of piston rings?

16. What are the essential maintenance tasks for the engine?

THE FUEL SYSTEM

CHAPTER 4

After reading this chapter, the reader should be able to:

- List the major components of the fuel system.
- List and describe the properties of gasoline related to engine performance.
- Describe the operation of the fuel delivery system.
- Explain the systems and functions of the emission control system.
- List the maintenance items of the fuel system.
- List some common problems in the fuel system.
- Describe the basic differences between gasoline and diesel engines.

AN OVERVIEW

For safety and performance, it pays to understand and maintain your car's fuel system. Most of the money you spend in operating your car probably goes into fuel. Therefore, the concepts in this chapter can pay real dividends in economy as well. To begin building your knowledge, this overview presents a brief discussion of the parts of a fuel system (Figure 4–1).

Fuel Tank

The **fuel tank** stores the gasoline for use by the engine. Modern fuel tanks have several safety features to prevent unburned fuel vapors—one of the major sources of pollution—from entering the atmosphere or air you breathe. A screen traps contaminant particles before they enter the fuel lines.

Fuel Pump

On most cars, the **fuel pump** is operated electrically and located near or within the fuel tank. The fuel pump draws fuel from the tank and forces it through **fuel lines** to a fuel delivery system known as the fuel injection system. Where fuel lines run under the car, they are made of steel tubing. At the ends of the fuel line, where connections are made and flexibility is necessary, special synthetic rubber or nylon hoses designed for use with gasoline are used.

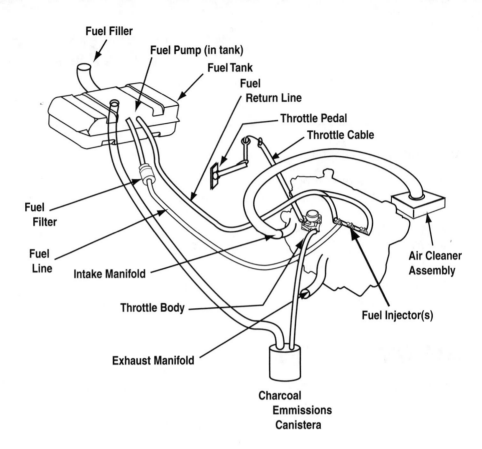

Figure 4–1: Major parts of a typical fuel system and their interconnections.

Fuel Filter

The function of the **fuel filter** is to remove contaminants, including dirt, rust particles, and water, from the fuel before it enters the fuel delivery system. The filter can be positioned at virtually any point in the fuel line. Most fuel filters are positioned either near the tank, along the frame, or in the engine compartment.

Air Cleaner Assembly

Just as there is a fuel filter in the fuel line, there is an **air filter** enclosed in an **air cleaner assembly.** The air filter traps and holds particles of dust carried by the incoming air. Thus, filtered air passes into the fuel delivery system. The air cleaner assembly also serves to muffle the noise of incoming air and may act as an antibackfire or flame-arresting device.

Fuel Delivery System

As air is drawn into the engine, it is mixed with a fine mist of fuel that is delivered by the fuel injection system. The amount of fuel delivered varies in response to both the pressure you put on the accelerator pedal and engine operating conditions.

At one time fuel was delivered to the engine by use of carburetors. These devices operated solely on mechanical and hydraulic (fluid) principles. In general, the greater the volume of air drawn through the carburetor and the greater the engine vacuum, the more fuel was delivered. Carburetors worked fairly well but could be mechanically complex. In addition, they were subject to wear and difficult to adjust precisely.

Today, fuel injection is controlled by a computer in response to the levels of oxygen in the exhaust of the engine. When oxygen levels change, either up or down, the computer signals the **fuel injectors** to increase or decrease fuel flow. By maintaining the air and fuel mixture at an optimal ratio, the best efficiency, the lowest pollution, and the most power can be obtained from the engine.

Electronic fuel injectors spray fuel into the intake manifold near the intake valve. Sensors on the accelerator linkage and in the engine, the exhaust system, transmission, and other parts of the vehicle provide input information for the computer. The computer provides output signals to increase or decrease the dwell, or "on" time, at which the injectors operate. Thus, the volume of fuel that is sprayed is adjusted to reflect driving conditions and air/fuel mixture requirements.

Intake Manifold

An **intake manifold** provides a series of hollow tubes, or passageways, through which air enters the cylinders. Gaskets are placed between the manifold and the cylinder head, and fasteners are used to secure the manifold tightly. If air leaks at the mating surfaces, an excessively lean mixture will be produced in one or more cylinders. The engine will run roughly. A simplified drawing of an intake manifold is shown in Figure 4–2.

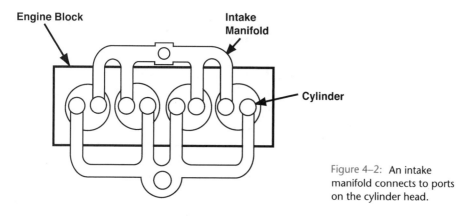

Figure 4–2: An intake manifold connects to ports on the cylinder head.

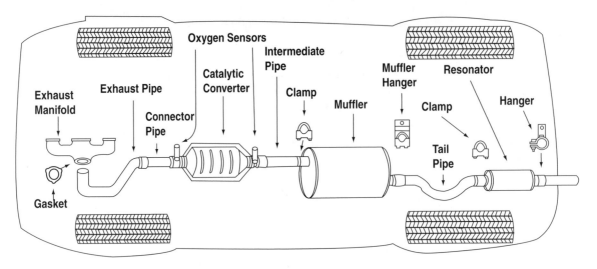

Figure 4–3: An exhaust system carries burned gases away from the engine.

Exhaust System

Exhaust gases pass out of the cylinder head into the exhaust system. The exhaust system, as shown in Figure 4–3, incorporates devices to substantially reduce the amount of pollutants from engine exhaust gasses. These include **oxygen sensors** and **catalytic converters.** The exhaust system also includes one or more **mufflers** to reduce the noise of the exhaust.

The basic purpose of the fuel system is to deliver air and fuel to the engine, provide varying air/fuel ratios to match engine operating conditions, and carry the exhaust gases away. Now it is time to take a look at the specifics of fuel systems.

GASOLINE

Just as all cars are not the same, all gasoline does not share the same performance qualities. To get the best performance and longest life from your engine, you should match the gasoline you buy to your car's needs. Gasoline is blended for different climates. Cars require unleaded fuel of a sufficient **octane rating** to prevent engine knocking and internal problems. Consult the owner's manual for the proper fuel to use. You need only use a gasoline that meets the manufacturer's recommended octane levels. Most cars will require the use of 87 or 88 octane gasoline, commonly referred to as *regular unleaded*. Higher altitudes may have 85 octane gasoline available. Another fuel widely used is *super unleaded*. This fuel is sometimes a blend of 90% regular unleaded and 10% **ethanol.** Ethanol may be blended into some or all of the gasoline available at the filling station depending on the emission requirements of the area. Cars may be operated with high-octane fuels and may see some slight mileage benefits, but is not necessary in most cases. Few cars will require the use of high-octane *premium unleaded* gasoline.

Additives

Gasoline has a variety of additives to enhance performance and fuel quality, and to maintain standards during distribution. These additives are extremely expensive and you don't have to worry about them being added in excess amounts. Detergents are added to eliminate or remove fuel system deposits, antioxidants are added to help prevent gum formation of stored gasoline, and corrosion inhibitors are added to prevent corrosion in the fuel system. There are many other additives used and Figure 4–4 shows some of the most common ones. Since gasoline is very carefully engineered and blended, it is not generally recommended that you use aftermarket additives in your gasoline. Using aftermarket additives can alter the composition and properties of the gasoline. While the additive may claim to fix one problem, it may cause another one such as valve deposits or other unwanted effects.

Gasoline Additives	
Additive	**Purpose**
Detergents/deposit-control additives*	Eliminate or remove fuel system deposits
Anti-icers	Prevent fuel-line freeze up
Fluidizer oils	Used with deposit control additives to control intake valve deposits.
Corrosion inhibitors	To minimize fuel system corrosion
Antioxidants	To minimize gum formation of stored gasoline
Metal deactivators	To minimize the effect of metal-based components that may occur in gasoline
Lead replacement additives	To minimize exhaust valve seat recession

* Deposit control additives can also control/reduce intake valve deposits.

Figure 4–4: Various gasoline additives are needed to enhance performance and fuel quality, and maintain standards during distribution.

Volatility

Volatility refers to gasoline's ability to vaporize. Gasoline can be refined chemically and blended to have different degrees of volatility. Thus, for different climates or seasons, fuels are produced that will vaporize properly in the temperatures to be encountered. In cold climates, a gasoline that is not volatile enough will not vaporize properly and the engine may run poorly.

Unleaded Fuel

The majority of cars on the road today use unleaded gasoline as their fuel. Since the mid-1970s, catalytic converters have been used in the exhaust systems of passenger cars. Unleaded fuel must be used in cars with catalytic converters. These vehicles have narrow fuel tank filler necks that will accommodate only the narrower nozzles of unleaded gasoline pumps. Catalytic converters and other parts of the exhaust system are described later in this chapter.

Octane Rating and Detonation

Detonation in a gasoline engine is a form of abnormal combustion commonly referred to as **knock** or **ping.** When gasoline detonates, a portion of the air/fuel mixture explodes before the piston reaches the top of its compression stroke. This causes extreme hammering pressure on the piston and rod bearings. Detonation that continues for more than a few seconds can cause severe engine damage.

Detonation occurs more frequently in older engines in which deposits have built up on the inside of the cylinder. These deposits increase the **compression ratio,** the degree to which an intake charge is compressed. With greater compression, more heat is built up, and the fuel is more likely to detonate.

As a general rule, if detonation can be heard during normal operation, you should try to use a higher-octane fuel. The lower the octane rating number, the more easily a gasoline detonates. Regular gasoline, leaded or unleaded, has a low octane rating. Super unleaded gasoline has a higher octane rating and higher resistance to detonation. The octane rating is posted on a small sticker attached to the gas pump. Today's gasoline octane ratings range from a low of approximately 85 to a high of approximately 93 in some states. A different brand of gasoline with the same octane rating also may reduce detonation in your engine.

The octane ratings found on gas pumps today are an average of the engine method, sometimes called the motor method, and the research method of determining octane ratings. The formula used to determine the average octane is shown on the pump (Figure 4–5). In the engine or motor

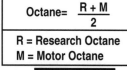

$$Octane = \frac{R + M}{2}$$

R = Research Octane
M = Motor Octane

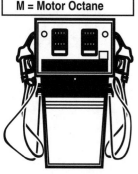

Figure 4–5: The octane rating for gasoline found on gas pumps today is the average of the research octane rating and the motor octane rating.

method, the fuel is tested in various engines and an octane rating determined. The research method tests the fuel under laboratory conditions and determines an octane rating. Since neither of these results will exactly match your engine or driving conditions, the best that can be done is to average the two methods to give the consumer a somewhat consistent number that can be used to show octane.

Alcohol Fuels

Many retail gasoline-fueling stations carry alcohol fuels containing ethanol and gasoline, usually a blend of 90% regular unleaded and 10% ethanol. Ethanol is an alcohol fuel made mainly from corn and is a renewable resource. Fuel delivery systems have been designed to use ethanol fuels. Use of these fuels will not damage fuel systems.

Alternative Fuels

There has been much research in the development and use of alternative fuels for cars. An alternative fuel is classified as any power source for an automobile that is not traditional gasoline or diesel fuel. Different fuels have been tried, but many have shortcomings that outweigh their benefits. Some hold great promise as an alternative to gasoline. The most promising fuels are E85 and compressed natural gas (CNG). E85 is a blend of 85% ethanol and 15% unleaded gasoline. This fuel would rely heavily on the production of ethanol, which is a renewable resource. Engines and fuel systems need few modifications in order to use E85 and may also use regular unleaded gasoline if E85 is unavailable. CNG is stored in special tanks at pressures approaching 3000 pounds per square inch (psi). Quite a number of modifications must be made to the fuel system in order to burn CNG because it is delivered to the fuel system as a high pressure vapor. Both E85 and CNG burn cleaner than gasoline and have many pollution-reducing benefits.

FUEL TANK

On earlier-model cars, gas tanks were vented. Unburned fuel vapors escaped directly into the atmosphere to cause air pollution. Modern fuel tanks have a system of **evaporative emission controls** designed to contain most of the vapors. A pressure-vented cap allows vapors to escape only if the build-up of pressure becomes great enough to damage the tank. In addition, the modern tank has a bulge at the top where vapors collect. It is important to fill the tank to a proper level and not overfill it. When the fuel nozzle shuts off, top off the tank by allowing the fuel nozzle to click off only a second or third time. Overfilling the fuel tank can flood the charcoal canister, and the excess fuel can cause a build up of pressure that can cause fuel spray and a potential fire. When the engine starts, the fuel vapors will travel through a vapor line and be drawn into the intake charge. They are then burned in the cylinders along with the normal air/fuel mixture.

The computer can also test to see if the fuel tank has a vapor leak. It does this by applying a slight vacuum to the tank and testing to see if it holds this vacuum. If the pressure increases in the tank, the computer may turn on a **MIL,** or malfunction indicator lamp, on the dash to alert the driver to a concern in one of the engine systems. Failure to properly tighten the gas cap after fueling a car can lead to a vapor leak.

FUEL PUMP

The fuel pump in your car is electrical (Figure 4–6). Its job is to move liquid gasoline from the tank to the fuel injection system. The pump is driven by a small electric motor and runs when the ignition switch is in the ON position. Most electric fuel pumps are located near or in the fuel tank. This location isolates the pump from the heat of the engine, which can shorten the life of the pump.

All modern fuel pumps are sealed units. They cannot be disassembled for repair. Instead, when a pump fails, you must replace the entire unit (see Chapter 12). If the pump is located in the fuel tank, then the fuel tank must be removed from the vehicle to replace the pump.

FUEL FILTER

A fuel filter contains a fine paper or metal mesh to remove impurities from the fuel. The filter is positioned between the fuel tank and the fuel delivery system. The fuel pump causes gasoline to flow through the filter.

The fuel filter element should be replaced approximately once every two to four years or every 30,000 to 60,000 miles, whichever comes first. You can ask to have the element replaced at tune-up time, or you can change it yourself. Follow the maintenance recommendations in your owner's manual. Should you notice that high-speed performance is lacking, replacing the fuel filter is a good place to start because a restricted filter will not allow the proper amount of fuel to flow to the engine. Details on changing a fuel filter are presented in Chapter 11. Figure 4–7 shows one type of replaceable fuel filter.

AIR CLEANER

The function of the air cleaner unit is to deliver clean, filtered air through the fuel delivery system and into the engine. Air enters the engine through the filter housing. The filter housing may be located on top of a throttle body or in a separate box connected to the fuel delivery system by hoses.

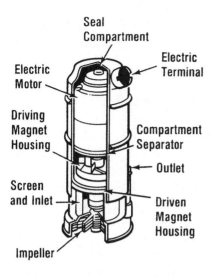

Figure 4–6: Electrical fuel pumps, generally located in the fuel tank, deliver gasoline by means of a rotary impeller.

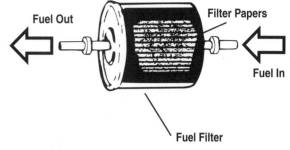

Figure 4–7: An in-line fuel filter traps dirt and small amounts of moisture before they can reach the fuel injectors.

On many new cars, a flexible hose connects the filter housing to an intake opening in the front part of the engine compartment. Cool air enters through the air horn. A replaceable filter takes up most of the space in the air filter housing (Figure 4–8). This filter permits the free flow of air, while it traps dust and other foreign matter. Most air filters are made of

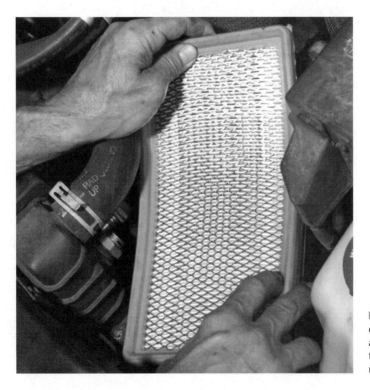

Figure 4–8: The air filter traps dust particles and allows clean air to pass through the throttle body and intake manifold.

pleated paper. Some are made of plastic foam or of a combination of foam and paper. The air filter should be replaced at regular intervals, usually every year or 15,000 to 20,000 miles.

FUEL DELIVERY SYSTEM

To cope with different driving conditions and driver demands, a fuel delivery system must provide the correct **ratio,** or proportion, of air and fuel in the intake mixture. The ratio is expressed as a set of numbers, such as 14.7:1. The number before the colon expresses the relative amount of air by weight measured in pounds, and the number after the colon, the relative amount of gasoline by weight also measured in pounds. The ideal mixture is approximately 14.7:1. This mixture provides the optimum amount of power, and tolerable exhaust emissions. This ideal ratio is known as **stoichiometric** and indicates the most efficient ratio for a given fuel. This ratio will vary with different gasoline fuels and the vehicle's computer automatically makes the necessary adjustment continuously.

However the stoichiometric ratio might not always be the desired or the most effective ratio (Figure 4–9). For acceleration and cold-engine operation, the fuel delivery system must provide a **rich** mixture, one with proportionally less air-to-fuel content. Under these conditions, an air/fuel ratio of as rich as 8:1 may be delivered. Or during acceleration the air/fuel ratios might be around 10:1. During deceleration the air/fuel mixture may go very **lean,** to as much as 20:1.

Carburetor

Older cars use a mechanical device, or carburetor, to mix air and fuel for delivery to the engine. A carburetor delivers fuel in proportion to the

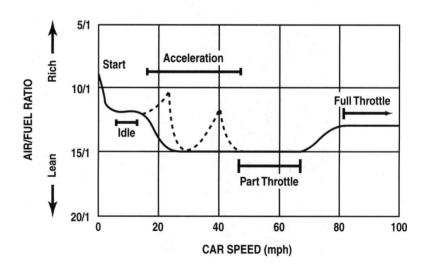

Figure 4–9: Air/fuel ratios will vary according to speed, load, and throttle opening.

amount of air flowing through it. As you press on the accelerator, the throttle valve opens wider to draw more air through the carburetor. The carburetor provides richer or leaner mixtures depending on a number of factors: engine speed, load, temperature, and throttle position. To meet complicated demands, a carburetor is a highly intricate device, with many internal passages and parts.

An automobile carburetor is designed with a **venturi** chamber. The venturi is simply a narrowed portion in the air passage. Air moving through the throat of the venturi speeds up as it travels through the narrowed passageway. The increased air speed through the venturi creates a low-pressure area.

Atmospheric pressure pushes down on a fuel reservoir within the carburetor known as the **float bowl.** Fuel is forced through a tube that extends into the low-pressure area of the venturi. Thus, fuel sprays through the end of the tube into the air system.

Electronic Fuel-Injection

At one time, all gasoline-engine vehicles were equipped with carburetors. Now fuel injection systems are the norm as they have many advantages over carburetors, especially with regard to fuel economy and lower exhaust emissions. Carburetors have difficulty providing the correct air/fuel mixture when the accelerator is released or pressed suddenly. Thus, the air/fuel ratio becomes incorrect for the immediate operating condition. This situation contributes to inefficient engine operation, lower fuel economy, and excessive exhaust emissions. Electronic fuel injection systems react much more rapidly by using electrical signals from sensors and computer processing units to control the air and fuel mixture.

A fuel injection system looks like a complicated method of fuel delivery, but really is rather simple (Figure 4–10). Sensors are mounted on the engine, in the exhaust manifold, and at other locations on the vehicle. These sensors provide information to the electronic control unit or computer. This computer is referred to as the powertrain control module (PCM). The PCM processes these inputs and provides output signals that cause the injectors to spray the proper amount of fuel. The injectors can be operated at more frequent or less frequent intervals, depending on a number of factors. The most common factors that affect fuel delivery are engine load, engine rpm, vehicle speed, oxygen levels in the exhaust, amount of air flowing into the engine, engine temperature, and air temperature conditions, as well as driver demands through the throttle pedal.

All fuel injection systems work in basically the same way; details of each car manufacturer's systems differ, although they are very similar in general terms. Two types of fuel injection are used:

1. Throttle-body injection (TBI)
2. Multiport fuel injection (so-called port-type injection)

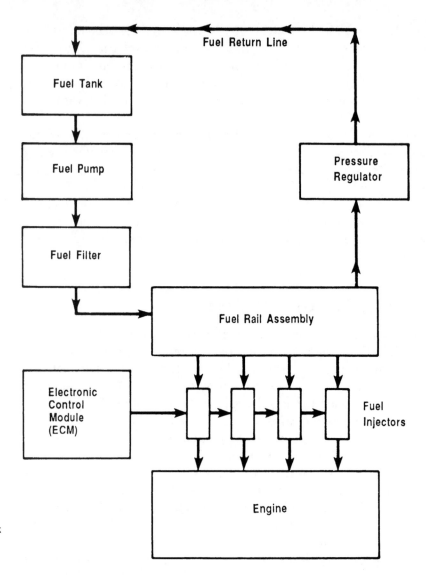

Figure 4–10: Simplified block diagram of a fuel injection system.

Throttle-Body Injection   A TBI system sprays fuel into a housing that contains a throttle valve, just as with the lower part of a carburetor. This assembly containing the throttle valve is called the **throttle body.** One or more fuel injectors are located above the throttle valve. The throttle body is mounted at the upper end of the intake manifold, in the same location that a carburetor would be mounted.

Multiport Fuel Injection   A port-type fuel injection system sprays fuel into each intake port, just in front of the intake valve to provide more efficient fuel delivery than a TBI system (Figure 4–11). The intake manifold is equipped with a throttle body to control the amount of air entering the engine. However, individual injectors must be provided for each cylinder (Figure 4–12). Thus, a port-type fuel injection system is typically more

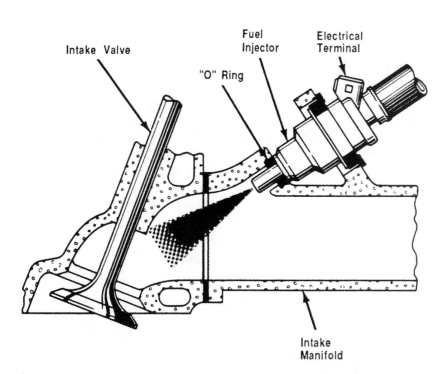

Figure 4–11: A port-type fuel injector sprays fuel directly into the intake manifold in front of an intake valve.

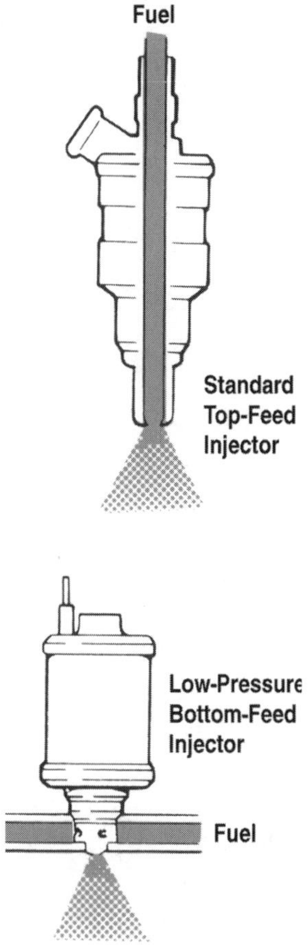

Figure 4–12: Either top-feed or bottom-feed injectors are used on port-type fuel injection systems.

complicated and expensive than a TBI system. But because each cylinder has its own injector, fuel delivery is more accurate and the fuel is mixed more precisely.

As you probably have guessed by now, you would be well advised to leave servicing of fuel injection systems to trained specialists only. Many of these systems operate with 40 psi or more on the fuel and require special procedures, equipment, and components.

Turbochargers and Superchargers

On most cars, air is drawn in through atmospheric pressure, as explained earlier. However, a **turbocharger** or supercharger can be added to force air in at higher than atmospheric pressures. Air contains oxygen, which makes burning possible. Thus, the greater the amount of air forced into the cylinder, the more oxygen will be present. If more fuel also is added, the burning process can produce much higher than normal pressures. Higher pressures exerted against the pistons produce more engine torque and power output.

The turbocharger contains two fan wheels connected by a metal shaft. One of the fan wheels, known as the *turbine*, is located in the exhaust manifold. The pressure of exhaust gases flowing out of the engine turns the turbine at high speeds. The other fan wheel, known as the *compressor*, is located in the intake system and turns with the turbine. The compressor wheel forces air under pressure into the intake manifold (Figure 4–13). Some turbocharged vehicles will have an **intercooler** between the turbocharger and the intake manifold. It looks like a large radiator and functions like one. When the air is compressed it heats up. The intercooler will help to cool the compressed air before it is delivered to the engine.

The principle behind a **supercharger** is the same as that of a turbocharger. However, a supercharger contains only a compressor wheel. The supercharger is mounted on the engine and driven by a belt, as with other engine accessories.

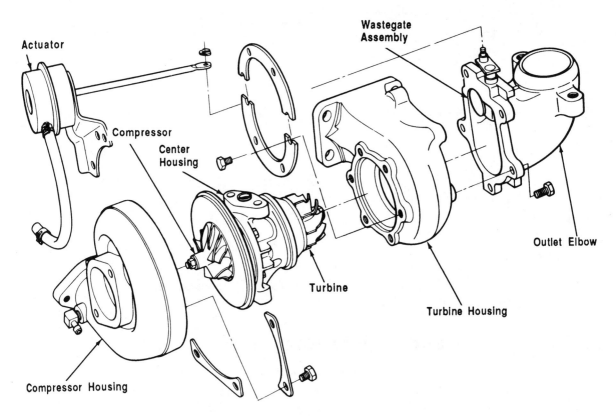

Figure 4–13: A turbocharger uses exhaust gas flow to force air into the intake manifold.

EMISSIONS CONTROL

Federal **Corporate Average Fuel Economy (CAFE) standards** mandated increasing average gas mileage standards beginning in 1979. In addition, the 1968 Federal Clean Air Act was passed to help reduce automotive emissions. Some of the continuing legislation related to the Clean Air Act required that parts with a direct effect on performance and emissions levels of the engine must be able to last at least 50,000 miles without servicing. Regular maintenance must still be performed, but the vehicle should require no other service work during this time.

Sources of air pollution from vehicles are illustrated in Figure 4–14. These sources include:

- Exhaust emissions
- Fuel system vapors
- Crankcase fumes

If the burning of hydrocarbon fuels inside your engine were complete, the products would be relatively harmless—water and **carbon dioxide (CO_2),** the gas exhaled in human breathing and absorbed by plants. However, for a number of complex reasons, combustion is never 100% complete. Pollutants from exhaust gases include unburned fuel, **carbon monoxide (CO),** and **oxides of nitrogen (NO_X).** Modern cars include many devices to help control the burning process, to make it more efficient, and to clean up the exhaust gases.

Earlier in this chapter parts of the fuel tank and related vapor control systems that prevent fuel vapors from escaping into the air were identified. In addition, catalytic converters and fuel injection systems help to reduce exhaust emissions. The specific systems you'll find in your car's total emissions control system include:

- Evaporative emission controls
- Catalytic converter
- Positive crankcase ventilation (PCV) systems

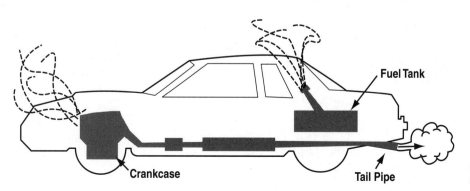

Figure 4–14: Sources of air pollution generated by passenger cars.

- Exhaust gas recirculation (EGR)
- Air injection

Specific emission control equipment as shown in Figure 4–15 is included as part of an overall, interrelated group of vehicle systems.

OBD-II

Computer systems are in most cars and light trucks on the road today. Manufacturers started using electronics to control engine functions in the 1970s and early 1980s primarily to help meet EPA emission standards. With a system on the vehicle that could control engine function, it only made sense to use the same system to alert the driver of a problem in the system. Thus began on-board diagnostics (OBD). This system could also be used to help diagnose engine problems. Through the late 1980s and the early 1990s, the OBD systems have become more sophisticated.

The Environmental Protection Agency has the responsibility for requiring manufacturers to build vehicles that must meet increasingly stiff emissions standards. The manufacturers must further maintain the emission standards of the cars for the useful life of the vehicle. So they developed a new system: **on-board diagnostics II (OBD-II).** Cars with the OBD-II system can detect component failures. Computer systems with OBD-II are capable of monitoring a number of systems and strategies used to maintain low emissions. Some of the things the OBD-II system can monitor are:

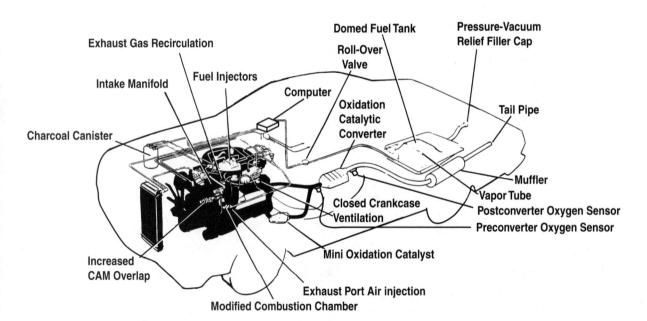

Figure 4–15: Key parts of an emissions control system and their linkage.

- Catalyst efficiency
- Engine misfire
- Fuel system
- Heated exhaust gas oxygen
- Exhaust gas recirculation
- Evaporative system
- Component function and effectiveness

All cars built since January 1, 1996, and sold in North America have OBD-II systems. Some manufacturers started incorporating OBD-II in various models as early as 1994. The main goal of OBD-II was to detect when engine or system wear or when component failure caused exhaust emissions to increase by 50% or more. OBD-II also calls for standard service procedures without the use of dedicated and specialized tools. The system will alert the driver through the MIL (sometimes called the "Check Engine Light"). If the MIL flashes occasionally, then an intermittent problem may be indicated. If the MIL is on continuously, then a failure has occurred somewhere in the system and the vehicle should be taken to a competent technician for diagnosis and repair.

Evaporative Emission Controls

The evaporative system consists of a special filler neck to help prevent overfilling of the fuel tank, a dome in the top of the fuel tank, a pressure/vacuum cap for the fuel tank, a vapor separator in the top of the tank, and a charcoal canister. Basically, fuel vapors from the tank are held in the tank and the charcoal canister. When the engine starts, the vapors are drawn into the intake manifold from the charcoal canister. Cars with OBD-II will monitor the evaporative system and alert the driver if there is a fault detected in the system.

Catalytic Converter

As already mentioned, the main pollutants in the exhaust gases are by-products of incomplete combustion: hydrocarbons (HC), carbon monoxide (CO), and oxides of nitrogen (NO_X). To help clean up the exhaust gases from a gasoline engine, a catalytic converter is included as part of the exhaust system. The catalytic converter contains special metals (platinum, palladium, and rhodium) that help to convert the HC, CO, and NO_X to water, carbon dioxide, and nitrogen (Figure 4–16). Carbon dioxide is the gas exhaled when you breathe; nitrogen gas makes up 78% of the air you breathe. Thus, a catalytic converter converts harmful gases into relatively harmless, more natural products. The catalytic converter operates solely from the heat of the exhaust gases and requires no special maintenance. However, an engine with a miss or a dead cylinder, or otherwise running

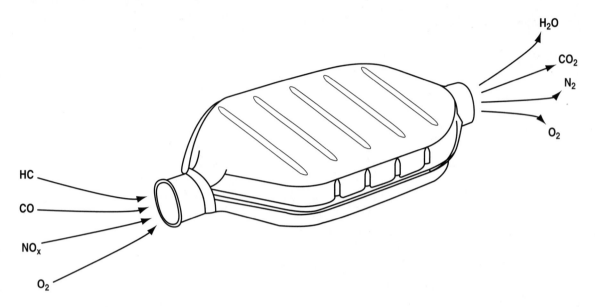

Figure 4–16: The catalytic converter contains catalysts to clean up an engine's emissions before they leave the exhaust system.

too rich can damage the catalytic converter. If there are large amounts of HC entering the converter, then excessive amounts of heat will be generated, possibly damaging the converter permanently. A properly running engine is the best way to maintain your catalytic converter. Vehicles with OBD-II monitor the functioning of the catalytic converter with exhaust oxygen sensors on each end of the converter.

Positive Crankcase Ventilation System

Blowby fumes from combustion and oil fumes combine in the crankcase area above the oil supply to produce vapors high in hydrocarbons. To keep these vapors from entering the atmosphere, a **positive crankcase ventilation (PCV)** valve system is used (Figure 4–17). The PCV system mixes the vapors with air and reintroduces them into the cylinders, where they are burned along with the intake charge. The idea is to burn up as many pollutants as possible.

Exhaust Gas Recirculation

To burn fuel in the intake charge as efficiently as possible, modern engines run at high temperatures. These high temperatures produce oxides of nitrogen, another air pollutant, in the exhaust gases. An exhaust gas recirculation (EGR) valve is used to reduce these emissions (Figure 4–18). The

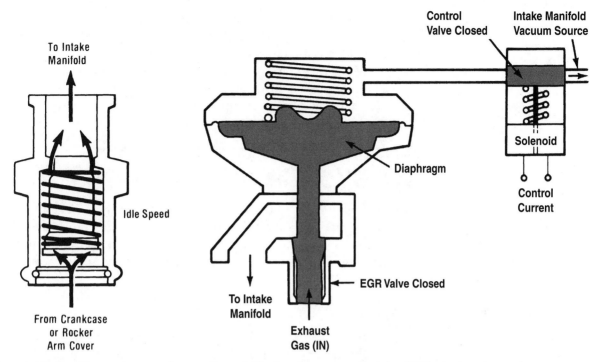

Figure 4–17: A positive crankcase ventilation (PCV) system recycles blowby gases from the crankcase into the intake charge.

Figure 4–18: An exhaust gas recirculation (EGR) valve introduces exhaust gases into the intake charge to lower burning temperatures.

EGR valve recirculates small amounts of exhaust gases into the engine cylinders to lower combustion temperatures. Even though the temperature of the exhaust gases will raise the air/fuel mixture temperature, the addition of the exhaust gases will lower the peak combustion temperature. Because the exhaust gases are mostly inert, they already have combusted; they have a quenching effect on combustion. This quenching effect serves to lower the temperature and pressure in the combustion chamber. Imagine sprinkling hot ashes over a campfire—the fire will be diminished proportional to the amount of inert material. Therefore, the lower the combustion temperature and pressure, the lower the amount of NO_X produced. Another benefit of lower combustion temperatures is the reduction of knocking or pinging in the engine. It's almost like a free octane boost for your car.

Air Injection

The hot exhaust gases contain unburned fuel. However, the oxygen they need to burn further has been used up in the combustion process. To help convert harmful exhaust substances such as carbon monoxide into relatively harmless carbon dioxide, oxygen is needed. To provide additional oxygen in the exhaust system, an auxiliary **air pump,** belt-driven or electric, may be used. Thus, additional burning takes place in the exhaust manifold. The air pump also is used to supply additional air to an **oxidizing catalytic converter.** Many cars may not need an air pump as fuel injection can precisely control the air/fuel mix, and there is sufficient oxygen present in the exhaust gas so that additional or supplemental oxygen is not needed.

SAFETY PRECAUTIONS FOR FUEL SYSTEM MAINTENANCE

Gasoline is highly flammable and gives off explosive vapors. A gasoline fire can produce severe burns and other injuries. In a vaporized form, one gallon of gasoline can produce the same explosive effect as 14 sticks of dynamite. That much explosive power can cause severe physical injuries, or worse.

CAUTION

Always have a fully charged fire extinguisher nearby that can extinguish class B (flammable liquid) fires before you work on any part of the fuel system.

CAUTION

Care must be exercised in storing and handling gasoline, especially in hot weather. Avoid getting gasoline on your skin or breathing its vapors. Don't use gasoline to clean parts. Don't attempt to siphon gasoline from a car's tank or other container by mouth because gasoline is poisonous and can cause serious injury if it is swallowed.

Backfires

A running engine may backfire through the throttle body at any time, without warning. In effect, a ball of fire explodes out of the top of the throttle body. Keep in mind that the danger is greatest when the air filter housing is off the throttle body. The air filter housing usually will contain and limit a small backfire if it occurs.

To avoid injuring your face and hands, do not look down the throat of a throttle body or hold your hands over the throttle body throat while the engine is running. Backfires through the throttle body can cause severe burns. Always shut off the engine before you look down the throttle body.

Engine Fires

Aside from protecting your face and hands, there is another danger from fires that begin in the throttle body you need to be alert to: an engine fire. After a backfire, flames may move from the throttle body toward the rest of the engine. In such circumstances, the first rule is: DON'T PANIC!

Extinguishing a fire promptly is your first priority. If the fire hasn't spread from the throttle body, you may be able to suck the flames back into the engine by pressing the accelerator while cranking the engine. However, if you encounter a fuel fire spreading from the throttle body, you can try to smother it with a blanket or jacket or use a class B fire extinguisher. While fire retardant material from the extinguisher may clog a throttle body and contaminate other components, you need to *put the fire out first* and worry about other problems afterward.

Fuel Odors

Suspect a fuel system problem—and a potential fire—if you smell gasoline fumes when the engine is running. If you notice strong gasoline fumes, pull over and turn off the engine. Open the hood immediately and look for leaks in the hoses leading to the fuel delivery system. Make sure these are repaired before you continue driving. As discussed in the following section, the checking and replacement of hoses is one of the main preventive maintenance steps that you should take to protect your car's fuel system.

FUEL SYSTEM MAINTENANCE

The fuel system is one place where your attention and periodic action can make a big difference in the economy, reliability, and emission levels of your car. Attention to just a few details can help assure trouble-free operation and savings.

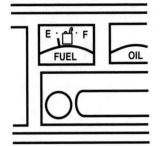

Figure 4–19: Fill the tank when the level drops to one-quarter full.

Fuel Tank Level

One of the important things you can do to keep your car running smoothly is to pay attention to your fuel gauge (Figure 4-19). It is not a good practice to let your car run out of gas. You should try to refill the gas tank soon after the gauge falls below the one-quarter level. Many owners' manuals will recommend that the fuel tank be filled when the level drops to the one-quarter level.

The fuel in most fuel injection systems is constantly circulated from the tank to the fuel injectors and back again to the tank. This action helps to keep cool fuel supplied for the fuel injectors at all times and minimizes the opportunity for the fuel to be heated while in the engine compartment. There almost always will be some water from condensation and dust or other impurities in the fuel. The lower you let the level of fuel in the tank become, the greater the probability that water or contaminants will be drawn into the fuel delivery system through the fuel line.

Hoses and Lines

The special synthetic rubber hoses in the fuel system should be inspected regularly. Fuel hoses connect the fuel tank to the metal lines (under the car) and the metal lines to the fuel injectors (in the engine compartment). Fuel hoses are also used on both ends of in-line fuel filters. Have any hose or line replaced at the first sign of cracking, crumbling, or leaks. Be sure all of the hoses and lines are in good condition and not leaking. Fuel leaks can lead to engine or vehicle fires.

Vacuum Hoses

Under the hood you should find a decal showing the routing of the vacuum hoses related to the emission control systems of the vehicle (Figure 4–20). This is called the **vehicle emission control information** decal (VECI). The information found on the VECI includes the following:

- Engine family number and displacement
- Manufacturer identification and trademark
- Tune-up information
- Emissions standards information
- Spark plug information
- Certification standards information
- OBD-II information.

Take a few minutes to look carefully at the small-diameter vacuum hoses in the engine compartment. These hoses are part of the emission control systems and may be included in other systems, such as windshield wipers and cruise control. Examine them for cracks, splits, or disconnections.

VEHICLE EMISSION CONTROL INFORMATION

	ENGINE FAMILY	EFN 2.GYBT2EA	OBD II CERTIFIED
A motor co, inc.	DISPLACEMENT	2.8 L	
	THIS VEHICLE CONFORMS TO THE U.S. EPA AND STATE OF CALIFORNIA REGULATIONS APPLICABLE TO 1997 MODEL YEAR NEW TLEV PASSENGER CARS.		

REFER TO SERVICE MANUAL FOR ADDITIONAL INFORMATION
TUNE UP CONDITIONS: NORMAL OPERATING ENGINE TEMPERATURE
ACCESSORIES OFF, COOLING FAN OFF, TRANSMISSION IN NEUTRAL.

EXHAUST EMISSION STANDARDS	STANDARD CATEGORY
CERTIFICATION	TLEV
IN-USE	TLEV INTERMEDIATE

| SPARK PLUG TYPE NGK BPREES-1F .GAP 1.1mn | CATALYST | EFN 2.8VBT2EA |

1. Engine family number and displacement

2. Manufacturer identification and trademark

3. Tune-up information

4. Emissions standards information

5. Spark plug information

6. Certification standards and information

7. OBD-II information

Figure 4–20: A sample VECI decal. Cars certified with this system will illuminate a dashboard "Check Engine" light when vehicle emissions exceed standards by 50%. These emissions can be caused by wear, malfunction, or defect. Your technician can use the OBD-II system to help diagnose and repair your car.

Vacuum hoses are used to control many systems that can affect engine operation. In addition, a disconnected vacuum hose creates a vacuum leak that can cause the air/fuel mixture to become lean. The computer can compensate for these small leaks and you might not even notice a problem, so checking them is one way to keep your car running well.

If only the end of a hose is cracked or expanded, and the hose is long enough, you can make a quick and simple repair by just snipping off the bad end of the hose with a pair of scissors or side cutters and pushing the remaining hose firmly onto the connection.

If the entire hose is cracked or brittle, vacuum tubing can be purchased by the foot or in six-foot lengths that can be cut to size. You can replace hoses and connectors yourself if you take the time to do the job carefully, *one hose at a time*. If you connect two or more vacuum hoses incorrectly, running problems or engine damage may result.

Further maintenance on the fuel system includes changing the air and fuel filters. These jobs are covered in Chapter 11.

FUEL SYSTEM PROBLEMS

Fuel system problems may be hard to identify because the same symptoms may result from a number of causes—even from causes not related to the fuel system. Some of the more common problems you may encounter—and troubleshooting techniques for dealing with them—are described in this section. If you encounter any of the problems described here, remember that your best approach is to think through the logical sequence of moving fuel from the gasoline tank, through the engine, and to and through the exhaust system.

Engine Cranks But Will Not Run

If your car won't start, think about what happened. Did the engine "crank," or turn over quickly and easily when you turned the ignition key? If so, your battery and starter system are all right. (Electrical system problems are discussed in Chapter 5.) Don't overlook the obvious. Before you go on to make more elaborate checks, look at the fuel gauge.

Check for Flooding If there is a strong odor of gasoline, the intake manifold may be flooded, which means that liquid gasoline is present instead of vaporized gasoline. Flooding usually occurs when the pedal is depressed on a car with fuel injection during cranking. The easiest and safest way to handle the situation is simply to turn off the ignition and wait for 10 or 15 minutes. The surplus fuel usually will evaporate during this time. When you try to start the car again, press the accelerator and hold it down to the floor. Don't pump the accelerator. Turn the ignition key and crank the engine until it starts, then release the accelerator pedal.

Check for Fuel Delivery If the car still won't start, listen to determine if the fuel pump is running when you turn the key ON. The fuel pump will run for a few seconds and then shut off. To listen again, just turn the key OFF, wait a few seconds, and then turn it back ON. If you have run out of fuel, fill the tank with some fuel and perform this procedure four to six times, and the fuel pump will deliver fuel to the injectors in the engine compartment. If the fuel pump is not running, then consult your owner's manual. Some cars have a fuel shutoff switch that will shut off power to the fuel pump in the event of an accident or vehicle rollover. If your car has one of these, simply locate the switch and press down on the red button of the switch to reset it. Sometimes a sharp jolt or severe pothole can trip this switch inadvertently.

The following checks should be made only while the engine is *off*. Locate the tail pipe of the car and smell for fuel. If you smell gasoline, then it is likely that the engine is at least getting fuel.

If you have TBI, then remove the air cleaner and crank the engine carefully looking at the bottom of the fuel injector. There should be a fine spray of fuel from the injector. It may help to have a flashlight to see this spray of fuel.

If your car has port fuel injection, then the fuel spray is not observable. However, you may still be able to hear the fuel injectors as they pulse to spray fuel. Locate a fuel injector and use a mechanic's stethoscope to listen for a clicking sound as the fuel injector pulses to deliver fuel (Figure 4–21). Place the tip of the stethoscope on the injector and as close to its axis as possible. Then crank the engine and listen for fuel injector clicking. Be sure to avoid moving and hot parts under the hood of the car and follow all the safety rules.

A sure way to test for fuel in the system is to use a fuel pressure gauge. This is a test best left to a trained professional because the fuel system operates under high pressure and even a pinhole leak can cause a dangerous situation and even a fire.

If the engine cranks but will not run, and the fuel system seems to be working properly, there may be an ignition system problem. If you can't verify fuel delivery, then get the car to a dealership service department or to a garage that specializes in the type of fuel injection system found on your car. The ignition system is discussed in Chapter 5.

Figure 4–21: **A mechanic's stethoscope is used to locate the source of engine and other noises.**

Engine Stalls Frequently

If your car stalls frequently, be aware that you may have ignition problems. These are covered in Chapter 5. However, there also is a chance that you have a fuel system problem. The fuel pump may not be delivering a strong, steady stream of fuel to the carburetor or fuel injection system. There can be several reasons for this condition. For example, the line leading from the tank to the fuel pump may be blocked, leaking fuel, or sucking air. To determine if the fuel pump is delivering enough fuel calls for the services of a trained technician.

Under normal driving conditions, a fuel pump should last 100,000 to 200,000 miles, often more. You will usually notice symptoms when the pump is about to wear out. These include hard starting, rough running, or a loss of power when going up hills. At the first signs of trouble, you should get your pump checked for both pressure and volume output.

Engine Runs Rough

If the engine is sputtering or "jerking" during idling or when you apply power, suspect loss of compression and consider the diagnostic tests described in Chapter 3. Also consider some fuel system possibilities. Run the same checks described in connection with stalling problems. Stalling and rough operation may stem from the same causes.

Consider Emission Control Equipment   Today, emission control systems are an integral part of your car. These systems and devices control and reduce specific pollutants and also are directly responsible for the continued proper operation of the fuel and ignition systems.

Thus, problems with emission control equipment can affect the way the engine runs. For example, if the PCV system is plugged, oil can be rerouted into the fuel delivery system intake. If the EGR valve malfunctions, the idle may be rough and may be serious enough for the engine to stall. This condition may also cause the engine to produce high emission levels.

These are examples of simple problems. Be aware that very complicated problems can be caused by the complex relationships between mechanical engine components and fuel, ignition, and emissions control systems.

Engine Idles Improperly

Idling too quickly or too slowly usually indicates a fuel system problem, emission system malfunction, or intake manifold problem. Idle speed is constantly checked and adjusted by the computer and is not adjustable. When the engine is cold, idle speed will be greater than normal. Only after the engine has reached full operating temperature will the engine rpm decrease to a "warm" idle. If the idle speed doesn't sound right or has changed markedly over a period of time, then it will need to be serviced by a trained technician.

Black Exhaust Is Coming from the Tailpipe

Black smoke is the symptom of an extremely rich air/fuel mixture. If you see black smoke coming out of the tail pipe continuously, a large quantity of fuel is passing through your engine unburned. A small amount of black smoke that lasts one or two seconds may be normal if you accelerate suddenly. If you see black smoke for more than a few seconds, you probably need some work on the fuel delivery system. Don't confuse this with the inside of the tailpipe appearing black in color. This black coating inside the tailpipe is normal and typical for the gasoline used in cars today.

DIESEL DIFFERENCES

If you have a diesel vehicle, the main differences between your engine and a gasoline engine lie in the fuel system. Unlike gasoline, diesel fuel is thick and not very volatile. Some diesel engines use a mechanical diesel injection pump that meters, times, and injects fuel into cylinders that already have a charge of highly compressed, heated air. Other modern diesel engines use a solenoid, similar to a fuel injector, that opens to allow oil pressure to enter a small chamber and a piston that in turn, fires a diesel fuel injector. These new style engines can benefit from a fuel system that is completely computer controlled. Temperature from heat of compression, rather than a spark plug, causes the diesel fuel to ignite.

The fuel tank and fuel line of a diesel vehicle are similar to that of a gasoline-powered vehicle, but this is virtually the only similarity. Even the emission control systems of diesels are different from those of gasoline engines.

Bear in mind that the fuel system is just one of several that have to function together to assure your car's smooth operation. Chapter 5 covers another operating element of your engine, the electrical system.

1. Name the common fuel system components of the automobile.
2. How is octane rating related to detonation?
3. What are some of the fuels available for cars today?
4. What are some of the common additives in gasoline?
5. What are alternative fuels and what alternative fuels are available?
6. What is an air/fuel ratio? Why does it need to vary?
7. Describe the basic operation of the fuel injection system
8. Where does pollution come from on a car?
9. What are some devices used to control emissions?
10. Which of the emission control devices do you think is the most significant control and why?
11. What is the minimum level of fuel you should keep in you fuel tank and why?
12. What are some common fuel system problems?
13. How do you check for flooding?
14. How do you check for fuel delivery?
15. What can indicate a rich air/fuel mixture?

THE ELECTRICAL SYSTEM

After reading this chapter, the reader should be able to:

- List the various types of electrical components.
- List the major components of the electrical system.
- List the precautions when working with the electrical system.
- Describe the importance of the battery and its maintenance.
- List the starting system components and describe their functions and maintenance.
- List the charging system components and describe their functions and maintenance.
- List the ignition system components and describe their functions and maintenance.
- Describe the various other electrical devices found in the automobile.
- Explain the basic operation of the computer and the devices it can control.

AN OVERVIEW

Electricity is essential for powering your car's starter motor and for operating a gasoline-powered engine. When the engine is running, electricity is used to provide a high-voltage electrical spark to ignite the air/fuel mixture. Lights, horns, windshield wipers, and other electrical accessories assure your safety, control over the vehicle, and physical comfort.

The **electrical system** of your car consists of an interrelated group of **circuits,** or complete paths through which electricity flows (Figure 5–1). An automotive electrical circuit has several essential components:

- The **battery** and alternator, which are sources of electrical power
- A **conductor,** usually a wire, to carry flowing electricity, or current
- An electrical **load,** which consumes electricity (bulbs, radios, motors, and so on)
- A **switch** to connect or interrupt the flow of electricity
- A **fuse** or **circuit breaker** that interrupts dangerously excessive electrical flow to protect the circuit and the electrical devices that form the load
- A **ground** to complete the circuit so that electricity can flow

For electricity to flow through a circuit and do work, the circuit must be complete, or uninterrupted. There must be connections to the battery and

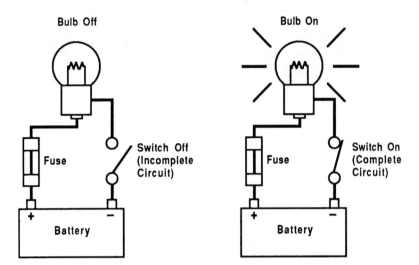

between all of the parts of the circuit. When the switch is on, the path for electrical flow is completed. Electricity flows from the battery to the electrical load, through the load, back to the battery, and out through the circuit again. This circular flow gives a circuit its name.

If, by some accident, wires from and to the electrical source are connected to each other without a load, a **short circuit** occurs. In effect, the wires become the electrical consumer. They heat up like the wires in a toaster, become red hot, or even explode. To prevent fires and damage from a short circuit or electrical overload, fuses or circuit breakers are included in the circuitry. More information on fuses is provided later in this chapter.

Each of your car's separate electrical circuits performs a specific function. All of the circuits are interconnected in a total system. The electrical system supports the continuing operation of the engine and accessories.

To understand the functioning of your car's electrical system, you should be aware of its major parts:

- The battery
- The starting system
- The charging system
- The ignition system
- Lights and accessories
- Electronic devices.
- Computers

THE BATTERY

The battery supplies power to start the engine. After the engine starts, the alternator provides the electricity used for igniting fuel and powering lights, motors, and all other accessories. But before all that can happen, the battery must produce enough electrical energy to start the car.

The battery produces electricity through chemical reaction. A cutaway drawing showing the internal parts of a typical automotive storage battery is presented as Figure 5–2. Based on the name, many people assume that a battery actually stores electricity. In reality, a battery stores chemicals that react on demand to produce electricity.

All electrical systems have two connections at the power source, negative and positive. To complete an electrical circuit, there must be connections to both the negative and positive terminals of an electrical source. The flow of electrons, the particles that make up part of an atom, create a current.

There are two theories about which terminal is the actual source of electricity. You may hear a technician refer to the **positive terminal** and connection to it as the "hot" side, or source, of electrical power. This way of looking at electricity is referred to as the "conventional" theory, which states that the current flows from positive terminal to the negative terminal.

However, for all scientific usage, the source of current is considered to be the **negative terminal.** Surplus, or extra, electrons originate from the negative terminal and are attracted toward a positive terminal, which lacks electrons. The flow of electrons creates an electrical current. This view is called the "electron" theory of electrical current. In this book, you will find explanations of the electrical system in terms of the electron theory.

Within the automotive battery, there are plates made from two dissimilar metals. Negative plates are made from sponge lead (Pb), or a porous form of lead, that has a surplus of electrons. Positive plates are made from

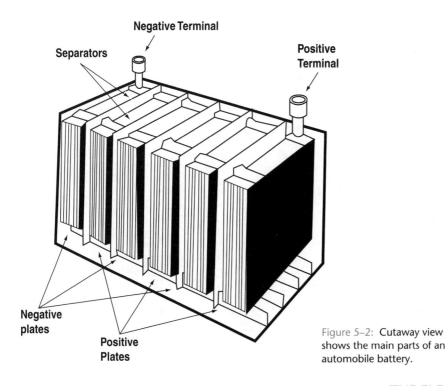

Figure 5–2: Cutaway view shows the main parts of an automobile battery.

lead peroxide (PbO_2) with a shortage of electrons so that it attracts and holds electrons.

Both the negative and positive plates are immersed in a liquid **electrolyte** solution that allows the electrons to flow from one set of plates to the other. In an automotive battery, the electrolyte solution is 40% sulfuric acid and 60% water.

Battery Charging

The battery provides power to the car during the times when the engine is not running. *Charging* refers to the process of preparing the materials in the battery to produce electricity. Thus, when a battery is charged, it can produce electricity on demand from the chemical energy stored in the plates and the electrolyte. In a discharged state, the battery lacks the energy stored in the chemicals and the negative plates are no longer able to supply electrons.

During normal operation of the vehicle, a **charging system** generates electrical current (a flow of electrons) and conducts them to the car's electrical system and the battery terminal. Thus, surplus electrons are returned to the negative plates and the battery becomes charged. The parts and operation of your car's charging system are discussed later in this chapter.

If a battery is not being charged by the car's charging system, an external battery charger, as shown in Figure 5–3, can be used. The battery charger converts normal home 110-volt AC electric current to the low-voltage 12-volt DC current required by the battery. However, the battery charger should not be used to supply power to other electrical systems of the car.

Delivering Current

The battery is a source of electrical power. To operate your car's electrical systems, the battery must be included in complete circuits. Wires connect

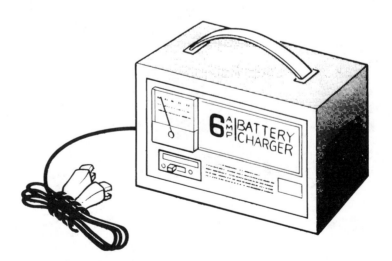

Figure 5–3: Small home-use battery chargers can help keep a battery in peak operating condition during winter cold.

electrical loads to the battery through switches and fuses. Typical loads in an automotive circuit include your car's starter, the radio, the horn, the lights, and so on—anything that uses electrical power.

Rather than using two separate wires to conduct electricity to and from each load, most cars have what is known as a "one wire" electrical system. One conductor connects the positive terminal of the battery to a load. The negative terminal of the battery is connected to the chassis and engine. Because both are made of metal, which conducts electricity, they become the *ground path* for electricity to flow from the battery's negative terminal. Some cars may use a "two wire" system and use a second conductor for the ground path of current. Typically, these cars have fiberglass or other composite body panels that don't conduct electricity well and require the use of the second wire.

Figure 5–4 illustrates typical automotive electrical circuits and their connection to the battery and alternator. The small inverted triangle symbols made of three parallel lines of different lengths indicate a connection to the car's ground path.

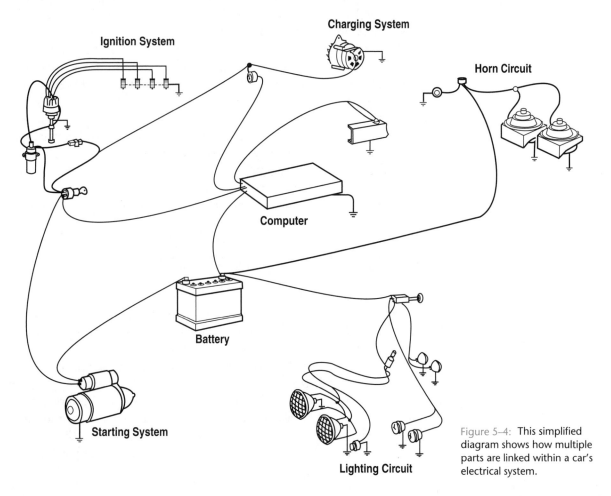

Figure 5–4: This simplified diagram shows how multiple parts are linked within a car's electrical system.

SAFETY PRECAUTIONS FOR BATTERY AND ELECTRICAL SYSTEM MAINTENANCE

Working around batteries or on electrical system parts can be dangerous. Batteries contain sulfuric acid, a powerful corrosive. When batteries charge, they release hydrogen gas, which is explosive. Metal objects and tools can conduct electricity and cause a short circuit or accidental ground that results in sparks and extreme heat. Always perform battery or electrical system work in accordance with the instructions printed in the car manufacturer's repair manual. Use the following list as a general guide to safe practices when you work on batteries and other electrical system parts:

- Wear eye protection at all times when you work on batteries. Battery explosions can splatter sulfuric acid into the eyes. Safety glasses or goggles can prevent eye injuries, infections, and possible blindness.
- Always disconnect the grounded, typically the negative (–), battery terminal when you work on the electrical system. This precaution will prevent sparks, fires, and damage to electrical parts.
- Remove all watches, rings, neck or ankle chains, and other jewelry from your body before you start work. Metal items can conduct electricity and cause a short circuit, sparks, burns, and possible battery explosions.
- Keep sparks, flames, and other ignition sources away from the battery. Sparks, flames, or smoking materials can ignite hydrogen gas produced by the battery and cause an explosion.
- Never connect or disconnect any electrical cables or connections while a battery charger is operating.
- Always clearly identify which terminals on the battery are positive (+ or POS) or (– or NEG)

BATTERY MAINTENANCE

Battery maintenance focuses on several key tasks to assure that the battery is charged and ready for use. Solid, clean connections must exist between the battery and the rest of the electrical system. There must be no electrolyte leaks or accidental discharges of current.

The battery is one of the most trouble-free pieces of equipment in the car. It requires very little maintenance compared to the work it performs and for its lifetime. Your first line of maintenance is to make sure that the battery terminals and cables are clean. Also inspect the battery hold down. A loose battery will bounce around and will fail much sooner than a properly secured battery. Another item to check is the electrolyte level to see if you have a serviceable battery. The procedures you follow will depend on the type of battery you have.

Electrolyte Level

Today's automobile battery has an outer case made of heavy plastic. On some "low maintenance" batteries, you will find removable caps at the top of the case, over the *cells*, or individual containers of electrolyte. You can lift these caps off to check the level of electrolyte (battery acid) in each cell. The electrolyte level is sufficient if liquid covers the plates and separators inside the battery and reaches almost to the bottom of the cell cover openings. The level should be equal in all cells. A battery with removable filling caps is shown in Figure 5–5.

Routine battery maintenance consists of adding water to the cells to bring the electrolyte up to the proper level. Distilled water is preferred because it does not contain the minerals and other impurities that can interfere with the normal chemical reactions within the battery. Water from a

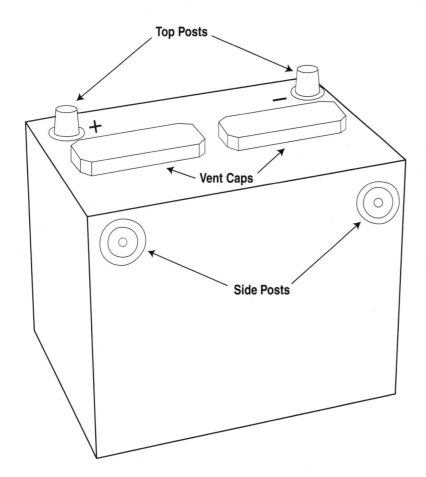

Figure 5–5: **Batteries have vent caps, some of which may be removed to check the electrolyte level, while others are sealed. Many replacement batteries may have both top and side posts.**

reverse osmosis purification system, used in some homes to purify water for drinking, may be used as well. If regular tap water is used in areas where the water is hard, or high in mineral content, the battery may become weakened. The contaminants in the water will interfere with the chemical processes within the battery. However, clean tap water can be used if distilled water is unavailable, but only use enough to cover the plates until distilled water can be obtained.

If you find that the battery needs water frequently, it's probably a sign that the charging system is overcharging, or providing too much electricity. Overcharging heats and evaporates the electrolyte. Internal battery problems also may cause the electrolyte to evaporate quickly.

Other types of storage batteries include the so-called maintenance-free units. To reduce the normal evaporative losses of electrolyte, special chemicals are added to the material of the plates. However, many of these batteries include filler caps under a protective plastic sheet. If the electrolyte level falls below the top of the plates, the plastic can be peeled up carefully and distilled water can be added through filler caps. Understand that this procedure may void a new battery warranty if the dealer from whom the battery was bought does not perform it.

There are also permanently sealed no-maintenance batteries. This type of battery is filled at the factory with electrolyte and sealed. The battery has no removable filler caps. Prying the top vents of the battery in an attempt to open it will simply puncture or crack the case and ruin it. If this type of battery fails because electrolyte levels become too low, it must be replaced.

Battery Charge

It's usually possible to check your battery's charge visually. Today, many batteries include a small plastic indicator window in the top (Figure 5–6). Instructions for checking the battery condition are printed near the window. Read the specific instructions on your battery and note the color or colors you can see at the bottom of the window opening. Typically, bright green, red, or blue colors indicate that the battery is fully charged and ready for use. For most commonly used permanently sealed batteries, black or dark colors indicate that the battery has lost its charge, but may be rechargeable.

If the battery has removable vent caps, you can use a battery **hydrometer** to check a battery's state of charge. There are many types of battery hydrometers, one of which is illustrated in Figure 5–7. Such a tester costs only a few dollars. A hydrometer operates on the principle that battery acid is denser than pure water. A precisely weighted bulb is contained within a tube. When liquid from a single battery cell is drawn into the tube, the bulb will float. A graduated scale is marked on the float to judge how much acid is present in the electrolyte. The level at which the upper surface of the electrolyte crosses the graduated scale indicates the acid content.

Figure 5–6: A maintenance-free battery is sealed. The condition of the battery is checked by noting the color in the small plastic window.

The charging action within the battery adds sulfur to the electrolyte, which combines with water to produce sulfuric acid. As the battery charges, the acid content increases and the water content decreases. Decreased acid content (in other words, more water) in the electrolyte indicates a loss of charge in the battery.

Realize that if you have just added water to the battery to bring the electrolyte level up, the acid in that cell will be diluted. In such cases, charge the battery by driving the car for a few days before you take hydrometer readings.

To use a hydrometer, squeeze the bulb tightly, insert the nozzle below the liquid level in a cell opening, and release the bulb slowly to draw electrolyte into the tube. When the bulb is released, bend over to view the graduated scale of the hydrometer at eye level. Note the numbered marking at which the top surface of the liquid crosses the scale. The reading on the hydrometer scale indicates the degree of charge for that individual cell only. A complete battery test requires that liquid be drawn from each cell individually and the readings compared.

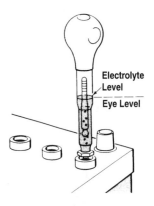

Figure 5–7: A battery hydrometer can be used to check the charge in each cell of an openable battery.

Because the acid concentration depends on the temperature of the electrolyte, good-quality hydrometers include a thermometer. At an electrolyte temperature of 80 degrees Fahrenheit (27 degrees Centigrade), hydrometer readings have the following indications:

- If the reading is 1.260, the cell is 100% charged.
- If the reading is 1.230, the cell is 75% charged.
- If the reading is 1.200, the cell is 50% charged.
- If the reading is 1.170, the cell is 25% charged.

At lower or higher electrolyte temperatures, the readings will be slightly different. A temperature scale typically is included on the hydrometer so that readings can be corrected for different electrolyte temperatures.

Note the readings from each cell for comparison. Significantly lower readings in one or two cells indicate that the battery may be near the end of its life. Assuming that water has not been added recently to some of the cells, all of the cells in a good battery should read at approximately the same level. If one is low, they should all be low. If there is a variation of more than 0.050 between individual cells, have a professional technician test the overall condition of the battery with a battery-starter tester.

No matter which type of battery you have, it should be kept fully charged in cold weather. When the temperature of the chemicals in the battery decreases, its ability to produce electricity is decreased. At the same time, an increased amount of electricity is needed to crank a cold engine and get it started. Also, as the level of charge decreases, protection against freezing decreases. Thus, a battery with a low charge can freeze and crack the case during severe winter months. A battery with a 25% charge can freeze when the temperature is 0 degrees Fahrenheit (–18 degrees Centigrade).

Battery Age-Dating

No matter what kind of battery you have, keep in mind its age as you check charge and electrolyte levels. Batteries typically have a prorated warranty for a certain amount of time. Don't count on getting much more practical use out of the battery than the length of its warranty. The warranty time usually indicates the life expectancy of the battery; it is surprisingly accurate. Batteries seem to be one of the few items that carry a warranty for the lifetime of the product. When you are near the end of its warranty period, it's best to install a new battery before the old one fails and leaves you stranded in some remote area or at an inconvenient time. The price of a service call may be twice the price of a new battery.

Terminals and Cables

Terminals are placed on either the top or the side of the battery case. You should take the time to locate and identify the positive and the negative battery terminals. Terminal identification is usually molded onto the case of the battery or stamped into the metal of the terminal itself. Markings will include a plus sign (+) or the abbreviation POS for the positive terminal. The negative terminal will be identified with a minus sign (–) or the abbreviation NEG. On batteries with top-mounted terminals, the markings frequently fade away. If the markings are unclear, you can identify top-mounted terminals because the positive terminal is always slightly larger in diameter than the negative terminal.

The vehicle's positive cable must be attached to the positive battery terminal and the negative cable to the negative terminal. If you reverse the connections, you will cause major, expensive damage to your electrical system (especially to electronic control units). You also may cause a battery explosion and be injured severely.

In use, corrosion builds up on the terminals because of the corrosive power of the acid in the electrolyte. Vents allow hydrogen gas to escape from the battery as it is charged. They also allow the passage of electrolyte vapors, which contain sulfuric acid. The result is corrosion on the battery

CAUTION

To prevent a short circuit and possible battery explosion, always remove the *negative* connection first. This action disconnects the vehicle's ground path. Thus, tools or other metal objects that might touch between the positive battery terminal and metal parts of the car cannot cause an accidental ground, or short circuit. When reconnecting the battery always connect the *negative* connection last.

terminal and its cable connection. As corrosion builds, battery performance declines. To clean terminals and clamps thoroughly, it is necessary to disconnect the cables from the battery.

There are two types of battery cable connections for a battery. The "top post" type has a small tapered post of lead on its top surface and uses a clamp style cable connection. The "side post" type has a small lead pad on the side, near the top surface of the battery. The cable is connected to the pad with a small battery bolt.

Battery cable connections can become badly corroded and difficult to remove from the battery terminals. Don't pry or twist against the battery terminal in your effort to loosen a cable. If you apply a lot of force, you can loosen or break the internal connection to the plates and ruin the battery. For a top post, make sure that you have loosened the cable connection bolt and nut fully. If the connection won't lift off easily, use a special battery terminal puller or spread the connector at the split with a screwdriver and lift the terminal off with a turning motion. For the side post, just loosen the small bolt on the side and the cable should easily detach from the battery terminal.

Once the connections have been removed, a battery terminal cleaning tool can be used to remove corrosion from the terminals and cable connections. The tool includes two shaped wire brushes. One fits inside the cable connection to brush and clean corrosion away. The other end fits over the battery terminal itself and is rotated to clean corrosion from the terminal surface. Brush the terminals briskly until they are clean and shiny-bright. When you are finished, wipe the battery case thoroughly with a paper towel dipped into a solution of baking soda and water. Rinse the case with fresh water and then dry the outside surfaces.

The battery cables link the battery to all of the circuits in your car's electrical system. Cables are usually heavy, braided wire covered with plastic insulation. When you are cleaning the cable connection, examine the insulation on both the positive and negative cables carefully. Also check the condition of the metal conductors in the cable. Make sure there are no breaks or cracks in the insulation of the positive cable. If the positive cable's metal conductor touches any metal part of the car, a direct short circuit across the battery is created at that point. The battery may explode or a fire may be started. Even if there is no contact between the bare cable and metal, cracks in the cable can cause corrosion problems. Corrosion in a battery cable causes resistance to electrical flow. Examine the cables for a consistent diameter. If the insulation on the cable changes diameter, such as a fatter spot in the cable, then it may have internal corrosion. This corrosion expands and stretches the insulation. If you find small cracks in the insulation, wrap the cable securely with electrical tape. If most of the insulation is missing from the cables, replace them with a new set.

Corrosion also should be removed from the battery carrier and hold-down brackets and clamps. For heavily corroded large areas, such as the battery tray, scrape the corrosion off with a putty knife or screwdriver blade. Use a large steel wire brush to remove corrosion down to the bare metal. Then wipe the metal with a paper towel dipped in a baking soda and water solution. Allow the area to dry before you reconnect the battery terminals.

After you have reconnected the battery terminals securely, apply some form of corrosion protection over the clean metal areas. A special silicone battery gel or battery spray works well. Grease can be used for protecting top post terminals. Coat the clamp and the post after they have been reconnected. The grease will keep the battery vent vapors and oxygen in the air from corroding the connections.

The Battery Case

While you are cleaning the terminals and performing other routine maintenance on the battery, also check the case. Look for cracks or wet spots. Make sure the battery is secured firmly in place by the hold-down brackets. If the battery is free to bounce, internal electrical connections may be loosened or broken. A loose battery may also cause a short to ground, resulting in battery damage or damage to nearby components.

If a battery case is cracked, don't attempt to patch the case with sealants. A cracked battery should be replaced as soon as possible.

Battery Recharging or Replacement

If tests show that the levels of charge in all of the cells in a battery are low—and all are about the same level—the battery needs recharging. People who drive short distances or who don't use their cars very often may expect to find a low battery charge. Most of the drain on the battery occurs when you start the car. The car will use power from the battery even if the car is not driven. Computers and other items in the car need electrical power for memory circuits such as computer memory, radio presets, remote or keyless entry, and the like. After the car is started, the charging system takes over and begins to replace the energy used in starting. In general, a minimum of approximately 100 miles of driving per week is necessary to keep a battery sufficiently charged for reliable starting.

If your typical driving pattern involves short trips and starting in cold weather, you may encounter a consistently low battery charge. In this case, a battery charger might be a good investment. Such units can represent real savings and convenience—as compared with having your car towed to a service garage and waiting for the battery to be recharged. Small trickle chargers are relatively inexpensive and are available at most auto supply or chain stores. A trickle charger produces a small current flow (a trickle of electricity) to recharge a battery overnight or within 24 hours. If it is used regularly, a trickle charger can maintain a seldom-used battery in top condition.

If you find that your battery has run down and you don't have a charger, an alternative is to go to a service garage. Commercial high-speed battery chargers can build up the charge sufficiently in an hour or so to permit you to start your car. While a high-speed charge may be more convenient, a trickle charger ensures longer battery life.

Assuming that the battery and your car's charging system are in good working order, simply driving the car will recharge the battery. Thus, if you can get your car started, the charging system should charge your battery. Drive the car for an hour or more to recharge the battery sufficiently for restarting.

If you car won't start, and another car with a charged battery is available, you can connect the two batteries with **jumper cables** to get your car going. A safe procedure for jump-starting is discussed later in this chapter.

If the battery keeps giving you trouble with low charges, the best course of action is to go to a service center, reputable dealer, or chain store. Ask to have a *charging system check* performed before you assume that the battery is the cause of the problem. Your problem might be caused by any of the following:

- The alternator may be defective.
- The voltage regulator may be defective (usually integral with the alternator).
- There may be a continuous low current drain, such as a trunk light that doesn't go out.
- The alternator belt may be loose.
- The starter may be drawing excess current.

When you need a battery, it is always a good policy to buy a unit with a slightly higher electrical capacity than the one that came with your car. Of course, the replacement battery should be of the same physical size as the original unit. Specific battery lengths, widths, heights, and terminal positions are coded as a **group number,** for example Group 24, Group 24F, Group 72, and so on. Many batteries will be equipped with both top-post and side-post terminals. The retail store will be able to look up the proper sizes available for your car. However, you can buy batteries of the same group number that have widely different electrical capacities (and different warranty periods).

Electrical capacity can be measured in several ways. Two of the most useful measurements for consumers are **cold cranking amps (CCA)** and **reserve capacity.** Cold cranking amps refers to the amount of current that the battery can produce when the electrolyte is cooled to 0 degrees Fahrenheit. Batteries with higher CCA numbers produce greater electrical output under severe cold weather conditions than low-rated batteries. Sometimes the cranking amp rating is given for a battery temperature of 32 degrees Fahrenheit (0 degrees Centigrade) and is referred to as *cranking amps* (CA). This rating will show a higher number and may confuse a buyer into thinking it is a more powerful battery. Be sure to locate the true CCA, which is given at 0 degrees Fahrenheit (–18 degrees Centigrade). It is the most important rating to consider when choosing a battery because it has the biggest impact on the battery's ability to crank the engine.

Reserve capacity is the length of time an output current of 25 amps can be maintained at an electrolyte temperature of 80 degrees Fahrenheit (27 degrees Centigrade) before the battery drops below 10.5 volts. Reserve capacity, stated in minutes, can be thought of as a rough measure of how long you could continue to drive if the alternator stops producing current.

When purchasing a battery, keep the receipt and warranty papers inside a sealed plastic bag in your glove compartment. Refer to these documents to determine the battery's age and the prorated warranty schedule. If the battery fails before the end of the warranty period, then an allowance for the number of months remaining will be applied toward a replacement battery.

THE STARTING SYSTEM

The function of the starting system is to turn or "crank" the engine crankshaft. As the crankshaft begins to rotate, pistons move, valves open and close, an air/fuel mixture is drawn in, and the spark plugs fire at the proper time. The four-stroke cycle is completed and the engine begins to run on its own. As shown in Figure 5–8, the parts of the starting system include:

- Battery
- Ignition switch
- Starter solenoid
- Starter motor
- Starter pinion drive gear
- Flywheel ring gear

The key-operated ignition switch on your steering column completes an electrical circuit that includes the starter motor. When the ignition key is turned to the START position, a small amount of electricity activates the starter **solenoid.** The starter solenoid performs two functions. First, it activates a pivoting mechanism that moves the pinion drive gear to mesh with the flywheel ring gear. Second, the starter solenoid **relay** operates to pro-

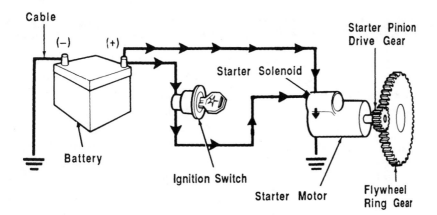

Figure 5–8: The starting system uses electrical current from the battery to turn the engine crankshaft by rotating the flywheel gear.

vide a large amount of current to the starter motor, which turns the flywheel and crankshaft.

A relay is a magnetically operated remote-control switch. Within the relay is a coil of electrical wire. When a small amount of electricity passes through the coil, a magnetic field is created. The magnetic field is used to move a spring mounted heavy-duty electrical switch contact. Thus, when the relay is activated, another circuit is switched on. This heavy-duty switch completes a circuit that includes the battery, battery cables, and starter motor. The starter motor is a powerful electric motor that provides enough torque to turn the engine crankshaft. A good deal of torque is required to turn the crankshaft against the resistance that results during each cylinder's compression stroke.

Once the engine begins to run, the driver allows the key to return to the ON position and the starting system is deactivated.

Although malfunctioning starter motors can be rebuilt, most people who repair their own cars don't have the time, tools, or expertise to rebuild their own units. A practical, quicker, and less expensive method is simply to exchange the old starter for a rebuilt unit.

The practice of exchanging worn parts for rebuilt ones is used for many types of auto repairs. Factories specialize in rebuilding many used auto parts, including starters. You simply go to an auto parts store, turn in the old part, and pay for a rebuilt exchange unit. The part you turn in is returned to the rebuilding facility. Eventually, your old part is cleaned, rebuilt with new parts, and returned to an auto parts store, where it is sold to someone else. If you don't turn in a rebuildable part in trade, you must pay a "core charge" that compensates the rebuilding facility for the loss of a rebuildable unit. When you return the core, the parts store will reimburse the price of the core charge. To buy a replacement starter motor, you need to know the year, make, model, and engine size of your car.

STARTING SYSTEM MAINTENANCE

Routine maintenance for the starting system is minimal. Simply make sure that all of the electrical connections at the battery, solenoid, and starter motor are tight and that the insulation on all of the exposed wires is intact. On many cars, the positive (+) battery cable is connected directly to a terminal bolt on the starter solenoid. When you perform other checks and routine maintenance under the hood or under the car, it is a good practice to check out electrical wire and battery cable connections at the solenoid.

In operation, never crank the starter for more than 30 seconds at a time. Continued cranking can cause the starter to overheat and the battery to become discharged. If the engine doesn't start after a few attempts, it likely will not start. Extended cranking will only serve to damage the starting system. As described in Chapter 4 and later in this chapter, perform a few checks to determine why the car is not starting.

THE CHARGING SYSTEM

The charging system's main function is to provide electrical power to the car while the engine is running. The charging system also provides current to recharge the battery. Unless your driving trips are very short, the charging system should keep the battery charged sufficiently to crank the engine quickly and easily.

A diagram of a charging system with an external voltage regulator is shown as Figure 5–9. The charging system includes:

- Alternator
- Battery
- Voltage regulator (usually integral with the alternator)
- Warning light or gauge.

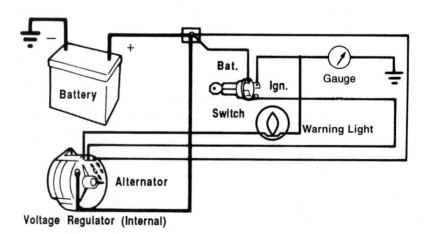

Figure 5–9: A typical charging system includes the alternator, voltage regulator, warning light or gauge, and battery.

Alternator

The term *alternator* is shorthand form for an alternating current (AC) generator. In reality it should be called a *generator*. A generator produces direct current (DC). The device in the car produces AC that is rectified or changed into DC for use in the battery and car's electrical systems. Many people refer to the device as an alternator because of the way it starts the production of electricity. So we may find that the alternator is sometimes called a generator. Either term is acceptable.

When your car is running, a belt that is connected to the crankshaft rotates the alternator mechanism. Thus, when the engine is operating, the alternator produces electricity to recharge the battery and to stabilize the voltage in the electrical system. If the alternator fails, the battery can supply power for the car until it becomes discharged.

Voltage Regulator

The voltage regulator controls the amount of electricity or voltage generated by the alternator to charge the battery. On older cars, the voltage regulator is a separate unit mounted in the engine compartment. On most vehicles, an electronic voltage regulator is built into the rear end of the alternator case. Thus, whenever you replace the alternator, you are also replacing the voltage regulator.

Warning Light or Gauge

Your car has a dashboard indicator showing the status of the charging system. This indicator may be a warning light (Figure 5–10), an ammeter, or a voltmeter gauge. Cars equipped with a light will either have an ALT lamp or a BAT lamp; the light on the dashboard should not stay on after the engine is started. A voltmeter indicates battery voltage. After starting and running, battery voltage should remain between 13.2 and 15.2 volts. If it drops below 13.2 volts, you probably have a charging system problem. If the voltage is above 15.2 volts, the regulator is defective. An ammeter

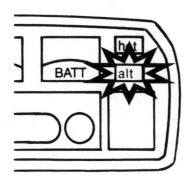

Figure 5–10: A dash warning light indicates that the battery is discharging and is not being recharged.

indicates whether the battery is being charged or discharged. The ammeter should indicate that the battery is being charged. After an extended period of running, it may indicate close to zero, which means that the battery is fully charged. But if the ammeter indicates negative amps, the battery is supplying power and the charging system is not functioning properly.

CHARGING SYSTEM MAINTENANCE

To keep the alternator operating, the drive belt must be in good condition and properly tightened. When you do your regular monthly electrical check, twist the alternator drive belt to view the driving surface, or underside, of the belt. If it is cracked or frayed, have it replaced as soon as possible. Minor cracks on a multigroove belt are acceptable. Procedures for replacing belts are covered in Chapter 11.

One method for checking belt tightness is to press down on the belt between the pulleys, or grooved wheels, into which the belt fits (Figure 5–11). If you can move the belt downward more than a half-inch per foot of span from its stationary position, it needs to be tightened.

If you have a multigroove belt with a self-tensioner, you may have a set of marks on the tensioner that indicate a belt stretched beyond its service limit. On occasion a tensioner will loose its ability to properly tension a belt. In this case a service center or dealership will need to test the belt tension and make a determination.

A loose alternator belt will slip on the pulleys and will not turn the alternator fast enough to keep the battery charged. Eventually, the belt can become glazed, that is, it develops a shiny driving surface that may squeal when the engine is first started or accelerated. The heat from slipping on

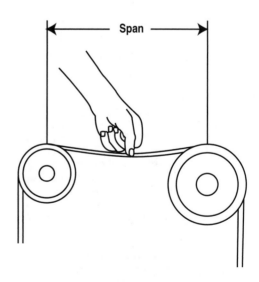

Figure 5–11: Belt tension can be checked by moving the belt downward with your thumb.

the pulleys dries the rubber, which promotes cracking and belt breakage. And while a loose alternator belt causes certain problems, a belt that is too tight will damage the bearings in the units turned by the belt, including the alternator or water pump.

For maximum reliability, it is recommended that you change all V-type belts every three years or 36,000 miles and serpentine belts every four to five years or 50,000 miles, whichever occurs first. Although belts may last longer than these suggested limits, changing them at the recommended interval will prevent most problems.

There are two types and more than 100 commonly used sizes of drive belts. If you take long trips, always carry spare belts for your car's alternator and coolant pump. Even if you don't want to replace these belts yourself, having the correct spare parts protects you against situations in which a service station does not have belts that fit your car.

During your regular electrical inspections, you also should check the wiring of the charging system. Make sure that all connections and plug-in connectors are tight.

Alternators, as with starters, usually are replaced as units rather than rebuilding. An alternator can be replaced with either a new unit or a rebuilt one. To buy the correct alternator for your car, you should know whether your car is specially equipped. For example, cars equipped for towing trailers usually have a heavy-duty electrical system with a high-output alternator. Today, with many additional electrical accessories and operating systems, manufacturers may specify high-output alternators as standard equipment. In any case, provide the auto parts store with as much information about your vehicle as possible. At minimum, you should be able to tell the counterperson the year, make, model, and engine size of your car. You also should know what other belt-driven accessories are on your car, such as air conditioning, power steering, an air pump, and so on.

THE IGNITION SYSTEM

The ignition system provides the spark needed to ignite fuel during the engine's operating cycle. The system has two basic functions. First, it must transform low-voltage current from the battery into sparks with voltage high enough to ignite the fuel mixture in the cylinders. Second, it must ignite the fuel in each cylinder at precisely the right time, when the piston is ready to start the power stroke.

The **ignition coil** produces high-voltage current. Some engines use a **distributor** to distribute the spark generated by the ignition coil to the individual **spark plugs.** The distributor is a mechanical device driven by the camshaft. Most engines do not use a distributor. Instead they use a **distributorless ignition system (DIS)** having a set of coils, known as a coil pack, for each pair of spark plugs (Figure 5–12). Other engines use individual coils

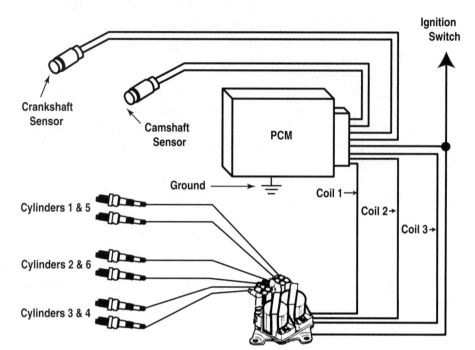

Crankshaft
Sensor

Camshaft
Sensor

PCM

Ground

Coil 1→

Coil 2→

Coil 3→

Cylinders 1 & 5

Cylinders 2 & 6

Figure 5–12: The parts of a
typical DIS ignition system. In
this system spark plugs are
fired in pairs.

Cylinders 3 & 4

for each spark plug. This arrangement is also a DIS and is known as a **coil-on-plug (COP)** ignition system or the coil-near-plug (CNP) system. The COP gets its name because the coil is mounted directly to the spark plug with no need for a spark-plug wire (Figure 5–13). The CNP has the coil very near the spark plug and is connected with a short spark-plug wire. These **electronic ignition systems** benefit from precise computer control just as fuel injection does.

Operation of the ignition system starts when you turn on the ignition switch. Electricity flows from the battery through the ignition coil, which is controlled electronically by a switching mechanism. The **coil** actually contains two windings of wire called the primary and secondary windings. The primary winding consists of relatively few turns of thick wire. The secondary winding consists of thousands of turns of extremely fine wire. In electrical terms, the coil functions as a transformer. The coil transforms 12-volt power from the battery into extremely high voltage—in the range of 20,000 to 70,000 volts.

The ignition system has two basic circuits that correspond to the two windings in the coil: primary and secondary. The primary circuit delivers low-voltage current from the battery through the ignition module to the primary coil, then to ground. Current flows in the primary winding circuit to build up a strong magnetic field. At the moment when a cylinder is nearly at the end of the compression stroke and is ready to begin the power

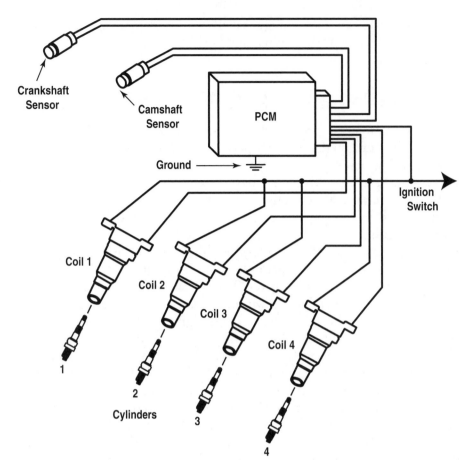

Crankshaft Sensor

Camshaft Sensor

PCM

Ground

Ignition Switch

Coil 1

Coil 2

Coil 3

Coil 4

Cylinders

1

2

3

4

Figure 5–13: In a typical COP ignition system, each cylinder has its own coil mounted directly to the spark plug.

stroke, the circuit is opened by the ignition module, which acts on a signal from the computer.

When the primary circuit is opened, the collapse of the coil's magnetic field creates high voltage in the secondary wiring. This high voltage is delivered to the center terminal of the distributor cap. This terminal then feeds the electricity through a rotor inside the distributor. The rotor turns continuously, aligning itself with terminals that deliver electricity to the spark plugs. As the rotor turns, the high-voltage electricity flows to each spark plug in turn and causes a spark to arc at the tip of each plug. This spark ignites the air/fuel mixture and keeps the engine running.

For a DIS, the coil operation is the same. The spark is delivered from the coil packs to the pair of spark plugs without the need for any high-voltage distributor components. With COP and CNP, the spark is delivered directly to the individual spark plugs.

The signaling apparatus within the distributor sends signals to the computer to indicate the relative position of the camshaft. With DIS, both

COP and CNP, camshaft and crankshaft position sensors are used by the computer to accurately determine their position and speed to make precise timing calculations.

IGNITION SYSTEM MAINTENANCE

The major maintenance required by the ignition system is a tune up. Simple parts of a tune up that you can perform are covered in Chapter 11.

Routine ignition system maintenance is much the same as for the charging system. Make sure all connections are secure and tight. Check all wires and make sure that the insulation is not cracked or frayed. If you see any loose or frayed wires, repair them or make sure that they are repaired as soon as possible.

A word of caution: Ignition cables, the thick wires leading from the distributor or coil packs to the spark plugs and the ignition coil, are delicate. COP does not use spark-plug wires. Though the ignition cables appear to be thick, the actual conductors within the insulated covers are relatively thin and fragile. The thick jacket of the ignition cables must insulate against many thousands of volts generated by the ignition system. In addition, the conductors generally are not made of metal wire. The conductors are fiberglass or other synthetic materials, impregnated with carbon to provide the proper electrical conductivity and reduce TV and radio interference. Do not pull on or twist the ignition cables. To connect or disconnect cables, grasp them only by the **boots,** or thick, molded portions at the spark-plug ends of the cables. First, twist the boot gently in a rotating motion to loosen it. Then remove the cable from the spark plug by pulling firmly on the boot. If the boot is cracked or the wire otherwise damaged, it will need to be replaced.

LIGHTS AND ACCESSORIES

There are enough electrical devices within a car that this portion of the electrical system could fill a book by itself. Only the most important items are covered here—those necessary to your safety and comfort.

Light and Accessory Circuits

Lights, motors, and other electrical devices operate on a series of individual circuits. Each of these circuits draws current from a source, contains switches and fuses, and supplies current to loads.

If you use lights or other accessories when the engine is not operating, power is drawn from the battery. Because the alternator does not operate unless the engine is running, the battery is not recharged by the charging system. Therefore, leaving any electrical device on when you turn your car off will discharge the battery.

Headlights

Two types of **headlights** are common on today's cars: sealed-beam head-lights and composite headlights (Figure 5–14). A **sealed-beam** headlight is, in effect, a very large bulb with a light source, reflector, and focusing lens all in a single unit. When a sealed-beam bulb burns out, the entire headlight is replaced. Sealed-beam bulbs may be either ordinary tungsten lights or **halogen** lights. A halogen bulb is filled with halogen gas, which helps to maintain the brightness of the light over long periods of time. Halogen headlights typically produce a more intense, whiter light than tung-sten lights. Composite headlights include a small, replaceable-bulb unit that fits inside a reflector that is mounted on the car. The composite head-lights are most commonly used today and allow manufacturers to make the headlight into any desired size and aerodynamic shape. These lights oper-ate at higher temperatures and may include a vent at the rear of the lamp assembly. When replacing the bulb, be sure not to touch the bulb as oils and salts from your skin will substantially shorten the life of the bulb.

Development of a third kind of headlight, called *high-intensity discharge* (HID), is underway. These lamps provide up to three times the light and

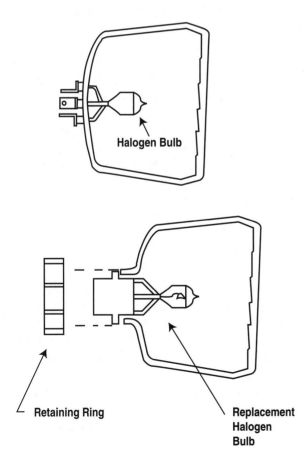

Figure 5–14: Sealed-beam headlights use halogen bulbs permanently mounted inside the headlight (*top*). Composite headlights (*bottom*) use a replaceable halogen bulb held in by a retaining ring.

illuminate a larger area than those of other systems. Although HID is still under development, it is now showing up on a few select vehicles.

Headlights can be either rectangular or round. Some systems have separate headlights for low and high beams. Others, typically sealed-beam units, use a single bulb to provide both low and high beams. When you operate your car's low- and high-beam switch, electricity flows to either one or both filaments within the bulb.

To replace the sealed-beam unit, you usually will have to remove two metal rings or moldings. The first molding is a decorative *bezel*. Under the bezel is a retaining ring that holds the sealed-beam unit in place. Sometimes, sealed-beam units are retained by spring mechanisms. On other units, you remove three screws to take off the retaining ring.

In removing a sealed-beam unit, be careful to loosen only the retaining-ring screws, which are identified easily because they pass through the ring itself. Do not turn or misadjust the aiming screws that will be near the sealed-beam unit you are replacing. Adjustments should be made only by a certified technician or other qualified repair facility.

On a composite headlight, unplug the wiring connector from the back of the bulb and remove the light-bulb retaining ring. Once the retaining ring is off, you can simply remove the old bulb and put in the new headlight. Reinstall the retaining ring, making sure it is properly seated. Then plug in the connector and test your headlights. Make sure that the new unit is the proper bulb for your car. Many composite headlight assemblies will have a vertical adjustment screw for headlight adjustment, and some may also have a horizontal adjustment. The vertical adjustment is more crucial because too high an adjustment could blind oncoming drivers, while too low an adjustment could impair driver vision on a dark surface. When checking the alignment, be sure the vehicle is on a level surface, the fuel tank is half full, and the vehicle is at normal ride height. The bubble in the sight window of the composite headlamps can confirm proper alignment. Otherwise, adjustments should be made by a certified technician or other qualified repair facility.

Other Lights

Different types of bulbs are used for other car lights (Figure 5–15). Bayonet bulbs are inserted by pressing them into a socket and then twisting. These bulbs are held in place by locating pins on the bulb base and matching slots within the sockets. Another type of bulb simply presses into a tight-fitting socket with two metal prongs. Pull the old bulb out and press the new bulb securely into place (Figure 5–16).

Bayonet or push-in bulbs of some exterior lights are reached by removing a decorative molding and lens (Figure 5–17). Some taillight and turn-signal light sockets are inside the car's body. You may have to reach under

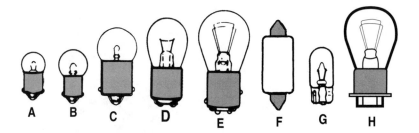

A, B   Miniature bayonet for indicator and instrument lights
C -   Single contact bayonet for licence and courtesy lights
D -   Double contact bayonet for trunk and underhood lights
E -   Double contact bayonet with staggerer indecing lights
       for stop, turn signals, and brake lights
F -   Cartrige type for dome lights
G -   Wedge base for instument lights
H -   Plug-in double contact for stop, turn signals, and brake lights

Figure 5–15: Common automotive bulbs.

a bumper. In other cases, you reach the sockets through the trunk. Or the lens assembly may need to be removed to gain access to the light bulb.

Changing bulbs is one of the easiest maintenance steps you can take. Simply remove the bulb and bring it with you to an auto supply store. Personnel will be able to look up the proper bulb if you provide them with the

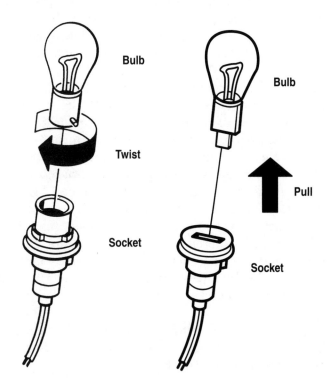

Figure 5–16: To replace a bayonet bulb, press down on the bulb and twist it. To replace a plug-in bulb, simply pull it straight out of the socket.

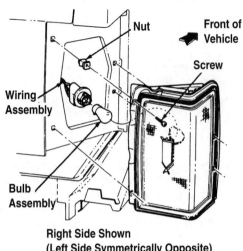

Nut

Front of Vehicle

Screw

Wiring Assembly

Bulb Assembly

Figure 5–17: One method to access a bulb.

**Right Side Shown
(Left Side Symmetrically Opposite)**

year, make, and model of the car. Most automotive bulbs are numbered on their bases. Make sure the number on the replacement bulb matches that of the old bulb.

Fuses

A fuse is placed within a circuit to protect against the possibility of electrical fires or burning out expensive devices. If there is a sharp increase in the amount of current—as happens, for example, when there is an accidental ground or short circuit—the material in the fuse burns out and breaks the circuit to protect electrical consumers, or loads, and conductors within the circuit.

Most electrical circuits within your vehicle are fused. (Some headlight, power seat, and other circuits may be protected by circuit breakers.) Typically, fuses for a number of circuits are grouped in one location for convenience—under the dash or in the engine compartment. Today's cars may have more than one fuse center, or grouping, with fuses for several circuits in each. These are sometimes called *fuse boxes, junction boxes, fuse panels,* or *power distribution centers.* Common types of automotive fuses are shown in Figure 5–18. Most popular are the mini fuse, ATC fuse, and the maxi fuse.

It is a good practice to identify the locations of key fuses. Fuse locations vary considerably with different makes and models of cars. Check your owner's manual and then take a few minutes to locate your car's fuses.

Fuses are either held between metal prongs, similar to some types of bulbs, or may plug into a socketlike holder. Late-model cars typically use plug-in type fuses.

Replacing a fuse doesn't automatically solve an electrical problem. When a fuse blows, something has caused excess current to pass through the circuit. A momentary surge of electricity, as when a switch is turned on,

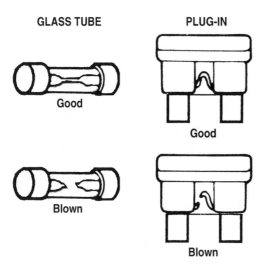

GLASS TUBE

Good

Blown

PLUG-IN

Good

Blown

Figure 5–18: Common types of automotive fuses.

can cause a fuse to blow. However, if a short circuit or accidental ground exists in a circuit, replacing a fuse will not fix the problem.

A handy way to test your fuses is to use a test light (Figure 5–19). A test light has a metal probe at the end of a handle with an encased light bulb at one end. The other end has a wire lead with a clip. To use the test light, attach the clip to a good ground. Make sure the fuse you are testing is in use; for example, if it is the park light fuse, the park lights will need to be on. Touch the probe to one end of the fuse. The test light should light. Now touch the other end of the fuse. The test light should again light. This indicates the fuse is okay. If only one end of the fuse lights and not the other, a blown fuse is indicated. If neither end of the fuse lights, the fuse is not in

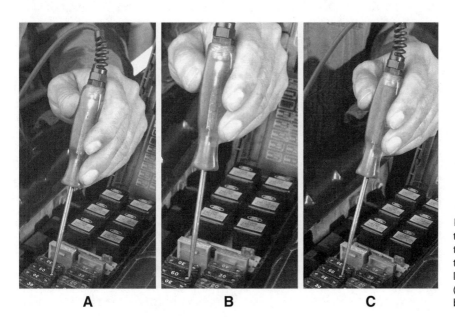

A   **B**   **C**

Figure 5–19: A test light used to test fuses. If both sides of the tester light (as in *A* and *B*), the fuse is okay. If one side lights and the other does not (as in *A* and *C*), the fuse is blown.

use, the test light is not properly grounded, or the bulb in the test light is broken.

CAUTION

When you replace a fuse, make sure the new one is of the same size and capacity as the one that has burned out. Avoid replacing blown fuses with new fuses that have a greater amperage rating. A larger-capacity fuse will overload the current-carrying capacity of the wires. The wires can overheat and possibly cause a fire or damage the wiring harness.

Windshield Wipers

Windshield wipers are operated by electrical motors on all modern cars. To be sure the wipers will work when you need them, test wiper operation from time to time. Clean the windshield of dirt and dust before you test wiper operation; otherwise, you may leave permanent scratches in the glass. This advice can be especially important if you live in a dry climate, such as the West or Southwest, where it may not rain for months at a time. If you don't check the wipers periodically, you can find either that the wiper blades have rotted or that dirt and corrosion has made the motor inoperable.

If the wipers don't work, check the fuse first. If you have to replace the fuse, and it blows again, see a certified technician. If the fuse is okay and the wipers still don't work, don't force the blades. Many people think they can get the wiper motor started by moving the blades back and forth across the windshield by hand. All that is accomplished is the breaking of moving parts or the stripping of gears within the drive system. If the windshield wiper motors don't start on their own, you won't do them any good by operating them manually.

MAINTENANCE FOR LIGHTS AND ACCESSORIES

At least once every two weeks, have someone sit in the driver's seat while you check the lights around your car. Ask the other person to operate the light switches while you check that the bulbs are working. Note that for many of these tests, the ignition switch must be in the ON position.

Exterior Lights

Ask the other person to step on the brake pedal when you are behind the car to check the brake lights. Ask the person to turn all of the light switches on and off, one at a time, and to operate the high-beam switch while you are in front of the car. Check the turn signals (front and back) with the headlights both on and off. Check the hazard flashers and the light for the rear license plate. Ask the person to move the shift lever to REVERSE to check the back-up lights.

If you find any lights that don't work during your inspection, check the fuses and change the bulbs as soon as possible.

Turn Signals

If the turn signals on your car don't operate properly, there are some easy diagnostics you can use. First, check the fuse center for a blown fuse.

If the turn signal indicator flashes for one direction but remains on for the other, the most probable cause is one burned-out turn signal bulb. Set the switch to signal a turn in the direction for which the flasher doesn't work. Then look at both the front and rear of the car. All bulbs must be operating for the turn signals to flash. Some vehicles have two or more lamps at the rear, so check them all.

If none of the turn-signal bulbs flash, and the hazard lights flash, the most likely cause is a defective turn-signal flasher relay. A flasher relay is a plug-in unit that can be removed and replaced easily. Check your owner's manual for the location of the flasher relay. It may be located at or near the fuse center. Some cars use a combination flasher relay that operates the turn signal and the hazard flashers.

ELECTRONIC DEVICES

Many electronic devices are used on modern automobiles. These devices can be broken down into two general groups: those that present information and those that control vehicle functions.

Warning and Indication Systems

Electronic circuits and display devices can be used in place of conventional gauges to present information. When information is presented as digits, or numbers on a display, it is known as *digital display*. Information displayed on a gauge is known as an *analog display*. Information also can be presented by a chime or in a spoken language. For example, when a door is ajar or the ignition key is left in the lock, a chime will sound or a verbal message is played to alert the driver.

Computers

A **computer** is a device that processes, or changes, raw data into useful information. The three main functions a computer performs are input,

Figure 5–20: **Many automotive operations are monitored and controlled by computers, also called control modules. These devices, like the one shown, are replaced as units.**

processing, and output. The data that computers receive and the information they deliver are in the form of electrical signals. Processing to convert data signals into information signals is directed by a set of instructions known as a **program.**

Computers can be used to control vehicle functions. The data input from sensors are processed and output signals are produced. The output signal is used to operate lights, relays, motors, solenoids, or other **output devices** to control vehicle functions. The main parts of a basic computer control system are:

- Input sensors
- Computer (PCM; Figure 5–20)
- Output devices
- Wiring connections

A full understanding of the inner parts and functions of a computer is not necessary. You, or your technician, need only to be able to determine if the unit is working properly.

In today's cars, the use of computers is being expanded constantly. A computer can control almost any mechanical or electrical function of an automobile, including:

- Electronic ignition
- Fuel delivery
- Emissions controls
- Cruise control
- Heating and air conditioning
- Automatic transmission
- Suspension
- Antilock brake system
- Power steering
- Antitheft system
- Traction control
- Memory settings for seat, mirror, and pedal position
- Air bag
- Speed-sensitive wipers and stereos
- Security systems
- Satellite navigation system
- Night vision
- Heads-up-display (HUD)
- Collision avoidance

Uses of computer controls now include "drive-by-wire" systems in which pedals are no longer mechanically linked to the systems they control. BMW produces a vehicle in which there is no cable or other mechanical link between the throttle pedal and the throttle body on the fuel injection system. This is a promising development with future innovations to include brake-by-wire and steer-by-wire systems.

Feedback Control   Computers, sensors, and output devices can be connected to monitor and adjust vehicle functions continuously. For example, a computer-controlled fuel injection system can maintain a 14.7:1 air/fuel ratio after the engine reaches normal operating temperature. An input signal from an exhaust gas oxygen sensor in the exhaust system is fed to the computer. When the oxygen sensor input signal indicates that the air/fuel mixture is varying from the ideal mixture, the computer produces an output signal that varies the time the fuel injectors are open. When the ideal air/fuel mixture is attained, the computer maintains a precise control over the injectors timing.

Closed-Loop Operation   Each computer receives a constant flow of data about the process it controls. The computer compares the input data to a set figure in its memory. If necessary, the computer changes its output. Whenever the process being controlled changes, the computer changes the output to bring everything back into line. This type of feedback control is known as a **closed-loop** mode of operation. In Figure 5–21, the line forms a closed, or continuous, loop.

Open-Loop Operation   For starting, warm-up, and hard acceleration (or wide-open throttle), the engine needs an air/fuel ratio richer than 14.7:1. Under these conditions, the computer can be directed to provide **open-loop** operation. In open-loop operation, the computer does not monitor and adjust the air/fuel ratio continually. Instead, the computer provides a richer mixture for smooth engine operation during warm-up

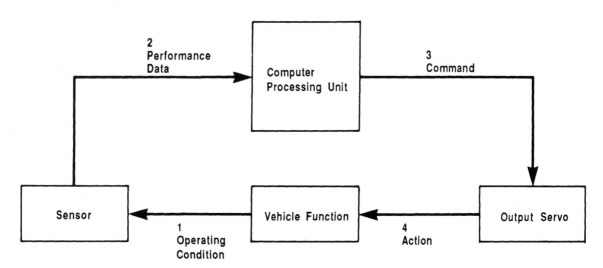

Figure 5–21: In closed-loop operation, a computer constantly monitors and adjusts a process to maintain a stable operating condition.

and wide-open throttle. After the engine warms up or the driver stops pushing hard on the accelerator pedal, the system will go into closed-loop operation.

MAINTENANCE FOR ELECTRONIC DEVICES

In most circumstances, no preventive maintenance is required for electronic sensors or computer modules. However, in areas of extreme humidity or dust, electrical connections can become degraded by oxidation or contaminated by dust. Poor connections can cause erratic operation or damage to sensitive electronic components.

Preventive maintenance in such areas might consist of disconnecting the battery negative terminal, then disconnecting and reconnecting electrical terminals to clean the contact surfaces. Mechanical output units operated by electronic devices may require periodic lubrication, adjustment, or other maintenance.

⚠ CAUTION

Always disconnect the battery negative terminal before disconnecting any electronic sensor or computer-controlled units. If the battery remains connected, voltage spikes may occur that will damage or destroy expensive computer control modules.

ELECTRICAL SYSTEM PROBLEMS

Pay attention to your battery in cold weather when the oil in the engine increases in **viscosity,** or thickness. Thick oil makes it harder for the engine crankshaft to turn. If the battery is not charged fully at all times during the winter, it can become discharged very quickly. Even in warm weather, problems can occur. Troubleshooting tips for some of these problems are described here.

Engine Won't Crank

Suppose you go out to start your car in the morning or after work. You put your key in the ignition switch and turn it to START, but the engine doesn't turn over. Make sure your transmission is in PARK or NEUTRAL, and if equipped with a clutch, have the clutch pedal depressed. Follow the manufacturer's recommended procedure for starting the vehicle.

Check for Battery Output If your car's engine won't turn over, try turning on the headlights. If the headlights are at full brightness, the battery is probably fully charged. If you have no headlights or they are very dim, the battery is very weak.

Check for Starter Current Draw If the headlights burn brightly, leave them on and try starting the engine. If the headlights remain bright and nothing happens, the problem is an inoperative starter circuit. The cause may be any of the following:

- Loose, broken, or corroded connections in the starter circuit
- Defective starter solenoid or relay
- Faulty ignition switch
- Faulty neutral safety switch or clutch switch

Feel the battery cables. If the engine won't turn over and the cables are warm or hot to the touch immediately after trying to start, the most likely cause is a short circuit or accidental ground in the starter circuit.

If the battery cables are cool, there is a chance that you may be able to jump-start the engine with booster cables and the energy from another car's battery. A safe procedure for jump-starting is described later in this chapter. If you try to jump-start your engine and it still won't turn over, call a professional technician.

If the car can be jump-started but will not start on its own again, have a technician check out both the charging system and the starter system.

If the headlights dim noticeably when you try to start the engine, there are several possible causes. Your battery may be weak. There may be enough current to turn the lights on, but not enough to carry the heavy drain of the starter, which can run as much as 150 to 300 amps.

Another possibility is that there may be a short circuit or an accidental ground in the starter circuit system that is draining off current quickly.

Listen for Noises In some cases, you may hear a rapid clicking sound as you try to start. The sound means the starter relay repeatedly is energizing and making contact, then deenergizing and releasing. This condition occurs when the starter draws a large amount of current from a weak battery.

You also may encounter a situation in which you hear one rather loud click, then nothing else. The click usually indicates a problem in the starter. The click you hear is the solenoid moving the pinion drive gear outward to mesh with the flywheel gear. However, the starter motor is defective and is not rotating the pinion drive gear.

If you hear a high-pitched whizzing or whining sound, the starter drive pinion is turning, but the drive pinion gear is not engaging the flywheel ring gear teeth. The most probable cause is a defective starter drive or solenoid.

An amperage draw test can be made to determine if the starter motor is malfunctioning. Other tests to check the starting system are possible, but should be performed by a professional technician.

Engine Cranks But Does Not Start

Suppose you turn the key in the ignition and the engine turns over rapidly as though it wants to start, but it never does start and run on its own. Remember, the job of the starting system is to crank the engine quickly. If cranking occurs normally, the battery and starter system are functioning properly.

As described in Chapter 3, an engine must have compression to run. Next, there must be a supply of fuel. Fuel delivery checks are described in Chapter 4. The next possibility is that there is no ignition spark.

Check for Ignition Spark You can check the ignition by using a screwdriver with an insulated handle. Remove one of the spark plug wires, making sure to handle it only by the boot. Insert the screwdriver point into the boot. Then hold the screwdriver so that the metal shaft is just above a grounded metal object, such as the engine block. Be careful of the spark and keep your hands away from the metal of the screwdriver. If you don't, you may get shocked. Another way to test for a spark is to use an old spark plug or a spark tester (Figure 5–22). A spark tester looks like a spark plug with a clip on it. Insert it into the end of the spark-plug wire and lay the body of the spark plug on a good ground, or clip the spark tester to a good metal ground. Have someone try to start the engine. If you get a sharp, blue-colored spark that

Spark Plug

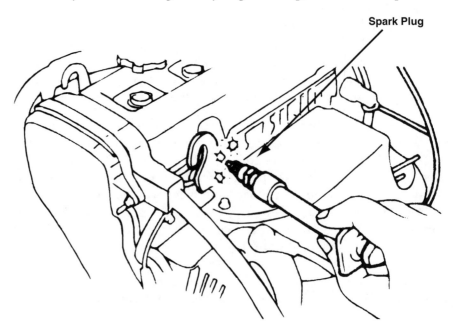

Figure 5–22: Spark can be checked by using a spark plug and holding the tip about ¼ inch away from a good metal ground and observing the spark while cranking.

arcs rapidly across at least one-quarter inch of space, the ignition system is functioning. If there is no spark, you may need ignition system work.

Consider Mechanical Problems   On engines with timing belts and more than 60,000 miles, the timing belt may be broken or may have skipped a tooth on the timing gears. Thus, the action of the valves and the action of the pistons are not properly synchronized. The engine will not run. Consult the owner's manual or service manual for recommended service intervals for your engine.

Other Problems

Occasionally, you will find a situation in which the engine will keep running as long as you have the key in the starting position. When you release the key, the engine sputters and dies. This problem could indicate an ignition system problem or a faulty ignition switch.

The best advice is to perform as many of the tests described here as you can. If you can't diagnose or repair the problem yourself, get help.

The Charging System Fails

If the engine is running and the alternator light comes on, or the voltmeter indicates less than 13.2 volts, the charging circuit is not operating properly.

Having an electrical system failure while you are driving may not seem as critical as if your engine overheats or loses oil pressure. The battery can continue to supply electricity to the ignition system for a short time. Barring other problems and use of other accessories, you could drive for a half an hour or more after the alternator stops charging the battery.

However, in many cars, the same belt that turns the alternator to produce electricity also turns the coolant pump (and possibly the fan) of the cooling system. While you can drive for a short time without a working alternator, you must have a working cooling system at all times. So, if you experience an electrical system failure, you must follow the same procedure as for other warning lights that come on suddenly: As soon as it is safe, pull off to the side of the road and shut off the engine. Then check to see if a fan/water pump/alternator belt has broken.

Even if the water pump or fan belt are intact and functioning, you should understand that if you keep driving after you experience an electrical failure, your car is running directly off the battery. The battery is not being recharged. So, when the battery drains, you won't be able to drive any more or start the car if it stalls. Depending on the condition of your battery at the time, you may be able to drive up to 50 miles or so before the engine dies. The best advice is to try to diagnose the problem immediately. If you see an obvious problem, such as a broken belt, you may try to replace it yourself. Otherwise, understand that you have a problem and get the car towed or fixed at your first opportunity.

Electronic Device Service

In many cases, computers store fault codes, or diagnostic trouble codes (DTC), that can be used to help diagnose problems in a computer-controlled system. The best way to retrieve these codes is through the use of a scan tool. The scan tool is used to communicate with the on-board computer and display to the technician the list of DTCs. The meanings of the DTCs for specific vehicles are included in the scan tool and the manufacturers' service manuals.

Computers are among the most reliable of automotive parts. A computer module that lasts the first 10,000 miles of use in a car will probably last for the life of the car. In most cases, computer modules are not serviceable. If a computer module is defective, it must be replaced as a unit.

Test procedures for particular computers processing units, or modules, are given in manufacturers' service manuals. With a given input signal, the computer should produce a certain output signal. If not, the unit is defective.

Because some technicians still are unfamiliar with computers, they may tend to blame a computer and replace it before checking other, basic systems. Manufacturers' surveys indicate that approximately 90% of automotive computers that have been replaced are functioning normally. However, as with other electrical parts, computers are not returnable. If you buy a replacement computer module, it cannot be returned for a refund or credit.

Jump-Starting

Earlier in this chapter, several situations are identified that call for connecting a dead battery to the battery of another car that is running. To perform a jump-start, you need a good-quality set of jumper cables. You can buy jumper cables in auto supply stores and department stores. It's a good idea to keep a set in the trunk.

Position a car with a good battery close to the disabled car. Don't let the two cars touch one anther or you may create an inadvertent path for current to flow. Connect the jumper cables in the number order shown in Figure 5–23. Use one cable to link the positive terminals of the two batteries. The other jumper cable goes from the negative terminal of the battery in the operating vehicle to a grounded metal part away from the disabled vehicle's battery. Make sure all cables are away from fans, belts, and pulleys on both cars.

⚠ CAUTION

Always connect the grounded cable clamp to an engine ground last. If the last connection is made at a battery terminal, a spark could ignite hydrogen gas from the battery and cause it to explode.

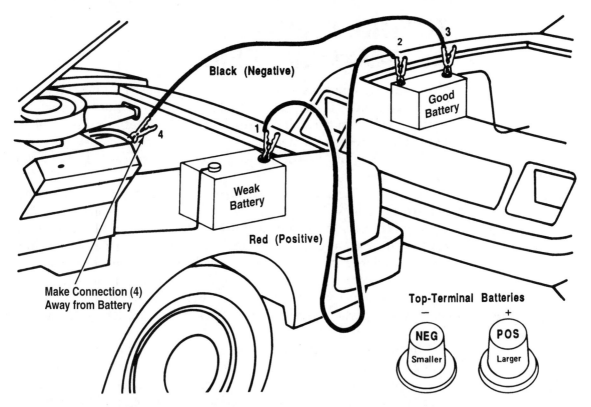

Black (Negative)

Good Battery

Weak Battery

Red (Positive)

Make Connection (4)
Away from Battery

Top-Terminal Batteries

NEG −
Smaller

POS +
Larger

Figure 5–23: Wear eye protection and follow the number sequence (1 through 4) to perform a safe jump-starting hookup. Make the last connection, 4, to a thick, grounded metal part as far away from the weak battery as possible.

Start the engine of the car with the good battery and accelerate it to a fast idle before you attempt to start the vehicle with the weak battery. Running the engine at a fast idle will ensure maximum output from the good battery. Wait a minute or two to allow the weak battery to accept a partial charge and help in starting the disabled vehicle. After a minute or two, start the disabled vehicle. Let the engine run for a minute or two before you disconnect the cables so that it will warm up and continue running on its own.

CAUTION

Avoid the fans, belts, and pulley when you disconnect the jumper cables. Severe injuries can result from being caught in moving machinery.

Disconnect the cables in the opposite order (4–3–2–1) shown in Figure 5–23. When you disconnect the cables, be careful not to touch the connectors (clamps) of the cables to each other while either cable is connected to a battery terminal. If the connectors touch, they will spark and may damage the battery or cause it to explode.

FUTURE ELECTRICAL SYSTEMS ARE HERE

Every year, automobiles require more electrical power for new vehicle features. Items such as electrically heated seats and entertainment, safety, and navigational systems are already reaching the limits of current 12-volt systems. The new 42-volt systems provide the extra power needed and also reduce the size of the electrical wiring, which makes a car lighter and more fuel-efficient.

Almost 5,000 feet of wiring will go into the cars of the near future. With a 42-volt system, wiring bulk is reduced because higher voltage lowers the electrical current, enabling smaller wiring. The 42-volt electrical system allows the manufacturer to satisfy customer demands for improved comfort and convenience items. A smaller 12-volt battery, roughly the size of a motorcycle battery, could be used supplementally to operate items such as the radio, emergency flashers, and other items such as cell-phone chargers, laptops, and portable gaming devices that would still require the use of a 12-volt power supply.

The high-voltage electrical system also paves the way for advanced technology items such as electromechanical valvetrain, electric superchargers, electric air conditioning, electric brakes and steering, and active suspension systems. With electronic valve actuation, valve timing could be easily programmed and controlled by the computer. The end result would be more horsepower from cleaner-running, smaller displacement, and more compact engines.

Next-generation vehicles will see higher levels of fuel economy and offer more comfort and convenience features by utilizing a technology that automatically shuts the engine off when the vehicle is stopped. The integrated starter-generator will allow the engine to automatically shut off when the vehicle is stopped, such as at a traffic signal, and then immediately start again when the need for power is anticipated. This system will require a more powerful electrical system that delivers 42-volts. A regenerative braking system that captures energy generated from the braking system to help recharge the vehicle's 42-volt battery could also be used.

Chapter 6 covers the lubrication and cooling systems, those parts and devices that keep the engine from overheating. These systems, as well as the fuel and electrical systems, are vital to the continued operation of your engine.

1. What are the essential components of the electrical system?
2. Describe a short circuit condition.
3. What does a battery store?
4. Describe the basic operation of the battery.
5. List the safety precautions when working on the battery.
6. What maintenance may be performed if the battery has removable vent caps?
7. Why might the battery require frequent addition of water?
8. What could cause a battery to be frequently undercharged?
9. List common starter system components.
10. How does the starting system operate?
11. What are the functions of the charging system?
12. What are the main components of the charging system?
13. What maintenance is required for the charging system?
14. What are the two functions of the ignition system?
15. How does the coil generate the high-voltage spark?
16. What are some common types of headlights?
17. What is the purpose of fuses?
18. Describe closed-loop and open-loop operation of the computer.
19. What are some common electrical system problems?
20. How can ignition spark be checked?
21. How can the charging system be checked?
22. Describe the proper jump-starting procedure.

THE LUBRICATING AND COOLING SYSTEMS

After reading this chapter, the reader should be able to:

- List the components and describe the function of the lubrication system.
- List the safety precautions for lubrication system maintenance.
- Describe the properties to look for in oil.
- List some common problems that could occur in the lubrication system.
- List the components and describe the function of the cooling system.
- List the safety precautions for cooling system maintenance.
- Describe the basic maintenance for the cooling system.
- Describe the different types of coolant available.
- List some common problems that could occur in the cooling system.

THE NEED FOR COOLING AND LUBRICATION

The burning of fuel within cylinders produces temperatures high enough to melt parts of your engine. In addition, metal parts rub against one another thousands of times each minute. Without the two systems discussed in this chapter, an engine would be badly damaged by heat and friction within minutes.

THE LUBRICATION SYSTEM

A **lubricant** is a material that reduces the effects of friction caused by contact of moving parts. The lubricant helps to protect, seal, and cool the internal moving parts of the engine. A system of parts and channels within the engine distributes the lubricant, which is motor **oil,** to perform five distinct functions:

1. Reduce friction.
2. Absorb heat and help in engine cooling.
3. Help to clean debris from between moving parts.
4. Form a film, or **oil seal,** between the piston and the cylinder wall, which lessens **blowby,** the leaking of combustion gases past the piston.
5. Cushion shock that results from the power stroke and movement of engine parts.

CHAPTER 6

Lubrication System Parts

To accomplish the functions listed, the lubrication system has a number of parts. There is a reservoir, or *oil pan*, to collect and store a sufficient amount of lubricating oil. The oil pickup and screen collects the oil in the oil pan and directs it to the oil pump. The *oil pump* forces the lubricant from the oil pan to the oil filter. The *oil filter* strains out particles that may be harmful to the moving parts. From there, the pressurized oil is distributed through a series of passageways, called *galleries*, throughout the engine. The lubricant circulates through the clearance, or gap, between moving parts to cushion shock and flush away debris. As the oil circulates, it picks up heat from the parts. Finally, the oil drains to the oil pan and is picked up by the pump again.

A typical lubrication system is shown in Figure 6–1. Key parts of the lubrication system are described in the following sections.

Oil Pan   Often called the *sump* or *reservoir*, the oil pan is a storage area for engine oil. It is typically made from sheet steel or cast aluminum. The oil pan is bolted to the bottom of the cylinder block just below the crankcase. A typical engine oil pan holds about 4 quarts of oil. There is approximately another quart in the oil filter. To determine an engine's oil capacity, refer to your owner's manual.

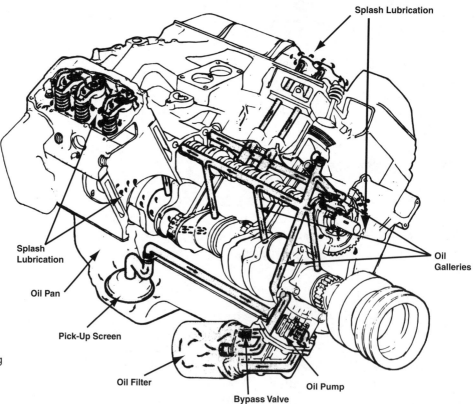

Figure 6–1: The lubricating system circulates oil under pressure to moving parts within an engine.

The shape of the oil pan varies, depending on how many chassis components it must clear. Also, to make sure that the oil pump pickup, or intake opening, can draw oil from the pan, some manufacturers install baffles, or barrier strips of metal, inside the pan. These baffles keep the oil from flowing away from the pickup area. Engines today use a "wet sump" oil pan, meaning that the oil is stored in the oil pan and the pump supplies the engine oil from the pan. Racecar engines may use a "dry sump" oil pan in which the oil is pumped out of the pan to a separate storage tank and the engine is supplied oil from the storage tank. The storage tank may be of any size and oil capacity and can be increased without the need for a larger oil pan under the car.

Heat from the engine will raise the oil's temperature. As the oil returns to the pan, this heat is transferred into the metal of the oil pan. Heat transfers from the pan to the air circulating around it. To enhance oil cooling, some automobiles have ribbed, or finned, oil pans. These ribs help dissipate more heat to the air that moves under the car. Some cars and trucks may have an added engine oil cooler. This is a small radiator-like device that can more effectively cool the oil for heavy-duty applications.

Oil Pickup and Screen Located in the bottom of the oil pan is the oil pickup and screen. This unit collects oil from the pan and directs it, via a tube, to the oil pump. The screen prevents any large particles of dirt from entering and damaging the pump.

It is important that enough oil be collected by the pickup and screen to feed the oil pump. Two common types of oil pickups are used in today's automobiles: the floating type, which is located just below the surface of the oil in the reservoir, and the fixed type, which does not move. The type of pickup used depends on the lubrication requirements and oil-pan design.

Oil Pump The oil pump does exactly what its name implies; it pumps oil through the engine. The oil pump is usually of the gear or rotor type (Figure 6–2). These are both *positive-displacement pumps;* that is, the faster the pumps are turned, the more oil is pumped into the engine and the greater the pressure produced.

At higher engine rpm, the oil pump might create too much pressure. To maintain the proper oil pressure, the oil pump has an oil pressure relief valve. This regulating valve contains a ball and a spring. The ball covers a passage that leads back to the oil pan. When pressure becomes excessive in the system, the ball is pushed off its seat. Thus, the oil is returned to the oil pan and the pressure within the lubrication system drops. In operation, the ball is unseated just enough to maintain a constant oil pressure at a given engine speed and temperature. The oil that continues through the pump passes on to the oil filter. The relief valve is not adjustable.

Oil Filter The oil filter removes contaminants from the oil before it is pumped through the oil galleries. Oil filtering systems used in vehicles are

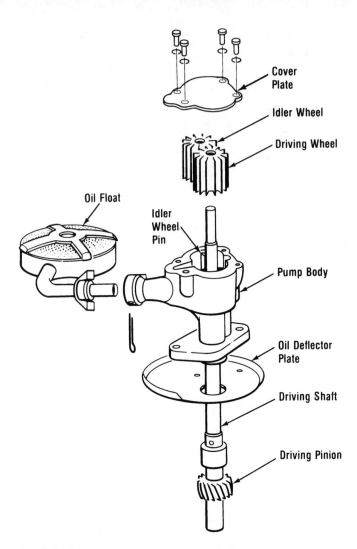

Cover Plate

Idler Wheel

Driving Wheel

Oil Float

Idler Wheel Pin

Pump Body

Oil Deflector Plate

Driving Shaft

Driving Pinion

Figure 6–2: Exploded view showing the internal parts of a gear-type oil pump.

the full-flow type with a bypass. Full-flow means that all the oil normally passes through the filter before it can flow through the galleries and to the moving parts of the engine. However, if the filter becomes clogged with contaminant particles, the oil pressure could drop, causing oil starvation to moving parts. To prevent this problem, a bypass valve, similar to the unit on the oil pump, is used to redirect unfiltered oil to the engine. Dirty oil is preferable to no oil at all.

The oil bypass valve may be located in the filter mount on the engine or within the filter itself. Never substitute another filter for your car without checking to see if it also meets the bypass requirements of your engine. Simply having a filter that fits doesn't mean it will work properly. You should always use the recommended filter.

Modern oil filters contain a pleated paper material designed to trap solid particles and small amounts of moisture. The filtering ability of these paper elements is measured in microns, or millionths of a meter. Thus, they remove even extremely small particles from the oil. The filter spins while installing it and is commonly called a *spin-on* filter. A filter in which only the paper element is replaced is called a *cartridge* type.

Oil Galleries As the crankshaft rotates within the oil in the oil pan, oil is splashed onto some of the moving parts of the engine. Modern overhead-valve and overhead-camshaft engine designs require a steady supply of oil to lubricate valvetrain parts within the cylinder head and bearings that support the crankshaft and camshaft. The pressurized lubrication systems force oil through a series of passageways—oil galleries—to deliver oil to parts within the block and upward to the cylinder head. The engine is crisscrossed with a network of galleries that direct oil to all of the moving parts of the engine. Oil galleries are either drilled or cast (formed when the molten metal is poured into a mold).

Splash lubrication is often used for timing chains, piston pins, and cylinder walls—oil is thrown off by the rotating crankshaft and splashes on these parts. Some engines have special holes in the connecting rods or oil jets in the galleries that squirt or spray oil to help lubricate and cool these components.

Whatever the method of lubrication, the oil still drips back into the oil pan and is recirculated by the pump.

SAFETY PRECAUTIONS FOR LUBRICATION SYSTEM MAINTENANCE

Observe the following safety precautions when you perform maintenance on the lubrication system:

- Wear eye protection whenever you work on cars. Safety glasses or goggles can prevent eye injuries, infections, and possible blindness.
- Position the car on a flat and level surface before you begin work. Never work on a car that is on an incline.
- Put the transmission in PARK (automatic transmissions) or in a gear (manual transmissions) and set the parking brake firmly. Place large wheel chocks or blocks both in front and in back of the wheels that remain on the ground to prevent the car from rolling or moving.
- Never put any part of your body under a car that is supported only by a jack. Always use safety stands or jack stands to support a car safely off the ground. A manufacturer's service manual will show you the proper locations on the car's underbody where you should place the stands. Alternately, use drive-up ramps to raise one end of the car safely. And remember to block the wheels that remain on the ground!

LUBRICATION SYSTEM MAINTENANCE

An engine must have an adequate supply of engine oil at all times to operate without damage to internal parts. The oil level must be checked regularly, and oil must be added if necessary. The most important maintenance for an engine's expensive internal parts is frequent engine oil and oil filter changes.

Many car owners are lax about checking oil and try to stretch intervals between changes. This short-term economy is an expensive mistake. You will shorten your engine's life severely if you let the oil level become too low or neglect to change the oil and oil filter on a regular basis.

Realize that doing any type of preventive maintenance more often than is recommended by the manufacturer cannot harm the engine or any other part of the car. However, you must try to balance the cost of the maintenance against the expected benefit. In the case of engine oil changes, you are balancing typically less than $25 for an oil change against $2,500 or more to rebuild an engine. Also realize that those fortunate individuals who set records of more than 200,000 miles on an engine without a rebuild, check and change their engine oil and filter much more often than is recommended in owners' manuals.

Choosing the Proper Lubricant

Three groups, the American Petroleum Institute (API), the Society of Automotive Engineers (SAE), and ILSAC (International Lubricant Standardization and Approval Committee), have established standards for engine oil. Your owner's manual will indicate what type and viscosity (weight) of oil is recommended for your engine. Quality oils will have API quality marks printed on the container (Figure 6–3). Look for the API quality marks every time you buy motor oil.

API Service Rating   When you buy oil, there will be a code printed on the container such as API Service SH, SJ, or something similar. The "S" indicates that the oil meets certain qualifications for spark-ignited (gasoline) engines. The coding system indicates grades from SA (nondetergent oil, not for use in engines built since 1963), through SL (detergent, highest grade, for use in all gasoline engines). Unlike grades in school, SL is the best grade of engine oil, and SA the worst. The scale is open-ended; that is, the next designation of higher-quality engine oil will be SM. Most oils available usually have the highest rating of SL.

If your car has a diesel engine, you should look for oil coded "CF-4," "CG-4," or "CH-4." The "C" indicates that oil is graded for compression-ignited engines, in other words, diesels. Some diesel engines can use either

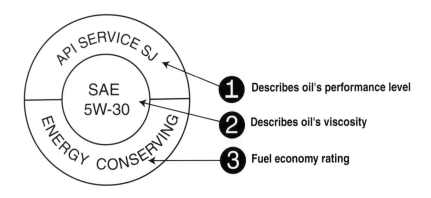

1. Describes oil's performance level

2. Describes oil's viscosity

3. Fuel economy rating

Figure 6–3: The API Service Symbol indicates an oil's service rating, viscosity, and fuel economy rating. The API Certification Mark, or starburst, indicates that the oil meets current standards of the International Lubricant Standardization and Approval Committee (ILSAC).

CF-4 or CG-4D oil. Others must use CH-4 only. Check your owner's manual or ask at your dealership. These oils are available at local auto supply shops or at commercial truck stops.

You may have heard of synthetic oil, made from chemicals other than petroleum. The synthetic oils produced by major oil companies are excellent products with antifriction characteristics superior to petroleum oils. When these oils were first marketed, there were claims that the oil change intervals could be increased to as much as 25,000 miles. However, no vehicle manufacturer or oil filter manufacturer recommends an oil and filter change interval longer than 10,000 miles or six months, whichever comes first. The oil and filter will become contaminated with dirt, water, and other combustion by-products and eventually require changing. The key to maintaining your vehicle warranty and to protecting your engine is in frequent, regular checking of oil levels and oil/filter changes.

SAE Number   The SAE number indicates the viscosity, or thickness, of oil. Cold tends to thicken oil. Therefore, thinner, or a lower viscosity, oil is needed for cold weather. Because heat will thin oil, thicker, or higher viscosity oil, is appropriate for warm climates.

For example, SAE 50 is a high viscosity oil. When the weather is extremely hot, the dense molecular structure of a high viscosity oil maintains an oil film that will protect engine parts against wear and scuffing. SAE 10, on the other hand, will flow easily in low temperatures. A "W" in an SAE number means that the oil has been tested for cold weather (winter) operation.

As you may know, oil manufacturers also produce **multiviscosity oils,** such as 5W30, 0W30, 10W-30, or 10W-40. The multiple numbers indicate that the oil has been tested for both cold weather and hot weather protective ability. Let's look at 10W-30 as an example. The first number, along with the "W," indicates that the oil viscosity is the equivalent to a single viscosity 10 weight oil at 0 degrees Fahrenheit (–180 degrees Centigrade). The second number, after the "W," shows the oil has the same viscosity as a 30-weight oil at 210 degrees Fahrenheit (99 degrees Centigrade). Multiviscosity oils do not really "change" weight or viscosity as they are heated. They still thin out from an increase in temperature, just not as fast as single viscosity oils. Multiviscosity can be thought of as a resistance to thinning out due to heat. These oils offer fuel economy advantages and improved protection from wear. For these reasons, multiviscosity oils are recommended in owner's manuals.

ILSAC   The International Lubricant Standardization and Approval Committee (ILSAC) in association with the API created a symbol for the front of the container that indicates the oil meets the latest performance category, SL, and the energy conserving standard. This symbol is called the API Certification Mark or "starburst." The starburst is designed to identify engine oils recommended for a specific application (such as gasoline service). An oil may be licensed to display the starburst only if the oil satisfies the most current requirements of the ILSAC minimum performance standard for this application, which is currently GF-3.

High-Mileage Oils   Many people are keeping their primary vehicle from five to seven years. And older vehicles don't fall apart after a few years either. Your car can run well beyond 100,000 miles if you maintain it on a regular basis. Oil companies are now marketing special motor oils for high-mileage cars. These new motor oils are formulated to condition seals, reduce oil consumption, reduce engine oil leaks, and improve engine performance—all bold statements. Reducing oil consumption is accomplished by manufacturing oil with a lower volatility base as its main ingredient. It is less prone to evaporate at operating temperatures than other oils. Conditioners are added that penetrate the pores of the seals to make them more effective. Better sealing of the valve seals reduces oil consumption past the valve guides and, hence, less oil means fewer deposits on the intake valves. Better sealing of crankshaft and camshaft seals reduces external oil leaks. High-mileage oils may also have more viscosity-index improvers added. They seal piston-to-cylinder clearances better and won't squeeze out as easily from the increased engine bearing clearances. They also may have more antiwear additives to slow the wear process.

Checking/Adding Oil

The oil level should be checked each time you stop for fuel. Regular checking of the oil level gives you an indication of whether the engine is using too much oil.

A few minutes after your engine is turned off, the oil drains down into the oil pan. At this time, you can measure the amount of oil by using the dipstick, which dips into the oil pan area (Figure 6–4). When you remove the dipstick from its position in the engine, markings on the metal shaft indicate how much oil is in the pan area.

You can recognize a dipstick by the characteristic loop or T-shaped handle at the top end. However, there may be more than one dipstick in the engine compartment. The engine oil dipstick's location will vary among different types of engines, but is typically on the side, rather than at the front or back, of the engine.

Wait a few minutes after the engine has been shut off. Pull straight up on the dipstick and remove it from its tube. Wipe the dipstick shaft with a rag or paper towel to remove oil that has splashed onto it while the engine was running. Note the markings on the end of the stick and any other

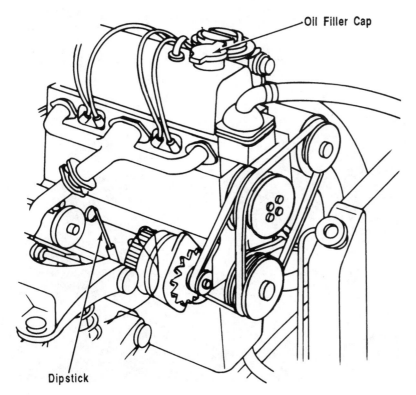

Oil Filler Cap

Dipstick

Figure 6–4: The oil dipstick extends downward to the oil supply in the bottom of the oil pan.

FULL

ADD

instructions stamped into the metal. Then reinsert the dipstick fully and remove it once more to check the oil level.

Figure 6–5 shows the markings at the end of a dipstick. The upper mark is the line at which the oil is at the FULL level. The lower line means ADD oil. If the oil level is below this mark, your engine may suffer extensive damage. When you reinsert the dipstick into the tube, make sure it is fully seated so that water and dirt cannot enter the engine through the tube.

If the oil level is between the FULL and ADD marks, there is enough oil in the engine for safe operation. If the oil level is at or below the ADD line, remove the oil filler cap, typically a large screw-on or push-on cap located near the top of the engine (Figure 6–6). Pour in about a half quart of oil. Allow a few minutes for the oil to drain down to the pan and recheck the level. Continue adding oil until you are at or near the FULL mark, but make sure not to overfill. Overfilling can cause frothing and foaming of the oil and lead to engine damage.

The newer plastic oil containers have funnel-shaped tops to help you pour the oil into the fill hole. However, on some engines you may need a long funnel to reach the fill hole without spilling oil (Figure 6–7). Always make sure that the area around the fill hole and the funnel are clean before use. Otherwise, you will be adding dirt to your engine in addition to oil.

Figure 6–5: The upper mark on a dipstick always indicates FULL and the lower mark always indicates level at which lubricant must be added to prevent engine damage. Some dipsticks do not have printed indications, just the lines.

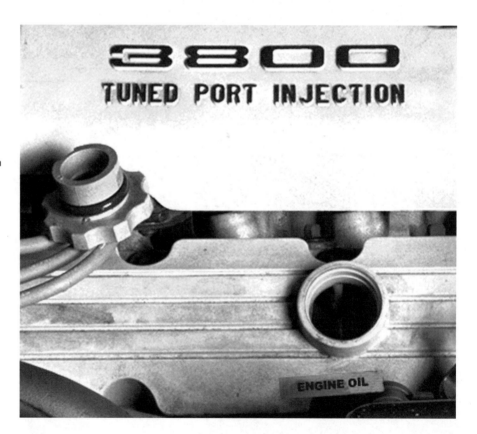

Figure 6–6: The oil filler cap usually is located on the valve cover. Note the dipstick nearby the filler hole.

Figure 6–7: **You may need a spout or funnel to add oil through the filler hole.**

You can save substantially on this type of routine car maintenance. For a quart of engine oil, service stations and repair shops typically charge double, or more, the price that you would pay in an auto parts store or supermarket.

If your car uses a quart of oil more than once every 1,000 miles, there may be problems. However, don't be upset if a new car uses a little oil during the **break-in** period. Typically, it may require a few thousand miles of driving for the piston rings to seat, or become matched, to the cylinder wall. Until the rings seat, they will not seal properly. After the break-in period, you may have to add a quart of oil between oil changes.

Also, in very hot weather or when driving at high speeds over long distances, oil consumption will be greater.

Changing the Oil and the Filter

Changing engine oil—and the oil filter at the same time—can be easy and inexpensive. Done at frequent intervals, this service will help your engine run efficiently for a long time. Automotive engines are designed for low emissions and require clean oil. Dirt, water vapor, and acid in the oil can seriously harm the moving parts.

You can save substantial amounts of money by changing your own oil and oil filter. If you are interested in doing this job yourself, procedures for oil changes are included in Chapter 11. For oil recycling information check out the parts store where you purchased your oil or the Internet. A good Web site to visit is http://www.earth911.org.

Checking for Leaks

There are only two ways to lose oil: Oil either leaks out or is burned in the engine. Check for leaks each time you return to your car. If you find a puddle of oil, it may be from your car.

Oil usually leaks from a **gasket,** which is a formed piece of cork, rubber, paper, or soft metal that is compressed to form a seal between parts. Gaskets are used to seal the oil pan, valve cover, oil filter, or any of a number of components. To determine the location of the leak, you must get under the car or look under the hood. The problem may be minor enough to correct by tightening a few bolts or nuts. However, cracked gaskets must be replaced to cure a leak. Chapter 11 provides details on changing a valve cover gasket in case you want to tackle the job yourself.

Oil may also leak from a seal. A seal is a metal housing with a rubber lip installed at the ends of the crankshaft and some camshafts. It allows the shaft to be connected to components outside of the lubrication areas of the engine while keeping the oil in the engine. It may not be uncommon for seals to leak after many miles or years.

Another common source of oil leaks is from an oil pressure sender, a unit that operates an oil pressure gauge or causes the OIL light to come on if the oil pressure is low. When an oil pressure sender fails, pressurized oil squirts out of the engine rapidly.

If you find a leak and are losing oil regularly, you should have the problem taken care of as soon as possible.

LUBRICATION SYSTEM PROBLEMS

There are many symptoms you can look for that indicate problems within the lubrication system. Checking oil levels and watching for leaks is only part of maintaining your car's lubrication system. The following are some simple checks to help determine what course of action to take.

Blue-White Exhaust

You can't always tell whether a smoking tail pipe indicates a problem unless you know what to look for. Blue-white smoke could be a sign of oil burning, as described in Chapter 3. If you suspect that your car is burning oil, just follow this procedure: While running at relatively high speed in a low-traffic situation, quickly lift your foot off the accelerator pedal. If the cylinders are not sealing properly, oil will be pulled past the piston rings or the valve guides as the engine is slowed abruptly. When you press the accelerator pedal again, the result is a puff of blue smoke. A friend following in another car will be able to see any puffs of smoke that occur.

If your car is putting out blue smoke regularly, you are probably violating emission standards. Aside from this consideration, whether you do anything about burning oil can be a matter of cost. How much do you spend on oil? Compare annual costs for oil with the expense of having the cylinder head reconditioned or the engine overhauled.

Check the tail pipe. If there is a smoking problem of any type, one simple diagnostic step is to inspect the tail pipe. If the engine is running normally, the tail pipe may have a light black coating at the end. If the tail pipe is sooty and is dark black, your engine may be burning too rich a fuel mixture. (Diesels typically have black tail pipes.) If it is black and has a coating of oil, you are burning oil.

Blowby

Blowby is the leakage of combustion gases past the piston rings and into the oil pan. Excessive blowby can create pressure on gaskets and seals at many locations within the engine, which causes gaskets and seals to fail. In addition, **oil pumping** occurs, so that oil is drawn up past the rings and burned in the combustion chamber.

Removing the oil filler cap when the engine is running can show excessive blowby. If you see a continual strong stream of oily smoke—not just a little vapor—you have found a symptom of excessive blowby in your engine. A more complete explanation of excess oil consumption is found in Chapter 3.

THE COOLING SYSTEM

The purpose of the cooling system is to rid the engine of excess heat. Only about one-third of the heat produced by burning fuel to create force to move the car is usable. Another third is wasted as exhaust heat. The final third must be dissipated by the cooling system to avoid damage to internal engine parts.

Cooling System Functions

To dissipate excess engine heat, the cooling system performs four functions:

1. Absorbs heat.
2. Circulates coolant.
3. Radiates heat.
4. Controls temperature.

The coolant absorbs heat as it moves through the engine block. Heat energy from the burning fuel in the cylinders passes into the cylinder walls and cylinder head. Liquid coolant circulates through hollow spaces within the engine block and head to absorb the heat from the metal parts of the engine. The hollow spaces are known as *water jackets*.

After absorbing the heat, the hot coolant passes out through the cylinder head and enters the radiator. As the coolant circulates through the radiator, it gives up its heat to the metal tubes of the radiator. The radiator is made of brass or aluminum, metals that conduct heat well. As air passes through the radiator fins and around the tubes, heat is transferred to the air.

However, if coolant circulated at all times from the engine to the radiator, the engine would run very cool on cold days. Remember that chemical reactions, including the burning of fuel, occur more efficiently at high temperature. Thus, for the engine to operate efficiently, there must be a temperature control mechanism. This mechanism is the **thermostat.** It regulates how much coolant is permitted to flow through the radiator. After you start the engine, it should heat to an efficient operating temperature as quickly as possible and maintain that temperature without overheating.

Parts of the Cooling System

The cooling system consists of several interdependent parts that function together to maintain proper engine temperatures (Figure 6–8). The parts include:

- Radiator
- Fan
- Coolant recovery system
- Pressure cap
- Coolant pump
- Water jackets
- Coolant
- Thermostat
- Soft plugs
- Hoses

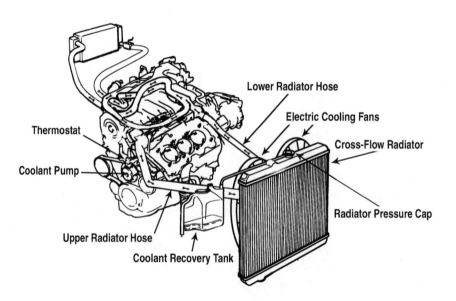

Figure 6–8: The main components of a cooling system.

Radiator   Radiator tubes may be arranged in a vertical (down-flow) or horizontal (cross-flow) pattern. Cross-flow radiators allow coolant to travel a longer distance, thus giving up more heat than down-flow radiators of comparable height. Also, the flat, horizontal shape of a cross-flow radiator can fit more easily into the front of vehicles with steeply sloped hoods.

The radiator consists of a network of tubes known as the **core.** A tank, or hollow half-round metal or plastic shell, is located at each end of the tubes. Thus, the tanks may be located at the top and bottom of the radiator, or at the sides depending on the radiator tube arrangement. The inlet side of the radiator includes a connection for the upper radiator hose and a neck where the radiator pressure cap fits.

Hot coolant from the engine enters near the top of the radiator. The coolant passes through finned tubes and releases its heat to the cooler air flowing around the core tubes. The fins increase the surface area of metal to which heat can be transferred.

The upper portion of the radiator removes any air bubbles that are formed as the coolant circulates through the engine. Air in the coolant can cause **cavitation** in the water pump. Cavitation is a condition under which a fluid pump attempts to pump air rather than liquid; it can cause wear to the water pump. Trapped air also can create hot spots in the engine block. Air bubbles that cling to the inside of water passages form an insulated area that does not allow the coolant to absorb engine heat.

By the time the coolant reaches the opposite end of the radiator, much of its heat has been dissipated into the airflow. If you car has an automatic transmission, the tank to which the lower radiator hose connects houses a transmission oil cooler, or heat exchanger. The transmission oil cooler is a sealed, hollow coil or chamber through which hot transmission fluid flows. The transmission fluid gives up its heat to the coolant in the outlet tank.

The lower-temperature coolant passes out through the radiator outlet and is drawn into the coolant pump and circulates through the engine again.

Fan   The fan is responsible for drawing air through the radiator to increase cooling. The fan is either electrically operated or driven by a belt.

On most cars, the fan operates electrically. A temperature sensor in one of the radiator tanks controls the operation of the electric motor that drives the fan. To improve fuel economy, the fan operates only when the liquid in the radiator is hot. The fan can either pull air through from the rear (engine side) of the radiator or push air from the front.

A belt-driven fan, used on some rear-wheel drive cars, is bolted to a pulley on the water pump and turns constantly with the engine. Thus, belt-driven fans always pull air through the radiator from the rear. The pulley on the crankshaft drives the belt.

Because a significant amount of engine power can be lost, most of today's belt-driven fans operate only when the engine and radiator heat up.

One such type of fan is the thermostatic fan. When the engine is cold, the fan freewheels and moves little air. When the engine warms up, the fan clutch engages to turn the fan.

Thermostatic fans operate either through a mechanical connection activated by a thermostatic spring or through a fluid coupling that uses a special silicone fluid. As with electric fans, thermostatic fans conserve power and contribute to fuel economy because they operate only when airflow is needed.

In heavy traffic when the weather is hot and the air conditioner is operating, you may notice a sudden rise in engine temperature. This increased temperature is normal for severe conditions. If the TEMP or HOT warning light comes on or the temperature gauge is getting near the top of the HOT range, use the following procedure. When it is safe, pull off the road out of traffic. Turn off the air conditioner and put the transmission in neutral. If the fan is belt driven, increase engine speed slightly. This procedure will speed up the fan, allowing it to draw more air through the radiator and increase cooling. If after a minute the temperature light does not turn off, shut the engine off, open the hood, and allow it to cool down.

A fan blade placed more than 3 inches from the **radiator core** becomes ineffective. It merely circulates the hot air around the fan blades. For this reason, some radiators are equipped with *shrouds*. A shroud is a large, circular piece of plastic or metal material extending outward from the radiator and enclosing the fan to increase its effectiveness. Do not permanently remove or cut the shroud.

Coolant Recovery System   The **coolant recovery system** saves coolant that ordinarily would be lost each time the engine overheats slightly. The system consists of:

- A special radiator pressure cap
- An overflow tube
- A tank or reservoir, most commonly a plastic tank

The coolant recovery system's operation can be seen after driving in heavy traffic or on a hot day. When you stop the engine, look under the hood. You will notice coolant flowing out of the radiator and into the reservoir. This overflow is caused by pressure buildup in the cooling system. As the coolant is heated, it expands and the excess is directed to the recovery tank.

The coolant recovery tank is located in the engine compartment near the radiator. A line runs from the overflow tube in the radiator neck into the recovery tank. When the engine cools off, the expanded liquid in the cooling system will contract. This contraction creates a low-pressure area in the radiator. Atmospheric pressure on the liquid in the recovery tank forces the liquid back through the tube and into the radiator and cooling system.

Some newer vehicles use a **pressurized coolant expansion tank.** Though not really a recovery tank, it does provide the same function. It allows for the expansion and contraction of the coolant. It also serves to eliminate air trapped in some parts of the cooling system. As its name implies, the tank is pressurized and the pressure cap is usually located on top.

Pressure Cap   The function of the **pressure cap** is to maintain pressure in the cooling system to help prevent the coolant from boiling (Figure 6–9).

The modern cooling system is very similar to a pressure cooker. Increased pressure raises the boiling point, allowing the coolant to be circulated at higher temperatures. For every 1 psi increase, the boiling point will increase 3.25 degrees Fahrenheit (1.80 degrees Centigrade). Thus, the coolant can absorb more heat, and there is less chance of boil-over. If the coolant boils, gases in the water jacket insulate the metal of the engine and the coolant cannot absorb the heat that is produced.

The radiator cap includes two valves: a pressure valve and a vacuum, or vent, valve. Under normal conditions, the pressure valve remains closed, completely sealing the cooling system. The large spring under the cap exerts pressure to maintain a sealed condition and constant pressure on the system. Because the system remains sealed, coolant loss is avoided. If the cooling system pressure exceeds the spring pressure of the cap, the spring is compressed and the pressure valve opens. Excess coolant then is pushed into the coolant recovery tank.

The vent valve allows coolant to be drawn back into the radiator from the coolant recovery tank. When the low pressure area forms within the radiator as the coolant loses heat and contracts, the vent valve opens.

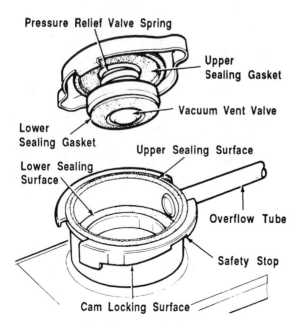

Figure 6–9: For the cooling system to operate properly, both the upper and lower sealing gaskets of the radiator cap must be in good condition.

Some cars have a pressure cap on the pressurized recovery tank instead of a radiator cap. In the case of the pressurized recovery reservoir, if the cooling system pressure exceeds the spring pressure of the cap, the pressurized air above the coolant is vented. The vent valve allows air to be drawn back into the tank as the coolant cools and contracts. This valve prevents the radiator tanks and other cooling system components from being crushed by atmospheric pressure.

Each pressure cap has a different rating, marked on the top of the cap. This rating determines the amount of pressure held in the cooling system by the spring. For domestic American cars, the pressure is stated in pounds per square inch (psi). For each psi stated on the cap, the system will raise the boiling level of the coolant approximately 3.25 degrees Fahrenheit (1.80 degrees Centigrade). Because the boiling point of water is 212 degrees Fahrenheit (100 degrees Centigrade), a 15-psi cap increases this to 206 degrees Fahrenheit (125 degrees Centigrade).

Some pressure caps typically have different markings. For example, many Japanese cars are marked "0.9"—meaning 0.9 times normal atmospheric pressure (14.7 psi)—or approximately 13.23 psi. European pressure caps may be marked "100," meaning 100% of atmospheric pressure, or 14.7 psi.

The coolant mixture of antifreeze and water also contributes to protection against boiling. As a result, most cars today run comfortably at between 220 and 230 degrees Fahrenheit (104.4 and 110 degrees Centigrade) without overheating.

Coolant Pump   The **coolant pump,** or water pump as it is often called, is responsible for circulating coolant. The coolant pump is mounted on the front of most engines and driven by a belt, but some may use a gear drive. In some cases, the water pump is located at the back of the engine and driven by the camshaft; its speed of operation, then, is controlled by engine speed. Fanlike blades, called the *impeller*, are enclosed within a housing. When the engine is operated, the engine causes the blades to spin and draw coolant in the inlet side and force it out of the outlet side of the pump.

The coolant pump draws coolant from the lower part of the radiator, brings it up through the lower hose, and pushes it through the engine. From there, the coolant passes through the thermostat, to an upper hose, then on to the upper part of the radiator. Figure 6–10 shows how coolant flows through the system.

Water Jackets   To transport coolant, water passages, or water jackets, are formed inside the block and the head when these parts are cast from molten metal. The passages are designed to deliver an even, unobstructed flow of coolant through the engine. Sharp bends or turns in the passages are avoided.

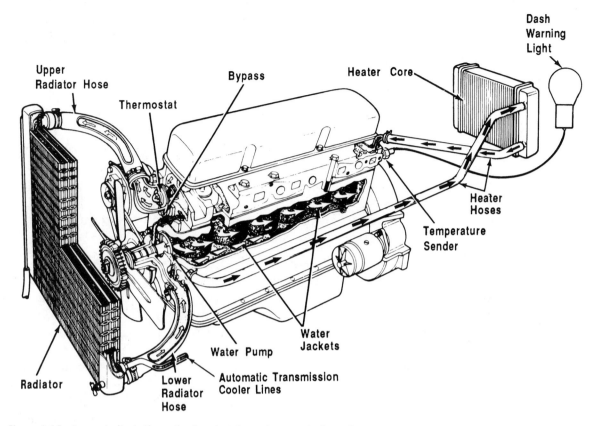

Figure 6–10: Arrows indicate the path of coolant through an engine's cooling system.

Coolant Coolant is a mixture of antifreeze and water. The proper mixture of antifreeze and water in the coolant will cut down on rust and lime deposits. These deposits insulate the walls of the water jacket and prevent the coolant from absorbing heat. Thus, hot spots that increase component wear and make overheating more likely are created in the engine. A 50/50 coolant mixture (50% antifreeze and 50% water) is the proper mixture to provide the best resistance to freezing and boiling. This mix will give a freeze protection of –34 degrees Fahrenheit (–37 degrees Centigrade).

There are several types of antifreeze from which to choose. The first is the traditional green antifreeze made with ethylene glycol (EG). This antifreeze is commonly used on many cars and has a two-year service rating; that is, the cooling system should be drained and flushed every two years.

EG smells and tastes sweet, thus making it easy to ingest, but it is toxic. When ingested, ethylene glycol converts to oxalic acid, which damages the kidneys and can cause kidney failure and death. Spills should always be cleaned up to prevent accidental ingestion by children and animals. Small amounts from a few teaspoons to several ounces can be fatal. More than

3,000 people a year accidentally ingest antifreeze, and over 90,000 pets die each year from antifreeze poisoning. Here are a few tips to follow when working with antifreeze:

- Wipe up and wash away any spills. When changing your car's antifreeze, keep your pets indoors.
- Keep stored antifreeze off the floor and away from pets.
- Keep new antifreeze in its original container.
- To store used antifreeze before disposal, use original containers and keep them sealed.
- Make sure your car or truck is in good working order and that the cooling system has no leaks.
- If your car's cooling system leaks, breaks down, or overheats, clean up spills thoroughly and dispose of rags carefully.

An alternative antifreeze is based on propylene glycol (PG). PG is much safer for the environment and is even found in many consumer products including cosmetics, pet food, and certain over-the-counter medications. This coolant also has a two-year service rating, which means that the cooling system should be drained and flushed every two years. PG coolants are also green and hence easily confused with traditional EG. Do not mix EG-based coolants and PG-based coolants, as they are not compatible. Some manufacturers do not recommend the use of PG-based coolants. They may even warn against the use of PG and state that its use could void the warranty. Be sure to consult the owner's manual for the recommended coolant. Complete and thorough flushing of the cooling system is recommended when changing to PG-based coolants.

Coolants blended from PG give better freeze protection but must be mixed at a different ratio. A 60/40 mixture will provide freeze protection of –54 degrees Fahrenheit, (–48 degrees Centigrade) and a 66/34 will provide freeze protection of –76 degrees Fahrenheit. PG costs more than EG, and the higher mixing ratios make it a more expensive coolant. Also, special antifreeze testers must be used because conventional testers are calibrated for EG. One popular brand of PG-based coolant is Sierra® and may be found in some parts stores and other retail outlets.

A third coolant emerged in 1995 as GM converted its new cars over to Dex-Cool®. This product is an organic acid technology (OAT) coolant designed to provide five-year or 100,000-mile service, whichever comes first. Known as a long-life or extended interval coolant, it is made from a traditional EG base and contains an OAT inhibitor to give it a longer service life. This antifreeze is sometimes referred to as EGOAT. Since it is EG-based, the precautions are the same as for traditional EG coolants.

The appeal of Dex-Cool is its extended service interval. It does cost a little more, but it has more than twice the life of traditional EG coolant. Dex-Cool is red in color and easily distinguished from its green coolant cousins. Mixing ratios, handling, service procedures, and testing of extended interval coolants is the same as for traditional EG products. If green coolant is added to a vehicle with red coolant, the cooling system must be treated as though it has EG as the coolant because any mixing of red and green coolants renders the OAT inhibitor ineffective.

If a vehicle was factory equipped with extended interval coolant, it is recommended that it be used throughout the life of the vehicle. If you want to change to extended interval coolant, a complete and thorough flushing of the cooling system is recommended before changing. The use of distilled water is also recommended. The coolant will be in service for five years or 100,000 miles and may react with contaminants in the water, particularly if the water contains calcium or is very hard.

Other manufacturers are also producing extended life coolants. A new technology from Zerex called G-05 is similar to Dex-Cool and uses hybrid organic acid technology to minimize inhibitor depletion. Zerex G-05 is approved by Ford and by DaimlerChrysler for worldwide applications, including all Mercedes engines.

Thermostat   When the engine is cool, the thermostat remains closed so that coolant cannot circulate through the radiator. Instead, coolant is circulated within the engine block and cylinder head until the coolant reaches a predetermined temperature. At that temperature, a waxlike pellet melts and expands inside the thermostat to open it and allow the coolant to flow (Figure 6–11). Today, most original-equipment thermostats open at approximately 195 to 215 degrees Fahrenheit (90 to 102 degrees Centigrade). In most cases, when the thermostat fails, it remains closed, and the engine overheats because coolant cannot flow into the radiator. Some people are tempted to install a lower temperature thermostat to keep the engine from running too hot. Always use the recommended temperature thermostat for your engine. Using a cooler thermostat will adversely affect engine operation and fuel economy, as the engine does not reach its designed normal operating temperature. A cooler thermostat is not a fix for an overheating problem.

Soft Plugs   Also called *core plugs* or *freeze plugs*, **soft plugs** are part of the engine block and cylinder head. These parts plug the holes that were used for casting water passages in the engine block and head. The term freeze plug comes from the fact that the holes they plugged were used to hold components in place as the molten metal cools into its solid form, or "freezes," as the engine block was cast. Some people think they will pop out to protect the engine block if the engine coolant freezes, which is not true. If plain water is used in the cooling system and it freezes, the engine will likely be ruined. The use of water or improperly mixed coolant may corrode, rust, and cause the soft plugs to leak. In some cases, the engine or manifolds must be removed to replace leaking soft plugs.

If the cooling system is not maintained properly, the engine soft plugs can rust through and leak. Rusted-through core plugs are an expensive and difficult replacement job. Though the price of a core plug is quite small,

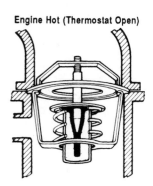

Engine Hot (Thermostat Open)

Engine Cool (Thermostat Closed)

Figure 6–11: The thermostat remains closed (*bottom*) until an engine has warmed to operating temperature, then opens (*top*) to allow coolant to flow through the radiator.

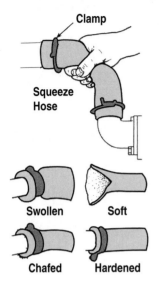

Clamp

Squeeze
Hose

Swollen   Soft

Chafed   Hardened

Figure 6–12: Common
defects in hoses.

labor is expensive. Standard core plugs are made of stamped steel and rust easily. It is best to have them replaced with brass or stainless steel plugs. However, if you maintain a good mixture of antifreeze and water in the cooling system and change it at regular intervals, you may never have to worry about core plugs rusting through.

Hoses   The hoses are the primary means of carrying the coolant to and from the radiator for the engine. They are also used to carry coolant to and from the **heater core.** Most every engine will use an upper and a lower radiator hose. These hoses are made from butyl or neoprene rubber and reinforced with nylon webbing or steel. Some hoses may be made from silicone. Silicone hoses are rather expensive but are very long lasting.

Hoses begin to deteriorate from the inside. It is very difficult to detect this deterioration but there are several signs to look and feel for (Figure 6–12). If the hose has any swelling, chafing, or bubbling, it needs replacement. When the engine is cool and the cooling system is not pressurized, gently squeeze the hose along its length. If any soft or hard spots are felt, replacement is needed. You may feel a wire or spring in the lower hose. This is normal and is used to prevent the lower radiator hose from collapsing. The lower hose should contain this spring.

Hoses generally last four years or 50,000 miles. Replacing hoses on a regular basis virtually eliminates radiator hose failure on the car. Be sure to properly install the hose as shown in Figure 6–13.

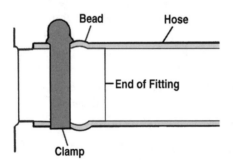

Bead   Hose

End of Fitting

Clamp

Figure 6–13: New clamps
should be installed immediately
after the bead of the fitting

SAFETY PRECAUTIONS FOR COOLING SYSTEM MAINTENANCE

As with all other systems on your car, there are hazards that can be avoided by using common sense. Observe the following specific safety precautions for maintenance of the cooling system:

- Whenever possible, avoid working on a running engine or one that is still hot. Serious burns can result from touching a hot cooling system or other engine parts. When possible, allow the engine to cool down for several hours before work is started.

- Do not open a radiator cap unless the metal of the radiator is comfortably cool to the touch. Removing the cap will lower the boiling point of the coolant and could allow the hot coolant to boil and gush out, causing severe burns.
- Avoid spilling antifreeze on hot exhaust system parts. Antifreeze can burst into flame if it is heated sufficiently. Severe burns can result.
- Shut off the engine before you check the fan, radiator, belts, or any part of the cooling system near them.
- Avoid working near electrically operated fans unless the battery is disconnected. Electrically operated fans can start at any time, even when the ignition switch is OFF.

COOLING SYSTEM MAINTENANCE

Checking and maintaining the cooling system can save you a lot of money—and a lot of grief. Overheating may lead only to an uncomfortable wait at roadside until the car cools, but results can be more severe. Overheating can damage the engine and automatic transmission and can lead to some very costly repair bills.

Check/Add Coolant

You must check the coolant level in both the radiator neck and in the recovery tank. If the radiator pressure cap is faulty, the level on the coolant recovery tank may indicate FULL when the radiator and cooling system are dangerously low. The radiator itself should be full to the neck. The coolant recovery tank is marked to indicate proper levels. If your car doesn't have a coolant recovery system, the level of the coolant in the radiator neck should be about one inch below the neck. Check the coolant level in the radiator neck only if the metal of the radiator itself is cool enough to be touched comfortably. On cars with a pressurized coolant reservoir, the level need only be checked at the reservoir tank as indicated by the marks.

CAUTION

Never take the pressure cap off the radiator of your car when the engine or radiator is hot!

Coolant heats as it is circulated through the engine to help control temperature. If you remove the pressure cap when the engine is hot, boiling coolant can spray out under pressure and scald you. Thus, the best and safest time to check and refill the coolant supply is when the engine is cool.

Some coolant is lost from evaporation, especially during very hot weather. This liquid must be replaced. If you notice that you are adding coolant more than once every few weeks, start looking for a leak.

Never add plain water to your system except in an emergency. Water causes rust and corrosion in the system. It's best to add a 50/50 mixture of antifreeze and water to your system. Antifreeze contains antirust and corrosion-inhibiting chemicals. But never use straight antifreeze. Straight antifreeze can cause overheating because it doesn't conduct heat as well as water, and it freezes at 10 degrees F (–12 degrees Centigrade). It is a strange curiosity of chemistry that mixing water that freezes at 32 degrees Fahrenheit (0 degrees Centigrade) and antifreeze that freezes at 10 degrees Fahrenheit (–12 degrees Centigrade) will yield a mixture that freezes at –34 degrees Fahrenheit (–37 degrees Centigrade) (50/50 mix). Straight antifreeze can also leave coking deposits, which will clog your cooling system.

The best mixture is one that follows the antifreeze manufacturer's recommendations. Those who drive in temperate climates should use a 50/50 mixture of coolant and water. Even if the car is driven in cold conditions, never use more than two-thirds coolant.

For convenience, it is best to premix your coolant in a clean plastic gallon bottle with a tight-fitting lid. Then, if you need to add coolant, just pour it in.

Check for Leaks

Under normal driving conditions, you should not be losing coolant. If you need to add coolant frequently, start looking for leaks. If you find large puddles of coolant, the most probable cause is a dried and cracked radiator or heater hose. The small leaks, though, are not always easy to spot. Coolant contains a coloring agent to help you find leaks. These coloring agents are usually red, orange, or green. Sometimes other colors may be used depending on the manufacturer. Just match the color of the leak to that of the coolant.

There are two kinds of leaks in your cooling system: a cold leak and a hot leak. You can spot a hot leak after you've been driving when the engine heats up and the cooling system builds up pressure. The cold leak doesn't begin to show up until the engine cools down.

The best time to check for a cold leak is in the morning, before you start the engine. Once the engine is started and the components expand, a cold leak will stop. There are certain telltale signs to look for—mainly coolant or rust stains. Check the hoses, especially at the ends. If you see evidence of a leak, tighten the hose clamps. Also, look for coolant or rust stains around the radiator. Check the inlet and outlet tanks, the seams, and the pressure cap.

If the pressure cap shows signs of leaking, it should be replaced. Look for cracks or worn gaskets. If you find any cracks in the gaskets, the entire cap must be replaced. The gaskets are not replaceable separately. A new cap usually costs five to ten dollars.

Leaks from the coolant pump usually are noticeable when the engine is hot. After the engine is shut off, you will find water dripping from the pump. If you have a leaking coolant pump, it is possible to come out in the morning and find almost all of the coolant on the ground. Use a mirror to check under and around the pump to determine if it actually is leaking. If you cannot replace the pump yourself, have someone else do it as soon as possible.

A radiator seldom leaks, but it is best to inspect it anyway. Age and repeated thermal cycling, heating and cooling, of the radiator can cause it to develop stress and cracks. Since it is out in front, debris from the road or other sources may have impacted the radiator and caused a leak.

If you suspect a leak but are not sure, you can take the car to a radiator shop and have a pressure test done. Pressure is applied to the entire cooling system. If the gauge shows a loss of pressure, there is a leak.

Clean Front of Radiator

Any restriction in the flow of air through the radiator can lead to overheating, so the front of the radiator must be kept clean. It is easy to accumulate bugs and debris as you drive. You can remove the accumulations by brushing them off with a paintbrush. Or, use a hose, flushing through the back of the radiator toward the grille.

Check Belts

As stated in previous chapters, if the belts show any signs of cracking or fraying, replace them. In the case of some new cars, only one belt may drive all of the systems.

To check belt tightness, simply press down on the belt. You should not be able to move it more than a half-inch per foot of span. If a belt moves farther, it needs to be tightened.

A loose belt will slip and may fail to turn the coolant pump and fan at full speed. Thus, overheating can occur. Too much slippage also will glaze, or polish, the belt. Excessive glazing can cause a belt to slip, even after it is tightened. A glazed belt should be replaced.

Check Fan Operation

Before you check the fan, be sure the ignition switch is off and the engine is not running. If the fan is powered by an electric motor, make sure to unhook the battery. Many serious accidents occur when an unsuspecting person sticks his or her hand into a moving fan or it starts suddenly.

On the belt-driven fan, just make sure the fan blades are not bent and are not hitting. If the fan seems noisy, grab the ends of the fan blades and wiggle the fan back and forth a few times. Looseness could indicate a bad coolant pump bearing. A bad bearing indicates that the coolant pump must be replaced.

The electric fan operation can usually be tested by starting the car and turning on the air conditioning. The cooling fans should start to run and pull air across the condensing coil located ahead of the radiator. If the fans don't start after a few seconds, there may be a faulty fan motor, relay, fuse, wiring, or computer problem. The computer is rarely at fault, so check the fuses and inspect the wiring first.

Temperature sensors that operate electric fans can fail and cause overheating. If the fan motor doesn't operate after the engine is warmed up, suspect the temperature sensor. If the temperature sensor fails completely, most cars will run the fan continuously and will turn on a light on the dash to alert the driver. If the fans run when the air conditioning is on and still don't run when the engine is warmed up, the temperature sensor may be slightly out of range. A technician may need a scan tool to retrieve diagnostic trouble codes and data from the computer. If the sensor is mounted in the lower part of the radiator, the radiator should be cooled and drained before you attempt to replace the unit.

If overheating problems occur, and your car is equipped with a belt-driven thermostatic fan, check the fan clutch. Since the thermostatic fan does not operate when it is cold, you should be able to spin the fan easily by hand. When the engine is warmed up, you should still be able to move the fan blades, but feel some resistance.

Some thermostatic fans on cars use a special silicone fluid to operate the fan clutch. If this fluid leaks out, the fan will not engage when the radiator heats up. In some cases, you can refill the fluid through a removable plug on the fan assembly. The special fluid is available at auto parts stores and dealership parts departments.

COOLING SYSTEM PROBLEMS

Cooling system problems can put your car out of business. Therefore, you should be able to understand some of the common troubles and be able to diagnose them.

Overheating

Overheating generally indicates a problem in the cooling system. A light or a temperature gauge on the dashboard will signal an overheating problem. When the engine overheats, the coolant may boil. You may hear the hissing of steam or see coolant coming from under the hood.

Again, as with most warning signals, pull over as soon as it is safe and shut off the engine. If you continue to drive, particularly at high speeds, your engine could seize, meaning that internal metal parts melt and stick to one another, ruining the engine.

If your car overheats because of cooling system problems, there are five common causes:

1. Loss of coolant. Coolant may have leaked out through a rotted or burst hose or from a leak in the radiator.

2. Improper circulation of coolant through the engine. Your cooling system may be full and your car may still overheat. Rust and corrosion may be blocking circulation in the system. Using plain water instead of properly mixed coolant in the radiator and coolant recovery tank typically causes this problem.

3. There may not be enough air circulating to keep the engine cool. Engine temperature is maintained by a total system, including airflow that carries heat away. If you are in heavy traffic, driving under stop-and-go conditions and using your air conditioner to keep yourself cool, the system can't transfer enough heat to the outside air, and the engine overheats temporarily.

   Air circulation also may be lost if the fan that circulates air through the radiator malfunctions. On older, rear-wheel-drive cars, this condition typically is caused by a loose or broken fan belt or defective fan drive mechanism. Airflow also may be reduced if the front of the radiator is coated with bugs, dirt, or debris. Air must flow freely through the radiator to allow the cooling system to give up heat from the engine. Many people put large bug screens in front of their radiators to keep debris and bugs from clogging the radiator fins. However, these screens generally decrease the circulation of air and increase the likelihood of overheating.

4. Loss in pressure. The pressure in the cooling system prevents the coolant from boiling until it reaches a temperature that is much higher than that encountered in normal driving. Any leakage in the hoses, radiator, or cap will cause a loss of pressurization. A defective pressure cap also can cause this problem. A cracked or broken rubber gasket on the pressure cap will allow the coolant to boil at a lower temperature than if the seal is intact. A defective cap can also allow the coolant to leak out and be lost.

5. Defective thermostat. The thermostat, which controls the flow of coolant, may malfunction or become stuck in the closed position. Thermostats are reliable and simple devices and use a wax pellet enclosed in a rubber jacket to open the thermostat. Thermostats can fail at any time. One symptom of a failing thermostat that is sometimes seen is an overheated condition indicated by a gauge on the dash when the car is first started and driven. Then the temperature gauge rapidly falls after a short time, followed by normal temperature operation. This symptom is caused by a sticking thermostat, which should be replaced.

In addition, there are many mechanical problems that can be related to engine overheating. These include:

- The water pump may be defective.
- The radiator hoses may be collapsed. The lower hose usually has a wire spring to prevent the suction of the water pump from collapsing the hose.
- A malfunction in the fuel system may lead to an incorrect fuel mixture. A malfunction indicator light on the dash should illuminate in this case.
- Ignition timing may be incorrect; early or late ignition of fuel can contribute to overheating.
- Improperly mixed coolant. Usually this is too little antifreeze and too much water.
- The automatic transmission may be overheating. Cars with automatic transmissions have a sealed heat exchanger located inside the radiator. If the automatic transmission malfunctions and overheats, it can cause the cooling system and engine to overheat.

Once you are in a safe location and the engine is off, lift the hood to improve air circulation and help cool the engine. Let the car cool by itself with the hood open for a half-hour or more before removing the radiator cap.

CAUTION

If the engine has overheated, do not attempt to open the radiator cap or locate the problem until the engine and radiator have cooled down safely. Serious burns can result if you attempt to work on an overheated cooling system.

You can perform a simple test to see if it is safe to remove the pressure cap. Wet your finger and touch the metal surface at the top of the radiator lightly. Be careful! The radiator might still be hot enough to burn you. Touch the metal the same way you would to test a hot iron. If you can touch the top surface of the radiator comfortably, it's safe to open the cap. Another test is to carefully squeeze the radiator hoses to determine if the cooling system is still pressurized. If the hoses are pressurized, then don't open the radiator cap.

When the engine and radiator have cooled to a safe temperature, look for the problem. Begin by looking for obvious leaks. At this point, the coolant system still may be under pressure, so leaks may be relatively easy to see. Coolant may be squirting out under pressure. If there is a leak, you can probably refill the radiator after the engine has cooled and drive slowly to a repair garage, keeping an eye on the temperature warning or gauge.

Coolant Leaks

The most common leaks come from hoses or the radiator. As explained earlier, check for rust or coolant stains around hose connections, the radiator, and soft plugs. Evidence of stains on parts or of coolant on the ground is a signal that further checking is needed. Avoid using stop-leak chemicals. Stop-leak chemicals are not a substitute for a proper repair, and, at best, should only be used in an emergency. These chemicals, though easy and tempting to use, can plug the radiator, heater core, and other cooling system passages.

A coolant pump leak is sometimes difficult to detect. A leak may not be visible when you are driving because the pump is pushing the coolant through the system. After the engine is stopped for a short period, however, the pressure built up in the system will make it leak.

Loss of large amounts of coolant without any visible signs of leakage can indicate a serious problem. When these conditions are accompanied by repeated overheating and large quantities of steam from the tailpipe, coolant may be leaking into the combustion chamber. At the very least, you can expect a leaking head gasket. Other possibilities include a cracked cylinder head or engine block. Should these symptoms exist, take your car to a professional technician. Coolant leaking into the combustion chamber is generally indicated by continuous white smoke from the exhaust pipe. More information on white exhaust is given in the following section.

To control a serious leak, you may be tempted to use some type of block-sealer chemical. This substance may stop a small seepage, but will not stop a sizable leak. As its name implies, it may also stop coolant flow through the radiator, heater core, and other cooling system passages. Again, these chemicals are no substitute for a proper repair. It is best to have the source of the problem fixed properly by having the block or head patched, welded, or replaced.

White Smoke from the Tail Pipe

White smoke (in reality, steam) from a tail pipe is normal in the morning when your engine is warming up. This phenomenon is especially noticeable in cold weather. Water vapor is a by-product of combustion. At cooler temperatures the water vapor condenses in the air and is visible as white smoke. As the exhaust systems warms up, the water vapor becomes invisible, but the water vapor is still present.

The water vapor mixes with other exhaust components from the burned gasoline to form corrosive compounds inside the muffler and tailpipe. If your car is driven for only short trips, the buildup of water and corrosive compounds eats away the muffler and tailpipe. Cars that are driven on long trips eventually evaporate the water buildup so that mufflers and tailpipes last longer.

If you notice large amounts of steam after the engine has warmed up or in hot weather, there could be a problem. For example, a defective head gasket or crack in the block or head will allow coolant to enter the cylinders. The heat of combustion will turn the coolant to steam. Check to see if you seem to be losing coolant, but are not leaking any onto the ground. The exhaust may exhibit a sweet smell. Keep in mind to limit your exposure to the exhaust as much as possible. There are simple tests that a professional technician can do to diagnose the cause. One simple check will test for the presence of exhaust fumes in the radiator or coolant recovery reservoir. If coolant is leaking out through the combustion chamber, it is very likely exhaust from the combustion chamber is leaking into the coolant.

Engine Doesn't Heat Up Fast Enough

If the engine always runs too cool, the thermostat has been removed, is stuck wide open, or has a temperature rating that is too low. In any case, the thermostat should be replaced.

There are a number of signs that indicate the thermostat is malfunctioning. You may notice that the temperature gauge does not reach the normal range after 10 minutes of driving. Or, if you have no gauge and it is wintertime, the heater will barely warm the air.

To check the thermostat, open the hood when the engine is cold. Start the car and immediately feel the upper hose. Avoid getting your hands or arms near the fan and belts. If you feel pressure when the engine is started, the thermostat is open. It should be closed to prevent coolant flow.

Allowing the engine to run at a cold temperature will contribute to poor fuel economy and engine damage. The fuel delivery system will provide excessive amounts of fuel, which also can wash lubricating oil from the cylinder walls and contribute to wear. In addition, acid and sludge will build up in the oil.

This chapter concludes an overview of the basic engine systems. Chapter 7 deals with the parts that control and transmit torque produced by the engine to the driving wheels—the drivetrain.

1. What is the function of the lubrication system?
2. Name the common components of the lubrication system
3. Name three groups that establish standards for oil.
4. How does a multiviscosity oil work?
5. Describe how to check engine oil.
6. What is the acceptable oil usage limit?
7. Where are common places for oil to leak?
8. What is the function of the cooling system?
9. Name the common components of the cooling system.
10. What component regulates how much heat is drawn away from the engine?
11. What are the different kinds of fans used in cooling systems?
12. What is the function of the coolant recovery tank?
13. What is the function of the pressure cap?
14. What is the proper mix of antifreeze and water when blending coolant?
15. Compare the different types of antifreeze.
16. Explain the operation of the thermostat.
17. What are the safety precautions when working on the cooling system?
18. When may coolant leaks be best checked for? Why?
19. How may the fan be tested?
20. What are some common cooling system problems?
21. What does white smoke from the tailpipe indicate?
22. What can cause insufficient engine warm up?

THE DRIVETRAIN

After reading this chapter, the reader should be able to:

- List the functions of the drivetrain.
- List the mechanisms that make up a drivetrain.
- List the sections and components that make up a manual transmission.
- Explain how a manual transmission operates.
- List the components that make up an automatic transmission.
- Explain how an automatic transmission operates.
- Describe the operation of the differential.
- Compare front-wheel drive to rear-wheel drive.
- List the safety precautions for drivetrain maintenance.
- List the basic maintenance for the drivetrain.
- List some common problems that could occur in the drivetrain.

DRIVETRAIN FUNCTIONS AND OPTIONS

The drivetrain transfers torque, or turning power, from the engine to the driving wheels. On most passenger vehicles, the drivetrain transfers torque from the engine crankshaft to either the front or rear wheels. On some passenger vehicles, trucks, and utility vehicles, the drivetrain can transfer torque to all four wheels.

Your car must be able to perform well under many different conditions. It must move easily from a stop and accelerate smoothly. It must be able to carry extra weight or passengers, go up or down steep grades, and travel in reverse. And it must go around corners smoothly and easily. Accomplishing these tasks is the function of the drivetrain. Mechanisms that make up the drivetrain include:

- Clutch or torque converter
- Transmission
- Differential
- Constant-velocity or universal joints
- Axles and drive shafts

Clutch or Torque Converter

To couple the engine to a manual transmission, a clutch is used. The clutch provides a smooth way to engage or disengage engine torque from the transmission. In cars with automatic transmissions, engine torque is coupled to the transmission by a torque converter.

Transmission

For your car to perform properly, varying amounts of engine torque must be applied to the drivetrain. More torque is required to accelerate or to climb a hill than to cruise. The engine produces torque over a relatively narrow range of rpm while vehicle speed (mph) and other loads can vary greatly. A transmission allows engine torque to be increased or decreased before it is transmitted to the rest of the drivetrain.

Differential

The differential is a set of gears located between two driving wheels. This mechanism has two functions. First, it receives torque flow from the transmission (either directly or through a drive shaft) and directs the torque to the driving wheels. Second, it allows the wheels to turn at different speeds. When a car turns, the outside wheels must cover a greater distance than the inside wheels. Therefore, the outer wheels must spin faster. If not for the differential, one wheel would be dragged, or the other would spin, and tire wear would be excessive.

Constant Velocity or Universal Joints

At appropriate locations within the drivetrain, flexible joints are used to transmit torque at varying angles as your car travels over bumps. Thus, torque can be transmitted constantly to turn the drive wheels without loss of power or smoothness.

Constant-velocity (CV) joints are used to transmit torque from a front-wheel drive transaxle to the driving wheels. (A transaxle combines the functions of a transmission and a differential.) Four-wheel drives with the front differential mounted to the frame also use CV joints. Rear-wheel drive vehicles use flexible joints known as **universal joints,** or U-joints, at either end of a long drive shaft that leads to the rear-mounted differential. On some rear-wheel drive cars, universal joints or constant-velocity joints are also used to transmit torque from the differential to axle shafts that turn the rear wheels.

Axles and Drive Shafts

The **axles** are shafts that transfer the power from the differential to the wheels. They may use CV joints and U-joints to allow movement of the wheel while turning the wheel to propel the vehicle forward. The wheels are attached to the outer end of the axle and fastened with lug nuts. The drive shaft is usually used to transfer power from one drivetrain component to another. In a rear-wheel drive vehicle, a drive shaft is used to transfer power from the transmission to the rear differential. A four-wheel drive may also use a front drive shaft to transfer power to the front differential.

TRANSMISSIONS

All transmissions must match engine torque and power to the load and speed of the vehicle. An easy way to view the transmission is in terms of three operating concepts:

1. The engine must be linked to the transmission gears or shifting mechanism.
2. There must be a set of gears or other means to vary and match engine torque.
3. There must be a method or system to change the gears.

The output of the transmission is then transmitted to the wheels through other components in the drivetrain.

There are two main types of transmissions: manual and automatic. The **manual transmission** consists of the clutch and the gearbox. The driver manually shifts from one gear to another gear. The **automatic transmission** consists of a torque converter and planetary gears. Shifting gears is automatic and is accomplished with the use of **servos** operated by either a valve body or, more commonly, a computer.

MANUAL TRANSMISSION CLUTCH

Proper use of the clutch smoothly engages and disengages the flow of torque from the engine to the transmission. The clutch mechanism, which allows you to shift gears easily, includes three basic sections:

1. Driving member
2. Driven member
3. Operating members

Use Figure 7–1 to help you understand the following discussion of clutch operation.

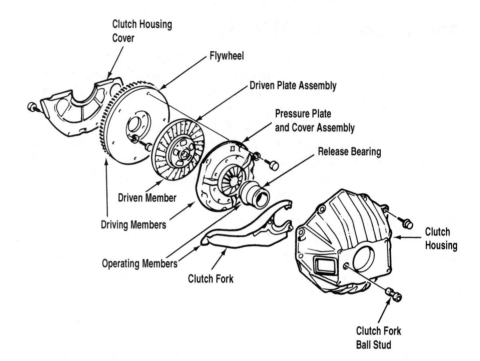

Figure 7–1: This exploded view shows major clutch components.

Driving Member

The driving member consists of two parts: the **flywheel** and the **pressure plate.** The flywheel is bolted directly to the engine crankshaft and rotates when the crankshaft turns. The pressure plate is bolted to the flywheel. The result is that both flywheel and pressure plate rotate together.

Driven Member

The driven member, or **clutch disc,** is located between the flywheel and pressure plate. The metal hub of the clutch disc has *splines,* or grooves, in its center so that it fits onto a splined transmission input shaft. The input shaft turns with the clutch disc. The clutch disc can move a short distance back and forth on the shaft.

When the clutch pedal is released, the pressure plate holds the clutch disc tightly against the flywheel. The clutch disc then turns with the flywheel. Thus, the transmission input shaft is coupled to the engine; the clutch is engaged. When you depress the clutch pedal, the clutch disc is no longer held against the flywheel. Then, the transmission is disengaged from the engine.

The clutch disc includes a frictional facing material (synthetic, organic, and metallic components are typically used to make the facing material, as with brake linings) that eventually wears out and must be replaced.

Operating Members

Operating members are the parts that release pressure from the clutch disc. They consist of the clutch pedal, clutch return spring, **clutch linkage,** clutch fork, and **release bearing.** The clutch linkage includes the clutch pedal and a mechanical or hydraulic system to move the other operating members.

When the clutch pedal is depressed, the clutch linkage operates the clutch fork. The clutch fork, or *release fork*, moves the release bearing against the pressure plate release levers. These levers then compress the pressure plate springs that normally hold the clutch disc tightly against the flywheel. This action allows the clutch disc to disengage.

At this point, the torque of the engine cannot turn the transmission input shaft. The gears in the transmission may be shifted or the vehicle can be brought to a full stop.

When the clutch pedal is released, the pressure plate forces the clutch disc against the flywheel. The clutch *return spring* helps raise the pedal.

Some cars use a mechanical linkage to operate the clutch. Other cars may use a cable similar to that used in the parking brake mechanism. And other cars may use a hydraulic clutch, which operates like the brake system except the driver depresses the clutch pedal, and the hydraulic fluid is pushed into a small cylinder. Brake fluid is normally used as the hydraulic fluid on these systems. This cylinder then operates the release fork. Figure 7–2 illustrates the mechanical, cable, and hydraulic clutch systems.

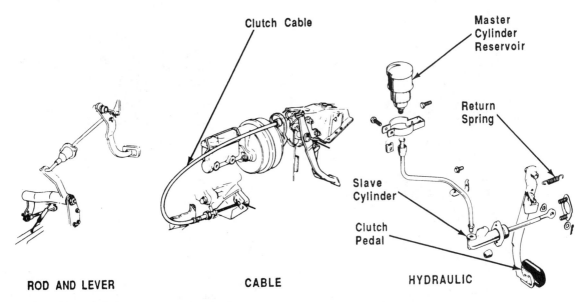

Figure 7–2: This exploded view shows major clutch components.

MANUAL TRANSMISSION

A transmission can have three, four, five, or more forward speeds, plus reverse. All manual transmissions work in basically the same manner. Almost all manual transmissions are "constant mesh." These transmissions have gears that are in constant contact with one another. These gears are then connected or coupled to the shafts for power output, usually through the use of a synchronizer. Most are fully synchronized, meaning that upshifts and downshifts through the forward gears can be made without any gear clash or noise. Upshifts occur when going from lower to higher gears as vehicle speed increases. Downshifts occur as the vehicle speed decreases. Figure 7–3 illustrates the flow of torque through a typical constant-mesh transmission.

The basic manual transmission consists of the following shafts:

- The input, or clutch, shaft
- The countershaft
- The output, or main, shaft
- The reverse idler shaft

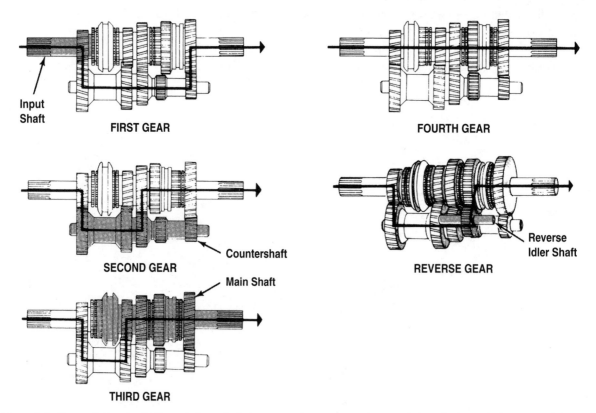

Figure 7–3: Torque is transferred through three shafts in a manual constant-mesh transmission: the input shaft, countershaft, and main shaft.

Each of these shafts has gears, or toothed wheels, that mesh, or fit together. When one meshed gear turns, the other gear is turned.

Input Shaft

The input, or clutch shaft is turned through its connection to the center of the turning clutch disc. At the transmission end of the input shaft is a gear called the *input gear*. As the input shaft turns, the input gear transfers engine torque to the countershaft and its gears.

Countershaft

The countershaft consists of a solid metal shaft with several gears. These countershaft gears, or "cluster gears," are part of the solid countershaft. Thus, when the countershaft turns, all these gears turn. The countershaft gears are in constant mesh with gears on the input shaft and the main shaft.

Main Shaft

If the shift lever is placed in neutral, all of the gears on the main shaft float or rotate freely on the shaft. When you move the lever to a gear position, one of the main-shaft gears becomes locked to the main shaft. Thus, the corresponding meshed countershaft gear turns the main shaft through the locked main-shaft gear.

Each set of meshed gears transmits a different amount of torque and speed to the main shaft. Thus, you can choose more speed and less torque (higher gear) or less speed and more torque (lower gear) to suit driving conditions.

In some transmissions, another method of transferring engine torque to the main shaft is used for high gear. The input shaft locks directly to the main shaft and turns it at the same speed as the engine crankshaft. This coupling action is known as *direct drive*. In many modern manual transaxles, all of the torque flows through the countershaft to turn the main shaft.

To produce lower engine speeds and better fuel economy under cruising conditions, the size of the gears can be modified so that the transmission output shaft turns *faster* than the engine. This gear arrangement is called *overdrive*. Some transmissions and transaxles will overdrive the top two gears. For example, fourth and fifth gears are both overdrive gears.

Reverse Idler Shaft

Moving the car backwards requires another gear in the transmission—a reverse idler gear. When you shift into reverse, the reverse idler gear meshes with gears on the countershaft and main shaft. The extra gear changes the direction of rotation. The main shaft, or output shaft, then turns in a reverse direction to cause your car to move backwards.

AUTOMATIC TRANSMISSION

As with a manual transmission, an automatic transmission contains sets of gears. The gears used within the automatic transmission are different from those found in a manual transmission. A clutch is not necessary, and the driver doesn't have to shift up or down through the forward gears. Rather, the driver selects the function for the transmission to perform and shifting occurs automatically.

The automatic transmission is coupled to the engine and operated hydraulically. Oil pressure is used to engage and disengage drive components within the transmission for shifting. The shifting functions operate automatically, in response to the speed of the vehicle and the load on the engine. The transmission is enclosed in a single housing just behind the engine.

Three basic systems make the automatic transmission function:

1. Torque converter
2. Planetary gearset
3. Shifting control system

The following description covers the highlights of automatic transmission operation. To help you understand the description, refer to Figure 7–4.

Torque Converter

The torque converter transfers engine torque to the transmission. Unlike a manual transmission's clutch, the torque converter allows the engine to idle when the car is stopped and still in gear. The torque converter also increases, or multiplies, engine torque. By multiplying torque, the torque converter produces better acceleration and minimizes the amount of shifting that is necessary.

The torque converter (Figure 7–5) uses a principle known as *fluid coupling* rather than a mechanical connection. To understand how fluid coupling works, imagine two electric fans placed face to face. If you switch on one of the fans, the airflow will cause the second fan's blades to turn.

The same principle is used in an automatic transmission fluid coupling. However, instead of air, **automatic transmission fluid (ATF)** is used to cause movement. One fanlike part, called the *impeller*, is attached to the engine and rotates when the engine crankshaft turns. A second fanlike part, called the *turbine*, is located close to and facing the impeller. The automatic transmission input shaft is connected to the turbine.

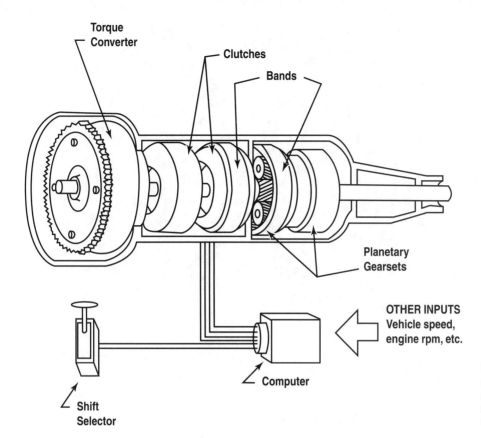

Torque Converter

Clutches

Bands

Planetary Gearsets

OTHER INPUTS
Vehicle speed,
engine rpm, etc.

Computer

Shift Selector

Figure 7–4: This simplified drawing shows the arrangement of major components of an automatic transmission.

The entire torque converter assembly is housed in a closed metal casing and is filled with automatic transmission fluid when the engine is running. The housing is connected to the flywheel, which is connected to the crankshaft. When the engine turns, the housing spins the impeller, which pumps fluid. The fluid flows and causes the turbine blades to spin. Torque is then transferred from the engine to the automatic transmission input shaft.

In a torque converter, an intermediate part, called the *stator,* is used between the impeller and the turbine. The stator has a series of blades that help direct the fluid flow from the turbine back to the impeller for efficient operation. Torque multiplication and efficiency are improved with the use of the stator.

Some automobiles have transmissions with *lockup torque converters;* the impeller and turbine connect mechanically at higher road speeds. The mechanical coupling eliminates slippage created by a fluid coupling. The result is improved fuel economy.

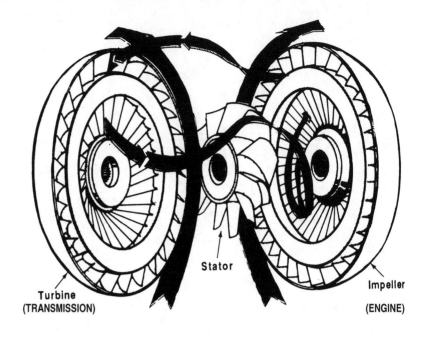

Figure 7–5: A torque converter consists of an engine-driven impeller, a stator, and a turbine.

Turbine
(TRANSMISSION)

Stator

Impeller

(ENGINE)

The input shaft of the automatic transmission, turned by the turbine, is connected to a **planetary gearset** within the transmission.

Planetary Gearset

Planetary gears (Figure 7–6) are used to change torque, change speed, and reverse direction of rotation. Planetary gears also provide a neutral position and act as a coupling for direct drive. Thus, planetary gears provide the same functions as manual transmission gears. The shifting of planetary gears is regulated by a hydraulic control system.

A few automatic transmissions use gears more similar to those found in a manual transmission. These transmissions use helical and square-cut gears instead of planetary gears. They shift by using servos located on shafts similar to those in a manual transmission. Power flow is comparable to that in a manual transmission.

Shifting Control System

A computer handles the shifting in most automatic transmissions, often the same computer that operates the engine. If a separate computer is used, it is known as a *transmission control module* (TCM). It is in constant communication with the main computer operating the engine. The computer receives inputs, processes the information using a shift schedule, and makes appropriate outputs to solenoids that, in turn, operate servos. The use of electronic shift control allows for improved performance, better fuel economy, lower exhaust emissions, and improved driver comfort.

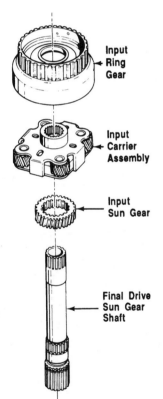

Input Ring Gear

Input Carrier Assembly

Input Sun Gear

Final Drive Sun Gear Shaft

Figure 7–6: Planetary gears rotate around a central, or sun, gear and mesh with an internal ring gear.

One of the most popularized options of these systems is the ability to manually shift the transmission. These systems allow the driver to manually upshift and downshift the transmission much like a manual transmission. Mostly found as a sport option on some cars, these are still automatic transmissions in which the computer controls the shifting, but shifts according to driver demand. The behavior of this option depends on the make and model of the vehicle. A performance decline might be noticed in manual mode as it takes an experienced driver to shift as accurately and efficiently as the computer can in automatic mode.

As a result of adaptive learning, the computer may change the behavior in which it operates in response to operating conditions and the habits of the driver. It is constantly learning about the vehicle and driver.

In some other and many older transmissions, the shifting is handled by a hydraulic control system. This system consists of a set of valves in the **valve body.** The valve body is the so-called brain of the automatic transmission. It can be thought of as a fluid computer. The valves direct hydraulic (ATF) pressure created by the spinning impeller through passages within the valve body. The amount of pressure controls the engagement of mechanical clutches and bands within the automatic transmission system. The bands and clutches shift the transmission's planetary gears.

There are three devices by which the valves are moved to open and close the passages: manual control linkage, a governor, or a vacuum modulator. Manual control linkage operates when you move the shift lever through the P-R-N-D-L positions. A unit known as a governor, which functions when the automatic transmission output shaft turns, can vary hydraulic pressure. Hydraulic pressure can be increased or decreased by a throttle valve, or vacuum modulator, which responds to pressure on the accelerator pedal.

There is a third type of transmission in limited production that is unlike anything previously available: the constantly variable transmission (CVT). The CVT uses a belt made of many laminations of steel. The transmission uses very few gears and instead relies on the belt to constantly change the ratio of the transmission for the needs of the vehicle. The idea is sound and makes sense to run the engine at its best power and economy speed while at the same time using an infinite number of "gear ratios" to vary the torque needed at the wheels. These transmissions are relatively new and if you have one, consult your owner's manual or a technician concerning maintenance and repairs.

DIFFERENTIAL

The function of the differential is to accept torque and redirect it to the driving wheel axle shafts. The differential also allows the drive wheels to turn at different speeds as the car travels around a corner.

In most front-wheel drive cars, the differential is located in the same housing as the transaxle gears. Thus, the mechanisms share the same lubricant supply.

Standard Differential

Differentials are of two types: standard and locking or limited-slip.

Figure 7–7 shows the internal gear arrangement of a typical standard differential, which is enclosed within a metal housing. The drive pinion transfers torque to the ring gear so that the ring gear spins in the same direction that the wheels must turn.

As shown, the ring gear is connected to the internal differential case, which is connected to the pinion gear shaft. As the drive pinion rotates, so do the ring gear, differential case, and pinion shaft. While the car is moving in a straight line, the pinion gears do not rotate on the pinion shaft. They act as a solid unit and move both axle gears at the same speed. When the car turns, the inside wheel resists rotation, causing the pinion gears to rotate on the shaft. As the pinion gears rotate, the outer wheel is driven at a higher speed. One wheel rotates faster than the ring gear, while the other rotates slower. Different amounts of torque are applied to the wheels.

Standard differentials are of the so-called open type. This is a differential in which most of the power may be sent to a single wheel, such as could occur if one wheel were on an icy patch and the other on dry pavement. The wheel on the icy patch would receive most of the power and spin helplessly. The car would not move forward very quickly, or the vehicle would get

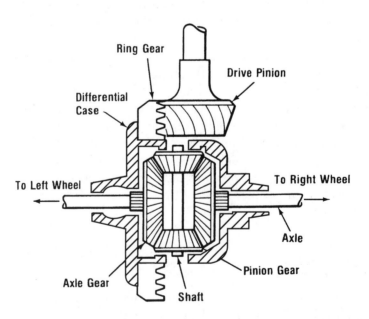

Figure 7–7: The internal gear arrangement of a standard differential.

stuck. The tire with traction would not receive much power. However, the tire with better traction tends to keep the vehicle on the road with a little more stability than it would with limited-slip-types of differentials.

Limited-Slip Differential

Limited-slip, or locking, differentials are used where it is more desirable to limit the power any one wheel can receive. A limited-slip differential operates in the same manner as a standard differential when the vehicle is moving straight or turning. However, if one wheel begins to spin, both rear axles lock together through a clutch or friction cone mechanism added to the differential assembly. Then both wheels deliver power.

The limited-slip differential once was used only as a high-performance option. These differentials are now found on some pickup trucks and sport utility vehicles, and a few cars. The unit can contribute significantly to vehicle safety on slippery surfaces and help you avoid getting stuck in mud and snow.

Note that limited-slip differentials require special lubricants to operate properly.

FRONT-WHEEL DRIVE

The front wheels drive most automobiles. Front-wheel drive (FWD) has advantages over rear-wheel drive. Because the engine and drivetrain components in a FWD vehicle are packaged compactly at the front of the car, the use of interior space is more efficient. Thus, a relatively small car can have a reasonable amount of room inside for passengers and luggage. In addition, the weight of the FWD powertrain is directly over the driving wheels. This positioning helps to increase traction on slippery surfaces.

The engines on most FWD cars are mounted sideways, or transversely, with the crankshaft parallel to the front axles, although some are positioned longitudinally, or front-to-back, as with rear-wheel drive vehicles. Regardless of its alignment, the engine still transfers torque through the clutch, or torque converter, to the transmission. After passing through the transmission, torque is transferred to the differential and driving axles.

These driving axles transfer torque from the differential to the front wheels. However, because the front wheels must travel up and down as well as steer left and right; extreme flexibility is necessary. To give these axles the necessary flexibility, constant-velocity (CV) and other types of flexible joints are used (Figure 7–8). One flexible joint is located at the transaxle end of the axle. Another joint is located where the axle is connected to the front wheel.

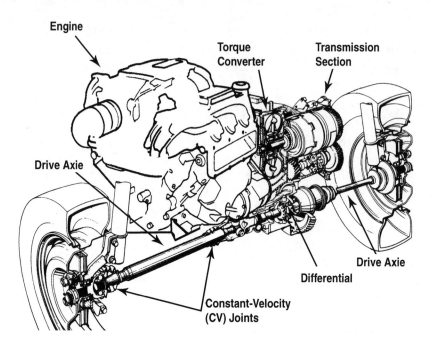

REAR-WHEEL DRIVE

Unlike a FWD car, the drivetrain parts of a typical rear-wheel drive (RWD) vehicle are separated. Torque from the engine and transmission, located near the front of the car, must be transferred all the way to the rear wheels. To transfer the torque, a **driveline** is required (Figure 7–9). The driveline transfers torque from the rear of the transmission to the front of a separate differential unit at the rear of the vehicle.

Drive Shaft

The drive shaft is a hollow metal tube. Since driveline rotation must be smooth and consistent, the drive shaft itself must be balanced. At the factory, small weights are welded to the drive shaft. The process is similar to the way weights are added to balance wheels, tires, and rims. The balancing weights eliminate vibration when the drive shaft turns.

Universal Joints

As the rear suspension moves up and down over bumps, the driveline must be able to flex and still continue to deliver torque. A solid, inflexible driveline between the transmission and differential would bend or break. U-joints, or Cardan joints, supply the needed flexibility (Figure 7–10).

A universal joint can transfer torque at an angle while it is rotating. The front U-joint accepts torque from the transmission and transmits the

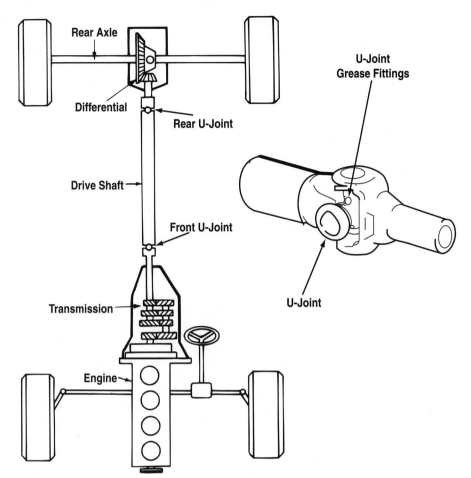

Figure 7–9: Flexible universal joints and a tubular drive shaft are used in the RWD and 4WD drivelines.

torque to the drive shaft. The rear U-joint accepts torque from the drive shaft and transmits it to the differential. Thus, the driveline can move as required by road and driving conditions and maintain a constant torque flow to the differential.

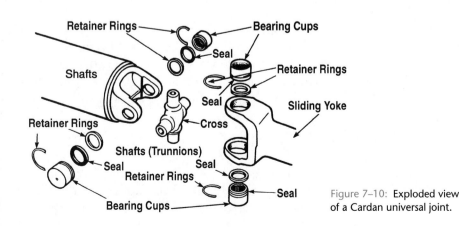

Figure 7–10: Exploded view of a Cardan universal joint.

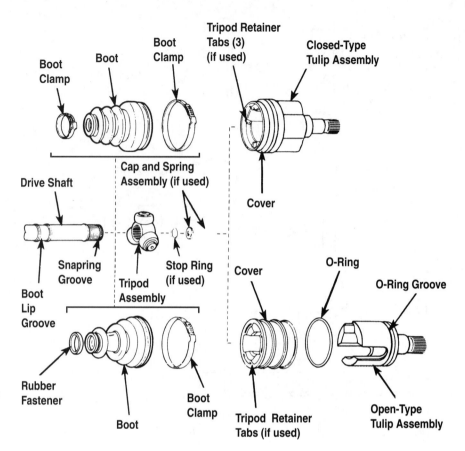

Figure 7–11: An exploded view of a CV joint, both open and closed types, and the boots used to protect the joint.

Each universal joint connects two yokes, or U-shaped castings that are welded to the driveshaft. Corresponding yokes are used at the transmission and differential ends. The transmission-end yoke is fitted to a hollow *slip joint*, which is splined to fit over the transmission output shaft. The two yokes are connected by a cross, or spider. At each end of the cross, a series of needle bearings is enclosed in a cap. The cap rotates inside a hole in the yoke.

Another type of joint more popularly used in place of the U-joint is the constant-velocity joint (CV joint, Figure 7–11). The CV joint is more complex to build than a U-joint but operates more smoothly and with less vibration. Also, it can transmit power over larger differences in angles between the transmission and wheels. CV joints function the same as U-joints.

FOUR-WHEEL DRIVE

As described in Chapter 1, front-wheel drive and rear-wheel drive can be combined on a single vehicle to drive all four wheels under power. This arrangement is known as four-wheel drive (4WD). Four-wheel drive turns all the vehicle's wheels to provide better traction on ice, snow, or mud, or under off-road conditions.

4WD Design Variations

Pickup trucks and sport utility vehicles are typically designed as rear-wheel drive vehicles. Then, to create four-wheel drive, a transfer case, front drive shaft, and front differential are added (Figure 7–12). Some smaller cars may be designed originally as front-wheel drive vehicles. A transfer case, rear driveshaft, and rear differential are added to send torque to the rear wheels. Thus, 4WD vehicles typically have two differentials and two sets of driving axles, and all the wheels are provided torque for improved traction.

Two important implementations of four-wheel drive must be noted. One less popular system is referred to as part-time 4WD. Vehicles with this system are not designed to be driven on paved roads while the four-wheel drive is engaged. Steering problems and excessive tire wear during turns will result if both front and rear wheels are driven at the same speed on solid surfaces.

The other system allows for driving in four-wheel drive mode while driving on dry surfaces or at highway speeds. It is known by several names including full time four wheel drive, shift on the fly, and all-wheel drive.

Full-time 4WD vehicles typically include a third differential (in addition to the front and rear differentials), mounted within the transfer case. In normal driving, this transfer case differential makes up for different distances traveled by the front-axle wheels and rear-axle wheels on turns. When maximum pulling power is required, the driver can lock the transfer case differential.

In recent years, full-time 4WD has been applied to high-performance sedans and sports cars as well as to trucks and utility vehicles.

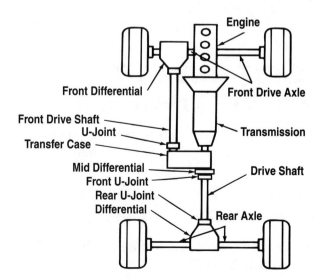

Figure 7–12: A 4WD is a combination of front- and rear-wheel drive with a transfer case.

Transfer Case

Many 4WD vehicles operate as two-wheel drive vehicles (either rear- or front-wheel drive) until a device known as a **transfer case,** attached to the vehicle's transmission, is engaged by moving a lever inside the vehicle. Gears or chains within the transfer case transfer torque from the transmission to an additional drive shaft and differential arrangement similar to that used on RWD vehicles.

Transfer cases can be shifted to produce two-wheel drive (2WD) or four-wheel drive. When it is shifted into a 4WD position, the transfer case sends torque from the transmission output shaft to all four wheels. All four wheels are turned under power to create four-wheel drive.

Transfer cases intended for use in truck and sport utility vehicles include both a 4WD HIGH (normal speed and torque) gear position and a 4WD LOW (half speed, twice the torque) gear position. For extreme conditions, the 4WD LOW range produces the most torque multiplication and pulling power.

In one design, the vehicle must be brought to a complete stop before the transfer case can be shifted. In addition, the transmission must be in NEUTRAL or PARK so that the transmission output shaft is not turning. On other designs, the driver can shift the transfer case into a 4WD position while the vehicle is still moving. This design is sometimes referred to as a "shift on the fly" capability.

Axle/Hub Engagement Systems

On many 4WD truck and utility vehicle designs, the front axles are engaged automatically to the transfer case so that the front wheels are turned by torque from the transfer case. Other designs may incorporate hub mechanisms that lock automatically when the transfer case begins to turn the front axles.

SAFETY PRECAUTIONS FOR DRIVETRAIN MAINTENANCE

To check or work on drivetrain parts, it's often necessary to get underneath the car. Observe the following safety precautions in performing maintenance on the drivetrain:

- Wear eye protection whenever you work under the vehicle. Safety glasses or goggles can prevent eye injuries, infections, and possible blindness.
- Position the car on a flat and level surface before you begin work. Never work on a car that is on an incline.

- Put the transmission in PARK (automatic transmissions) or in a gear (manual transmissions) and set the parking brake firmly. Place large wheel chocks or blocks both in front and in back of the wheels that remain on the ground to prevent the car from rolling or moving.
- Never put any part of your body under a car that is supported only by a jack. Always use safety stands, or jack stands, to support a car safely off the ground. A manufacturer's service manual will show you the proper locations on the car's underbody where you should place the stands. As an alternative, use drive-up ramps to raise one end of the car safely. And remember to block the wheels that remain on the ground!

ROUTINE CHECKS AND LUBRICATION

The drivetrain is a mechanical system. As with all other mechanical systems, wear occurs. The maintenance procedures described here can help to avoid breakdowns.

Check for Leaks

Each time you return to your car, look underneath for evidence of leaks. If you're sure the leaks are from your car, you can get underneath the car for a closer look.

Check for drivetrain leaks from the transaxle/transmission and, on RWD vehicles, from the rear-mounted differential housing and axle tubes. Leaks of differential lubricant may also be noticed from the rear brake units. The most likely cause of drivetrain leaks is a worn or defective oil seal around a rotating shaft.

Inspect the boots of CV joints for cracks, splits, tears, and lubricant leaks. If the boot is not in proper condition, the lubricant will leak out, and dirt and other contaminants can get in and greatly shorten the life of the unit.

If you find leaks at these locations under the car, have a competent technician correct the problem. Most of these jobs require experience and tools that the average car owner is not likely to possess. Don't ignore drivetrain leaks. If significant amounts of lubricant are lost, an expensive system may be ruined.

Check/Add Automatic Transmission Fluid

To check automatic transaxle/transmission fluid levels, locate the automatic transmission dipstick *before* you start the engine. This dipstick is located in a tube that leads to the automatic transmission at the rear of the engine. Some FWD vehicles have two dipsticks for the automatic transmission. Check your owner's manual for this procedure. Some cars do not have an automatic transmission dipstick, most notably GM cars in recent years. GM concluded that the dipstick tube was a major source of contamination. These cars require a technician to check the level of the fluid.

⚠ CAUTION

Before you check the automatic transmission fluid level, apply the parking brake securely. Put the transmission in PARK.

VEHICLE LEVEL

ENG IDLING

TRANS HOT

CHECK IN PARK

FULL HOT

ADD 1 PT

Figure 7–13: Automatic transmission dipsticks may include information on conditions for checking fluid level.

The best time to check transmission fluid levels is after normal driving, when the engine and transmission are at normal operating temperature.

On almost every automatic transmission—with the exception of some Honda automobiles—the engine must be *running* to check the fluid level correctly. If the engine is cold, start the engine and let it idle for about two to three minutes. As the engine is warming up, hold your foot on the brake and move the shift lever through each of the different gear positions. This step assures that all pistons and passages within the hydraulic control unit are filled with transmission fluid.

Refer to your owner's manual for proper procedures to check your automatic transmission fluid level. On most automatic transmissions, the fluid level is checked with the engine running and the transmission selector in the PARK position.

After you remove the dipstick, wipe it on a white rag or white paper towel. If the fluid is brownish or blackish in color and has a burned smell, the transmission has been overheated severely. The fluid, filter, and transmission pan gasket should be changed. The burned smell and dark particles in the oil are caused by overheating and wear of the clutches and bands. If these clutches and bands become excessively worn, the transmission may need to be overhauled.

If the fluid is normal, that is, a transparent red, wipe off the dipstick, replace it in the tube, and make sure it seats completely. Then remove the dipstick once more and take your reading. A typical transmission dipstick is shown in Figure 7–13.

The most common type of automatic transmission fluid for passenger cars is Dexron III and it is recommended by nearly all manufacturers. Dexron III replaces all other Dexron variations. Other manufacturers may recommend the use of Mercon. In some of these vehicles, Dexron III may be used as a secondary choice. However, always refer to your owner's manual or call a dealership if you aren't sure if this is the proper ATF to use. Older vehicles may use type A, CJ, F, or H fluid. Use of the wrong fluid can void manufacturers' warranties, cause erratic shifting, or damage internal transmission parts.

The reading on the dipstick should be close to the FULL mark. If fluid is needed, add it through the automatic transmission dipstick tube. It usually takes a pint, or half quart, to raise the fluid from the ADD mark to the FULL mark. To pour the fluid in, you will need a long-necked transmission funnel, available at auto parts stores.

As with engine oil, ATF is sold in quart containers. Let the engine idle and add small amounts of fluid. Allow time for the fluid to drain to the bottom of the transmission then recheck the level. Avoid overfilling the transmission. If you overfill, the fluid will be whipped into a foam, and the transmission may be damaged severely. If your transmission requires frequent filling, check for leaks.

Check Manual Transmission Hydraulic Clutch Fluid

Many vehicles are equipped with a hydraulic clutch linkage system (Figure 7–14). When you depress the clutch pedal, a clutch master cylinder, similar to the brake master cylinder, forces brake fluid to move through a rubber tube. At the transmission end, a clutch slave cylinder provides force to move the operating members. Check and add brake fluid if necessary. However, the need to add fluid typically indicates that the system is leaking and requires repair.

If the clutch does not operate after adding fluid, an air bubble may be present in the fluid line. If you are unfamiliar with bleeding procedures, refer to the manufacturer's service manual or have the job done by a professional technician.

Check Clutch Pedal Free Play

The clutch should engage and disengage fully at a point approximately halfway to two-thirds of the way down to the floor. If you are not sure that the clutch is engaging at the proper place, make the following check: Depress the pedal with your fingers. There should be some "free play," or **free travel,** before you begin to feel resistance. Use a ruler to measure how much free play exists before you feel resistance. Refer to your owner's manual or a manufacturer's service manual for the correct amount of free

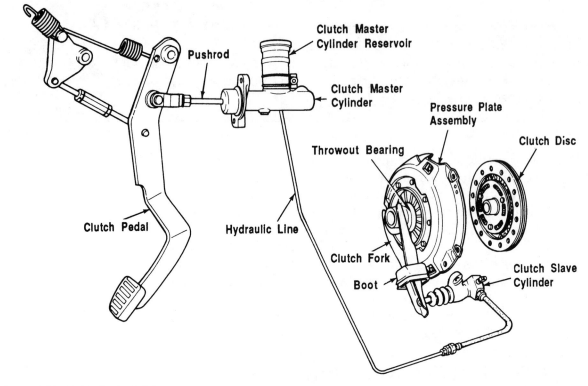

Figure 7–14: Hydraulic clutch linkage.

play. If the free play is greater or less than specified, the clutch linkage needs adjustment.

As the clutch wears, free play will diminish. If there is no free play, your clutch may start to slip, or you may not be able to shift gears. The clutch simply may require adjustment; however, if it is badly worn, the clutch plate must be replaced. Clutch adjustments vary considerably from car to car. Some vehicles have clutches that are self-adjusting or require no adjustments. If you want to attempt a clutch adjustment yourself, follow the instructions in a manufacturer's service manual. Otherwise, have the work done by a competent, professional technician.

Check CV or U-joints

The CV joints on a FWD powertrain's driving axles are enclosed in protective rubber boots. The boots keep water and dirt out of the joints, while keeping in the lubricant. If the boots become cracked, the joint may become ruined by contaminants. Check for cracks, tears, and leaking lubricant.

Damaged boots must be replaced before costly CV joint repairs become necessary. Some replacement boots are split down one side, and can be slipped over the joint easily without removing the axle shafts from the car.

Other replacement boots are solid and require that the axle shafts be removed for boot replacement. Replacing the entire shaft with a remanufactured unit may often be an economical alternative, especially on cars that are a few years old or older. Leave the last two types of repair to a competent technician.

Attempt to shake and rattle each CV joint. Back-and-forth movement along the length of the shaft is normal. However, you should not hear any rattling or feel any movement perpendicular to the length of the shaft.

Many RWD cars use the Cardan joint, or U-joint, in the driveline, though there is a trend toward using CV joints in this area. To check a car's U-joints, grasp the driveshaft and attempt to shake and rattle the joints. There should be no noticeable rattling or movement in any direction. Rusty powder around the joints indicates that the needle bearings have failed from lack of lubrication.

In normal service, CV joints are not lubricated unless a protective boot fails. If the joint still is serviceable, it is cleaned and replaced with special grease when the new boot is installed. If the joint is not serviceable, it is replaced.

On many RWD and 4WD vehicles, original-equipment U-joints are sealed and cannot be lubricated. Replacement U-joints may include grease fittings. If there are grease fittings on U-joints, wipe off the fitting with a rag. Pump grease into the fitting until grease appears at the seals.

Replacing CV joints or U-joints requires special tools and expertise. Leave such work to professional technicians.

Check/Add Manual Transaxle/Transmission Lubricant

A FWD manual transaxle typically includes a filler/check plug located on the side, toward the middle of the housing. Some FWD manual transaxles may include a dipstick to check lubricant level. Sometimes checking the lubricant level can be done from under the hood rather than from under the car.

Checking a manual transmission or transaxle lubricant may require raising the vehicle until it is level. A transaxle or transmission filler/check plug must be unscrewed with a wrench or socket. Remove the plug and insert your finger into the hole (Figure 7–15). The proper lubrication level is at the lower edge of the plug hole or within a quarter-inch of the opening of the plug. If the liquid is below this level, the proper lubricant must be added.

Manual transaxles and transmissions may require extreme pressure (EP) gear lubricant, engine oil, automatic transmission fluid, or special mineral oils. For this reason, check your owner's manual or a manufacturer's service manual to identify the recommended lubricant.

Transmission

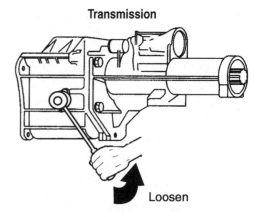

Loosen

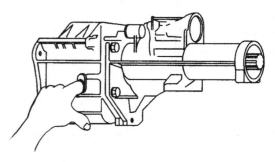

Figure 7–15: Remove the transmission's plug to inspect fluid level and condition. Differentials have similar plugs.

Filling a FWD car's manual transaxle may require a long-necked funnel, similar to that used for filling an automatic transmission. Filling a RWD car's transmission may require a suction pump to draw the lubricant up and force it into the filler opening.

Check/Add RWD or 4WD Differential Lubricant

Checking a separate differential unit is much the same as checking a manual transmission. A plug is located on the front or rear of the differential. A special, large Allen wrench, open-end wrench, or a ratchet and extension bar may be needed to remove the plug.

As with a manual transmission, the lubricant level should be near or even with the bottom of the plug hole. Separate differentials typically require 90-wt EP, 80W90, or synthetic 75W90 gear lubricants. Check your owner's manual or a manufacturer's service manual for the proper lubricant. Remember that limited-slip differentials require special lubricants and may require the use of special additives. If the lubricant is low, add enough lubricant to fill the differential to the level of the plug opening.

Check/Add 4WD Transfer Case Lubricant

Checking or adding lubricant to a 4WD vehicle's transfer case is essentially the same procedure as checking a manual transmission or differential. Remove the check plug on the side of the case and check the level of the lubricant. Add lubricant if necessary. Transfer cases may use a gear lubricant or Dexron III. Adding the wrong lubricant in the transfer case can destroy it in a short time. Check your owner's manual or a manufacturer's service manual for the recommended lubricant.

DRIVETRAIN PREVENTIVE MAINTENANCE

For maximum life and reliability, transaxles, transmissions, separate differentials, and transfer cases should be drained and refilled with fresh lubricant at regular intervals. Follow the recommended interval for your vehicle. It may be every two years or 24,000 miles, or it may have a longer interval. If your vehicle is used for extensive stop-and-go driving, trailer towing, or off-road driving, the maintenance should be done more frequently as these types of driving place additional stress on the fluid.

Automatic Transaxle/Transmission Fluid and Filter Change

To drain the ATF and change the filter, the oil pan must be removed on most vehicles. While most automatic transaxles and transmissions typically do not have a drain plug, there are a few that do provide a drain plug and use an external filter similar to the filter used on the engine for the lubrication system (Figure 7–16). On cars equipped this way the procedure to drain and refill will be the same as it is for changing engine oil.

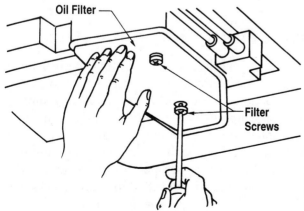

Figure 7–16: Replace the filter and oil at the recommended intervals. Always replace the filter and gasket in the correct and proper positions.

On cars without a drain plug, a large drain pan is needed to catch the ATF as it drains when the pan is removed. When the transmission pan is removed, the pan's interior surface should be examined for heavy accumulations of black or brownish deposits. Such deposits indicate that the internal bands or clutches have worn excessively and that the transmission needs to be overhauled.

If there are no heavy deposits, the inner filter or screens should be examined for evidence of metal shavings and other deposits. Large amounts of metal shavings can indicate severe wear to transmission internal parts.

Used filters must be replaced; screens may be cleaned in solvent and reused if they are not torn or broken. Some vehicles have a reuseable pan gasket. If it is not reusable, then the pan gasket is replaced and the transmission is refilled with fresh ATF of the correct type.

Many service centers have special equipment that can be attached to the car's automatic transmission cooling lines. Many shops will remove the pan and change the filter before exchanging the fluid, and then there are some shops that will exchange the fluid only. While exchanging fluid only is a convenient and quicker method for replacing your automatic transmission fluid, it is important to note that the filter cannot be examined, and you do not have the opportunity to inspect for deposits in the pan. While this may sound like a good option for a newer car, for proper maintenance, the pan should be removed and the filter changed.

Manual Transaxle/Transmission Fluid Change

Manual transaxles and transmissions typically include a drain plug on the bottom of the gearbox housing. The plug is removed to drain the lubricant. If large amounts of metal shavings are found in the lubricant, significant wear has occurred to the gears, bearings, shafts, or case. Small amounts of metal shavings may be normal. After the lubricant has drained, the drain plug is replaced and the unit refilled with the proper lubricant up to the bottom of the fill plug hole.

RWD Differential Fluid Change

Some differentials have drain plugs; others require removing a rear metal cover to drain the lubricant. Or the lubricant may be removed using a tube or small hose and a fluid evacuation tool or squeeze bulb. Small amounts of metal shavings are normal; large amounts of shavings indicate wear to the differential gears, bearings, and shafts. Evacuating the fluid may not bring out the metal particles with the oil. If a metal cover must be removed to drain the lubricant, the old gasket must be scraped off the cover and differential housing, and a new gasket must be installed to prevent leaks. Refill with a lubricant that is approved by the vehicle manufacturer.

Clutch Adjustment

As the clutch lining wears, due to friction between the flywheel and pressure plate, adjustments to clutch linkage are necessary. Symptoms of improper clutch adjustment include:

- The clutch engages and disengages near the very bottom or very top of the pedal travel.
- The clutch won't disengage, and it is difficult or impossible to shift gears.
- The gears grind and clash when you upshift or downshift, or when you try to shift into reverse.

Each car manufacturer may have different procedures for adjusting clutch linkage. Basically, clutch linkage adjustment restores the proper amount of free play to the operating members. If clutch adjustment is improper and is not corrected, the clutch plate and operating member parts may be damaged. Refer to the manufacturer's service manual for proper free play measurements and clutch linkage adjustment procedures.

DRIVETRAIN PROBLEMS

One indication of a drivetrain problem is excessive noise in the drivetrain. However, noises can be transmitted through the drivetrain so that the actual source of the noise is difficult to find. Loud, unusual noises from a transaxle, transmission, differential, or transfer case usually indicate serious problems. The causes of noises do not go away; they just get worse. If servicing is delayed, a noise may indicate a major problem.

Driveline Noises

Noises can occur if driveline CV or universal joints are worn or defective. Use the following information to help diagnose the causes of driveline noise.

CV Joint Noises   Clicking or grinding sounds may be heard from a FWD vehicle's CV joints, especially when accelerating the vehicle with the wheels turned. Typically, going around a corner puts more strain on the *outer* joints near the wheels. Thus, a noise heard during left turns may be caused by the right side CV joint. The left side CV joint may cause a noise heard during right turns. This noise may be noticed as a humming in the early stages of joint failure. More severe damage can cause a clicking or popping noise.

U-Joint Noises   When a RWD vehicle's universal joints begin to fail, you may hear some warning noises. As you shift an automatic transmission from PARK to REVERSE or DRIVE, you may hear a distinct clanging or ringing noise. In cars with both automatic and manual transmissions, you may hear screeching or chirping noises as you start from rest and accelerate. Or, when you back up, you may hear a continuous clicking sound.

Axle Bearing Noises Axle bearings are used on both the driving and nondriving wheels of both FWD and RWD cars to allow the wheels to turn freely. When the bearings fail, they make noise whenever the car moves.

As with CV joints, you can often determine whether a noise is from the left or right side of the car by driving through a turn. If you turn right, you are putting pressure on the left side bearings. If the noise intensifies, the left bearing is at fault. Then try driving through a left turn to check right side bearings. Noise will become progressively worse as you drive. Badly failed bearings may make a crunching or grinding noise.

Transaxle, Transmission, Differential, and Transfer Case Noises Noises from major drivetrain components may be gear noises or bearing noises. Gear problems usually are heard as howling or whining. Sharp knocking or clunking also may be heard if internal gear shafts and parts are loose and hitting against one another. Defective bearings may make whining or deep, rumbling noises, or may produce a clicking sound that varies with road speed. Any of these noises may occur only during acceleration or deceleration. In other cases, the noise may occur whenever the vehicle is moving. It will be helpful to note when these noises occur and whether they occur at a particular speed, during accelerating, cruising, braking, or turning, or under other conditions.

If there is any doubt about which drivetrain part is causing the noise, you may be able to locate the source by checking under the car. Raise and support the vehicle properly on safety stands or jack stands. Inspect each component, checking for evidence of looseness or damage. Never operate the vehicle while it is supported.

If the suspected source of the noise is an internal transaxle, transmission, differential, or transfer case problem, have the car checked by a competent professional technician.

Clutch Problems

Common clutch problems include slipping, noises, and grabbing. The following information will help you to determine the cause and cure of clutch problems.

Slippage You are experiencing clutch slippage if the following occurs: You are in gear and the clutch is released, but the engine races and the car barely moves. This situation may occur only briefly during acceleration, but it is a sure sign that something needs attention.

Slippage occurs when the clutch does not transfer the torque of the engine flywheel to the transmission input shaft effectively. The clutch disc and driving members may be glazed, meaning that the surfaces become polished through heat and friction and slip against each other.

Improper use of the clutch is one of the main causes of clutch glazing and slippage. "Riding the clutch" occurs when you rest your foot on the pedal, keeping it partly engaged when it should be fully engaged or fully released. Do not use the clutch pedal as a footrest.

Improper clutch linkage adjustment also can cause glazing. If there is not enough clutch free play, the release bearing presses against the release levers and releases pressure on the clutch disc. The result is that you do not get a positive lockup between the pressure plate, the disc, and the flywheel.

Another cause of clutch slippage is oil on the friction surfaces. A defect in engine or transmission seals can allow oil to contaminate the clutch disc.

If a clutch linkage adjustment doesn't stop the slippage, the clutch disc, pressure plate, and flywheel must be serviced. To service these parts, the manual transaxle or transmission must be removed from the car. It is a difficult and expensive procedure.

To maintain a clutch in good condition, proper driving habits should be observed and clutch adjustments must be made promptly when necessary. If you pay attention to the way you use your clutch, and to clutch pedal free play, you can save yourself some money and benefit from extended clutch life.

Noises   Any noise that comes from the clutch warrants prompt attention. A noisy clutch may result from any of the following:

- A clutch fork may be loose and is hitting the pressure plate.
- A spring from the pressure plate may be loose, broken, or disconnected.
- A loose or broken pressure plate release lever may be hitting a driven member.
- A defective release bearing will make noise when you depress the clutch pedal. As you depress the pedal, the release bearing begins to turn.
- A return spring is weak. The spring is not returning the clutch pedal to its full upright position. This failure for a complete return will release pressure from the clutch disc and cause it to make a clicking sound.

In some cases, a clutch linkage adjustment may take care of the noise. However, most clutch noises indicate that a part is defective and must be replaced. Have your clutch checked by a service center that specializes in clutch and transmission work.

Grabbing   The symptoms of a grabbing clutch are a sudden jerking, or bucking, when the clutch is engaged. Typically, grabbing occurs only when you accelerate from rest.

Causes of a grabbing clutch include:

- The clutch disc may be contaminated with oil or may be oil-glazed due to defective transmission or engine seals.
- The metal splines on the input shaft or on the clutch disc may be worn.
- The pressure plate or the flywheel may be warped.
- The engine, transmission, or transaxle mounts may be loose or broken.

The remedy for a grabbing clutch usually involves replacing the clutch plate and other worn parts. Refer such servicing to competent, professional technicians.

SHIFTING PROBLEMS

Shifting problems can occur with either manual or automatic transaxles and transmissions. The following information may help you to diagnose shifting problems.

Automatic Transaxle/Transmission Shifting or Slipping Problems

If your automatic transaxle or transmission does not shift properly or the car moves erratically, check the ATF fluid level. If the fluid is low, add fluid to the proper level and then take a test drive to see if the problem still exists.

Some older automatic transmissions with a valve body use a vacuum modulator to help shift the transmission. If the transmission shifts erratically, and the ATF level drops constantly, the diaphragm in the modulator may be ruptured. When the diaphragm breaks, ATF can be drawn into the engine intake manifold and burned in the cylinder. A symptom is excessive blue-white smoke from the tailpipe. To check a modulator, pull off the rubber vacuum hose attached to the unit. If ATF drips from the hose or modulator, the diaphragm is leaking and the modulator unit must be replaced.

If you have corrected the fluid level, and there are no leaks but the problem still exists, the transmission pan should be removed to check for filter condition and deposits. If no indications of problems are found and a transmission lubricant and filter change doesn't correct a shifting or slipping problem, see a transmission specialist.

Manual Transaxle/Transmission Gear Shifting Problems

You may find that a manual transmission won't shift. It may be locked in neutral, or in a gear, or the gears will clash when you try to shift. The most common causes of this condition are improper clutch adjustment and linkage problems. This condition could also mean that the clutch disc is warped. Any of these conditions may prevent the clutch from disengaging fully, causing the grinding sound.

Don't try to force the shift lever. Something may break. If the gears won't shift with a normal amount of pressure, see a competent technician.

This concludes the discussion of the drivetrain. Chapters 8 and 9 discuss the systems and parts that have a direct effect on your car's safety: tires, brakes, suspension, and steering.

1. What are the mechanisms that make up the drivetrain?

2. What is the main purpose of each of the mechanisms that make up the drivetrain?

3. What are the main components of a manual clutch?

4. What are the three functions of a transmission?

5. What are the three shafts in a manual transmission?

6. Describe the operation of the torque converter.

7. Describe shifting control for an automatic transmission.

8. What is the function of the differential?

9. List the types of differentials available and describe their basic functions.

10. What additional components must be added to create an AWD/4WD from either a FWD or RWD vehicle?

11. What precautions should be followed when checking the ATF in an automatic transmission?

12. What are some signs of U-joint failure?

13. What are some signs of CV joint failure?

14. List common clutch problems.

15. Why does riding the clutch shorten clutch life?

16. What should be done if an automatic transmission does not shift properly or slips?

TIRES AND BRAKES

After reading this chapter, the reader should be able to:

- List the systems that support the car's weight, control its direction, and bring it to a stop.
- List the different types of tire construction.
- Explain the coding system used for tires.
- Describe the maintenance for tires.
- Explain how power brakes operate.
- Describe the two types of brake mechanisms.
- List the components of the antilock brake system and explain its operation.
- List the safety precautions for brake system maintenance.
- List the basic maintenance for the brake system.

The parts and systems in the vehicle that support your car's weight, control its direction, and bring it to a stop include:

- Tires
- Brake system
- Suspension system
- Steering system

Chapter 8 focuses on the tires and the brake system and Chapter 9 focuses on the suspension and steering systems.

TIRES

A tire and its wheel, the circular component on which the tire is mounted, forms the "tire and wheel assembly," commonly called the *wheel*. (Wheels may be made of stamped sections of steel fastened together or from aluminum that is either cast or forged.) The most important component of the wheel is the tire. The entire weight of your automobile rests on four "contact patches" formed by the bottom surface of the tires. Each of these areas is about the size of your outstretched hand. Your car's ability to remain stable on straight roads and in curves and when you brake depends on the traction provided by the tires.

For maximum traction and safety, tires should be checked frequently and inflated to the manufacturer's specifications. If tires aren't inflated properly, their useful lives can be shortened by one-half or more. Also, underinflated tires may blow out suddenly due to increased flexing and heat buildup. Finally, underinflated tires require more energy to roll down the road, which results in lowered gas mileage.

Tires eventually wear out and need to be replaced. If your tires wear out due to underinflation and other poor maintenance practices, you are wasting money and risking your life and the lives of others. In an effort to save money, many drivers delay tire purchases until long after their tires reach a dangerous condition. Don't be one of them.

Proper tire maintenance and good driving habits will help to keep your tires in a safe condition and extend their useful life. Procedures for routine tire maintenance are covered in Chapter 11.

Tire Construction

Tires are constructed to combine the properties of a comfortable ride, good traction, and durability. The basic materials used in tire construction are rubber, fabric, and synthetic or steel reinforcing strands. The construction of a tire is basically a sandwich made of layers of rubber and fabric. Each layer of rubber and fabric is known as a *ply*.

When fabric is woven, the direction in which the vertical strands of thread run can be referred to as the "grain." As with wood, the fabric has greater strength and stability in this direction. When the layers of fabric and rubber are molded into the tire, the way that the fabric is arranged will have a direct effect on the flexibility, strength, and performance of the tire.

No matter what size, style, or brand of tire you buy, there are only three basic types of tire construction according to the arrangement of the plies:

1. Radial ply
2. Bias ply
3. Bias belted

Refer to Figure 8–1 as you read the following explanations of how tire plies are arranged.

Radial The plies on a radial tire extend around the tire from bead to bead. (The bead is the molded edge of the tire that fits into the wheel rim.) This radial, or circular, arrangement creates a supple sidewall that can flex to keep the tread in contact with the road surface. To reinforce and strengthen the tire, additional *belted*, or woven reinforcing materials, are placed directly under the tread area. The belts encircle the tire under the tread, just as a belt goes around your waist. Tread-reinforcing belts are woven from steel strands, fiberglass, or strong synthetic fibers.

Radial tires are standard equipment on nearly every new passenger car. The radial tire has a firm ride, but it offers the advantages of superior traction in cornering and braking, as well as long tire life. Radials also offer modest improvements in fuel mileage for most vehicles as compared with other tires due to decreased rolling resistance.

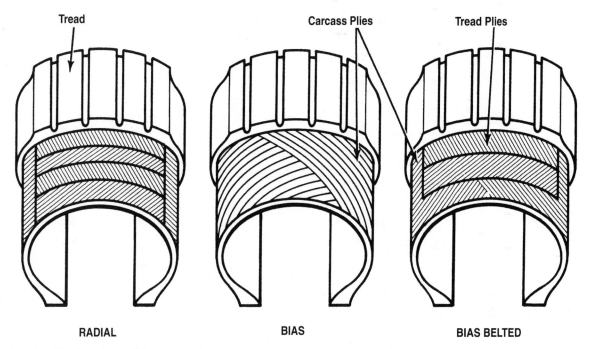

Tread   **Carcass Plies**   **Tread Plies**

RADIAL   BIAS   BIAS BELTED

Figure 8–1: The three types of tire construction.

Bias Ply   In a bias-ply tire, the plies of material crisscross from bead to bead. To create a relatively stiff tread area, the sidewall also must be stiff. During hard cornering, the stiff sidewall tends to tilt with the rest of the car. This tilting shifts the vehicle's weight from the full tread area to the outside edge of the tire, which results in decreased traction.

Bias Belted   In a bias-belted tire, the plies crisscross at an angle, or on a bias, from bead to bead around the tire carcass. To create a flexible sidewall, relatively few bias plies are used. Belts are used to stiffen the tread, as in a radial tire. This type of construction is considered a compromise between a true radial-ply tire and the oldest design of tire, bias ply.

Bias-belted tires perform better than bias ply. Radial tires are best. For safety, it is recommended that you do not intermix radials with bias-belted or bias-ply tires. All of the tires on the car should be of the same type. If the different types are mixed, vehicle instability could be the result.

Bead and Tread Design

In addition to the basic design of the tire, elements of tire construction include bead and tread design.

Bead   The bead is the portion of the tire that is held firmly against the metal wheel rim to form an airtight seal and hold the tire securely in place. The tire takes a great deal of strain when it runs over bumps or during high-speed cornering. For added strength, steel wire cables are molded into the rubber in the bead area. In addition, the rim has a special lip built into the bead. This "safety rim" is designed to hold the bead securely.

Tread Design   The tread is the portion of the tire that contacts the ground. It comes in a variety of patterns. Some treads work better on dry pavement; others are best when it is wet. Most are a compromise for both dry and wet road conditions and are known as "all season" tires.

There are several types of special-use tires. These include snow tires with deeply ribbed treads for traction under winter conditions. Snow tire treads do not dissipate heat as quickly as the treads on normal tires. Therefore, snow tires will wear rapidly at high speeds on dry pavement. Snow tires with steel studs embedded in the rubber tread are effective on ice. However, studded tires are illegal for use in some states because they cause serious damage to pavement. The use of chains on the car's driving wheels also is an effective way to deal with icy conditions. Consult with local authorities for the rules and regulations concerning studded tires or chain usage.

Increasing numbers of people are becoming involved in off-road recreational activities. Special off-road tires have extra-wide, extra-deep treads. As with snow tires, off-road tires tend to wear rapidly on pavement at highway speeds. Just like anything else you buy, you must base tire purchase decisions on your personal driving needs.

Tire Sizes

Your owner's manual states the size of tires that were installed as original equipment for your car. Over the years, different systems of tire sizing have been used. Tire stores maintain charts that list former size codes and current equivalent tire sizes. Improperly sized tires—too large or too small—can affect the safety and normal operation of steering and suspension parts.

However, in many cases it may be possible to improve traction and braking performance by replacing narrow tires with slightly wider tires. Also, larger diameter wheels that will accept larger tires may be compatible with your car. But, you must check that the new tires or tire/wheel combinations don't interfere with steering or suspension functions. On most cars, front and rear tires are the same width. Using one or more tires of a different size on these cars will affect handling adversely.

Coding marks embossed on the tire's sidewall describe each tire and its performance ratings. The systems presently in use combine metric and U.S. standard measurements. If your car has older tires, the coding may be different. Check with a tire store or car dealership service department to determine the proper replacement size under the newer system.

Metric Tire Sizing   The numbers on a typical new tire intended for an American car might be: P225/75R-15. The *P* stands for passenger car. If the tire were intended for use as a temporary spare only, the first letter would be a *T*. The number 225 represents the tire width, from sidewall to sidewall, in millimeters. Thus, this tire is 225 millimeters (8.9 inches) wide at its maximum width.

The number after the slash or dash, in this case 75, indicates the aspect ratio, or profile. The *aspect ratio* is a comparison between the width of the tire and its height from bead to tread. In this case, the tire's height is 75% of its width. (The smaller the aspect ratio number, the shorter the tire will be in comparison to its width. Thus, a 70-series tread is shorter than a 75-series tread, a 60-series is shorter than a 70-series; and so on.)

The next letter, *R*, stands for a radial-ply tire. (*B* would indicate bias belted and *D* would mean bias ply.)

The last number, 15, indicates the diameter, in inches, of the wheel rim on which the tire will be mounted.

Truck Tire Sizes   For 4WD and other trucks, another tire-sizing system may be used. An example is 30×9.50R-16. The first number, 30, indicates the total height of the rim and tire combination in inches. The second number, 9.50, indicates that the tire is 9.5 inches wide, sidewall to sidewall. The *R* indicates a radial tire. The last number, 16, means that the tire is to be mounted on a 16 inch diameter rim.

Uniform Tire Quality Grading

Since 1981, tires have been subject to a government grading process to help consumers determine quality. A typical rating under the Uniform Tire Quality Grading (UTQG) system might look like this: 100 AB. The first part of the rating code, the number 100, indicates a measure of the tire's anticipated tread wear. For example, a tire graded 150 should last approximately 1.5 times as long as a tire graded 100. Because driving conditions vary, the tread wear rating is a comparative measure only and does not indicate any specific number of miles that the tire is guaranteed to last.

The next letter, *A*, stands for the tire's traction rating. An A grade is the best, representing the tire's ability to stop on wet pavement of asphalt or concrete, as measured on a government test track. Possible grades are A, B, and C. A tire marked C has only average traction performance.

The last letter in the 100 AB rating, *B*, indicates resistance to problems caused by heat buildup within the tire, for example, blowouts. Again, an A grade would be best, followed by B and C. All tires are required to have a minimum C rating on temperature resistance.

Speed Rating

A tire may also carry a speed rating for high-performance use. The rating indicates the maximum speed at which a tire can safely operate. An S rating indicates that the tire is for use on vehicles that can attain speeds of up to 112 mph. An H rating is for a maximum speed of 130 mph, and a V rating is for speeds over 130 mph. Although it is illegal to drive at any of these speeds in the United States, the construction and testing of speed-rated tires are assurances that they will stand up to severe usage. Domestic tires only need pass a test that runs them at 85 mph for 30 minutes. However, some domestic tire manufacturers now build tires that meet the H or V speed-rating requirements.

Figure 8–2 shows the kinds of information that can be found on the sidewall of a tire. It is to your advantage to buy tires that have the highest possible grades for tread wear, traction, and temperature resistance. Naturally, top-grade tires will cost more. You should buy the best tires your budget can afford, taking into consideration the kind of driving you do. A good tire is low-cost insurance for your safety.

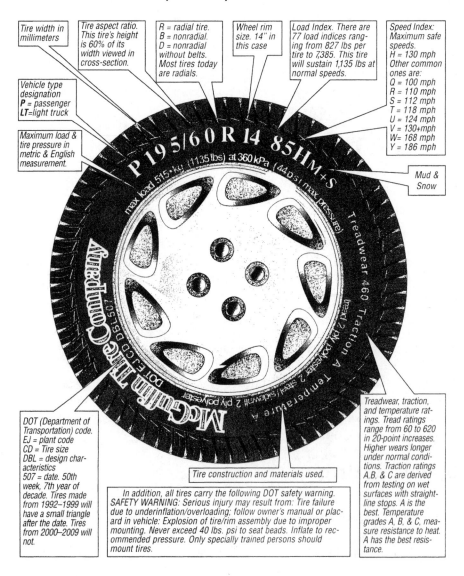

Tire width in millimeters

Tire aspect ratio. This tire's height is 60% of its width viewed in cross-section.

R = radial tire.
B = nonradial.
D = nonradial without belts. Most tires today are radials.

Wheel rim size. 14" in this case

Load Index. There are 77 load indices ranging from 827 lbs per tire to 7,385. This tire will sustain 1,135 lbs at normal speeds.

Speed Index: Maximum safe speeds.
H = 130 mph
Other common ones are:
Q = 100 mph
R = 110 mph
S = 112 mph
T = 118 mph
U = 124 mph
V = 130+mph
W = 168 mph
Y = 186 mph

Vehicle type designation
P = passenger
LT = light truck

Maximum load & tire pressure in metric & English measurement.

Mud & Snow

DOT (Department of Transportation) code.
EJ = plant code
CD = Tire size
DBL = design characteristics
507 = date. 50th week, 7th year of decade. Tires made from 1992–1999 will have a small triangle after the date. Tires from 2000–2009 will not.

Tire construction and materials used.

In addition, all tires carry the following DOT safety warning. SAFETY WARNING: Serious injury may result from: Tire failure due to underinflation/overloading; follow owner's manual or placard in vehicle. Explosion of tire/rim assembly due to improper mounting. Never exceed 40 lbs. psi to seat beads. Inflate to recommended pressure. Only specially trained persons should mount tires.

Treadwear, traction, and temperature ratings. Tread ratings range from 60 to 620 in 20-point increases. Higher wears longer under normal conditions. Traction ratings A,B, & C are derived from testing on wet surfaces with straight-line stops. A is the best. Temperature grades A, B, & C, measure resistance to heat. A has the best resistance.

Figure 8–2: Tire size, tread wear, heat rating, and other information can be found on the sidewall of the tire.

Run-Flat Tires

It all started in the early 1990s when a few tire companies began marketing their so-called runflat and zero-pressure tires. These tires use an extremely stiff sidewall. This stiffer tire can support vehicle weight without any air pressure, allowing drivers to continue driving for limited speeds up to 55 mph and at range up to 100 miles after the tire goes flat. A pressure monitoring system is used to alert drivers when a tire is losing air. Without it, drivers would likely not be aware of the flat tire condition and continue driving right through the 55 mph 100-mile safety limit.

These tires can eliminate the need to have to change a tire on the side of a busy road or in an unfamiliar part of town. They may also eliminate the need for a spare tire and jack. In a small car with limited storage space, not having a jack and spare tire can mean carrying a few more items. It also means less weight to lug around and a little improvement in fuel mileage. Run-flat technology isn't the final answer to safe travel, but it does offer added safety when driving. We may yet see the spare tire and jack become a part of automotive history.

Temporary Spare Tires

Many new cars have a temporary spare tire in the trunk. These special tire/wheel combinations are typically smaller in diameter and narrower than the normal tires and wheels on the car (Figure 8–3).

These tires are designed for temporary use only. Typically, the maximum speed at which they are to be driven is 50 mph. Special care must be taken when installing the temporary spare on the drive axles (front axle of a FWD or the rear axle of a RWD). Since the tire is smaller, it will spin faster to keep up with the larger tire on the other side, thus working the

Figure 8–3: Many late-model cars have special, undersized spare tires that are meant for temporary use only.

differential more than normal. If a vehicle has limited slip and the tire on the drive axle is flat, the temporary spare should be mounted on the non-drive axle and a regular tire moved to the drive axle. If you are not sure if you have a limited-slip differential, then place the temporary spare on the nondrive axle and transfer the regular tire to the drive axle just to be on the safe side. In addition, after one hour of driving. you must stop and allow the tire to cool.

Because the temporary spare is smaller than the normal tire, the car will tilt to the side on which it is mounted. This tilt will remind you that maximum speed should be reduced and that the normal tire should be repaired and replaced as soon as possible.

Typically, the temporary spare is a high-pressure tire that should be kept inflated to 60 psi, much higher than usual tire pressures. This extra pressure is required so that the temporary spare can support the weight of the vehicle. Check your owner's manual.

In past years, other types of temporary spare tires have been used. Another type is the collapsible spare. This tire is deflated, or collapsed, to save space. A special canister of compressed air must be used to fill the tire and expand it to normal size before use.

Full-Size Spare

Larger vehicles that have room may use a full-size spare. Though they take more space to store, when installed they allow the vehicle to operate as though it has the regular tire mounted. Many AWD and 4WD vehicles use a full-size spare. Vehicles with a limited-slip differential should also use a full size spare.

ROUTINE TIRE MAINTENANCE

Train yourself to look for possible trouble areas and keep your tires in the best possible condition. Remember, the tires are the car's only contact with the pavement, and you don't want to jeopardize that contact.

A tire maintenance kit costs only a few dollars. Such a kit contains tire pressure and tread depth gauges. In today's age of self-service gasoline stations, it is important to be able to check tire pressures and wear patterns yourself.

Check Air Pressure

Proper tire inflation is vital to the safety and economical operation of your car. However, tires lose air pressure over time because the materials used in a tire are porous. In addition, nails or other punctures can cause a tire to lose air. Thus, tires must be checked frequently and air must be added as necessary. It is a good habit to check the tires whenever you stop to buy gasoline.

The information on the tire sidewall lists the *maximum* safe pressure—not the *recommended* pressure for your specific vehicle. The same tire may be used on many different vehicles, each of which may require a different pressure for safety, comfort, and wear considerations.

The recommended tire pressures for your car are typically printed on a sticker, or placard, that is attached to the vehicle. The most common location is on or near the driver's door (Figure 8–4). Unfortunately, there is no standard location for this informational sticker. Common locations include:

- The front driver's or passenger's side door edge or doorjamb
- The inner surface of the glove box door or floor console door
- The trunk or by the spare tire
- The gas tank filler door

If you cannot locate the sticker, check your owner's manual or ask at a service center or tire store for the recommended pressure.

Recommended inflation pressures are typically listed in pounds per square inch (psi). Most tires for passenger cars will typically be 32 psi. Different pressures may be listed on the sticker for different tires that may be fitted to the car, or for different load conditions.

In many cases, different pressures are recommended for front and rear tires. Because of the different amounts of weight on the front and rear of vehicles, tire pressures may be different to keep your automobile stable and safe under all driving conditions.

The normal loss of tire pressure in a typical tire ranges from $\frac{1}{2}$ psi to 3 psi per month. If you haven't checked your tires recently—especially the spare—there may not be enough air pressure to support the car safely.

Always check tire pressures when the tires are cold. The best time is first thing in the morning. Make a note of how much additional air each tire needs. Air expands and air pressure increases when the tire is heated, as from driving. So, if you must drive to a service station or garage to add air, the tire

Figure 8–4: A tire placard on a doorjamb.

pressure may rise during the trip. Simply add the amount of air that you noted was needed. When the tire cools again, the pressure will be correct.

On long trips and in hot climates, you should never let air out of a hot tire to bring the pressure down to the recommended level because you'll end up with an underinflated tire. Remember, the recommendation is for cold tires.

Some select vehicles may use on-board air pressure sensors in each wheel to alert the driver to an underinflated tire. These systems can be expensive, but they continuously monitor the pressure for the driver and alert the driver should a low-pressure condition occur. Another system in development gauges changes in wheel speed, as an underinflated tire will rotate faster than normal. This system could be easily added to antilock brake systems (ABS), but that would mean the vehicle would need to be equipped with ABS.

Underinflated tires wear quickly, decrease gas mileage, increase stopping distances, and produce excess heat that can cause a blowout. A 5 psi drop in pressure could mean a drop of up to 10% in fuel mileage. Checking tire pressure saves money and excess tire wear, and is a safe thing to do.

Check Tread Wear

Your tires should wear flat and level across the tread surface. If they are wearing unevenly, you could have problems with inflation pressures, suspension or steering parts, or possibly poor driving habits. Problems associated with uneven tire wear are discussed further under suspension system problems in Chapter 9.

Tire wear is usually measured in increments of $1/32$ of an inch. Your tires should have more than $2/32$ inch ($1/16$ inch) of tread, the minimum safe and legal tread depth. You can check tread depth with a tread gauge or a Lincoln penny. Insert the penny in the tire's tread with Lincoln's head toward the tire. If you can see the top of Lincoln's head above the tread, the tire is dangerously worn and should be replaced. Most tires feature a tread depth indicator built in during manufacture. Visible wear indicator bars are molded into the tread of the tire to alert you when it is time to buy new tires. These bars are $2/32$ inch high. Once the tread wears to the same level of these bars, a bare spot will appear across the tread surface (Figure 8–5). If the tread is dangerously worn, you will notice these bare spots, or bare strips, across the tread from side to side.

Inspect for Defects

In addition to tread wear, other defects can cause a tire to be unsafe. Inspect the sidewalls for wear, cracks, cuts, bubbles, or any other irregularities. If you find deep cracks, cuts, or punctures in the sidewall, the tire must be replaced. Because the sidewall must flex, these types of defects will eventually cause a blowout.

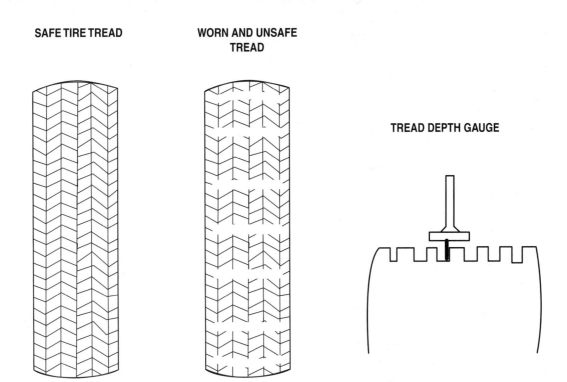

SAFE TIRE TREAD   **WORN AND UNSAFE TREAD**   **TREAD DEPTH GAUGE**

Figure 8–5: Methods of determining safe tire tread depth include wear indicator bars and a tread depth gauge.

Check the valves and valve stems, too. A valve stem that is beginning to crack should be replaced immediately. Valve stems should always be replaced when new tires are installed.

Look for any nails, glass, or other objects lodged in the tread. If you find a nail in a tire, the safest way to proceed is to drive to a tire service center or service station as soon as possible. Remember, when the nail is pulled, the tire will go flat. As long as the nail is in place, the air pressure will be held. However, if a nail or other object is left in a tire over a long period of time, it may enlarge the hole and weaken the tire carcass so that the tire cannot be repaired.

Rotate the Tires

Rotating the tires simply means putting a rear tire on the front and a front tire on the rear to even out the wear. On most cars—especially FWD vehicles—the front tires wear more rapidly than the rear tires. For this reason, it is best to rotate your tires every 6 months or 6,000 miles, whichever occurs first.

Some procedures for tire rotation are covered in Chapter 11.

Check Wheel Balance

Proper wheel balance allows the tire to roll straight without excess wear to the tires. But an out-of-balance condition can subject the suspension system to increased wear of joints, shocks, and other suspension components. Tires don't wear perfectly evenly. Some small variations in wear will occur and may put the tire out of balance. Or the tire may have struck a pothole or other obstacle to cause a slight shift in the tire and throw it out of balance. Should a vibration, shimmy, or **wheel tramp** show up while driving the car, a wheel balance is necessary. Tire balancing and perhaps a wheel alignment may be needed if abnormal or uneven tire wear is indicated.

Wheel balancing distributes wheel weights along the rim to counterbalance heavy spots in the tire. Wheel weights are clipped or adhered to the rim. Two types of balancing can be used: *static* and *dynamic*.

Static balancing uses a bubble in a round sight glass. The tire is placed on a flange and the wheel weights are placed on the side of the rim that the bubble floats toward. Weights are divided between the front and the back of the rim. While a simple method and better than no balancing at all, static balancing is not effective in most cases.

Dynamic balancing uses a machine, a wheel balancer, to spin the wheel and a computer to measure the dynamic forces created during spinning (Figure 8–6). These dynamic, or centrifugal forces, allow the computer to

Figure 8–6: A computerized spinning balancer called a "dynamic" wheel balancer is used to balance wheels.

accurately indicate to the technician where to place the weights along the rim. Static balancing may not counterbalance a heavy spot on a tire, especially if the heavy spot is on one edge of the tread. Dynamic balancing will counterbalance these out-of-balance conditions. It is the preferred type of balancing. A tire that is dynamically balanced will also be statically balanced.

THE BRAKE SYSTEM

The function of the brake system is to slow and stop your car safely. The brake system creates friction between **brake linings** and disc or drum brake units, as described in Chapter 1. The friction changes the kinetic (moving) energy of your car into heat energy, which then is dissipated to the air.

Through the hydraulic pressure of the **brake fluid,** force is transmitted from your foot on the brake pedal to the disc or drum brake units located at each of your wheels. When you press the brake pedal, you are forcing a plunger pump within the **master cylinder** to operate (Figure 8–7). This action creates pressure within the **brake lines.**

At the end of the brake lines, the pressure operates *wheel cylinders* that create force to move disc brake pads or drum brake shoes into contact with friction surfaces. The hydraulic system is designed so that it multiplies the force you apply to the brake pedal.

Two kinds of hydraulic brake systems and the major parts common to both of them are covered in the following sections.

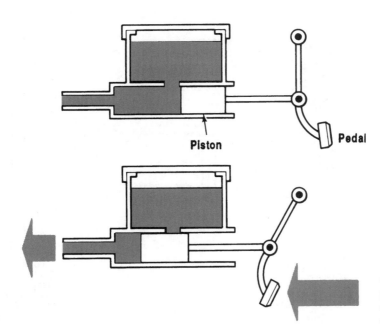

Piston

Pedal

Figure 8–7: When you depress the brake pedal, a plunger pump in the master cylinder applies pressure to operate the brakes.

Non-Power Brakes

Though few cars today have non-power brakes, the principles by which the brakes operate remain the same. Non-power brakes rely on the driver's foot pressure to operate. The system multiplies the force exerted on the pedal through hydraulic means and mechanical leverage. A non-power brake system is illustrated in Figure 8–8.

The brake pedal is linked mechanically to the master cylinder, located beneath the brake fluid reservoir. As the pedal is depressed, the master cylinder transfers brake fluid through the brake lines to each wheel. At the wheel, a wheel cylinder expands, applying pressure to brake shoes, or a piston is pushed outward to apply pressure to the disc brake pads. The pressure forces brake linings into contact with friction surfaces.

When the brake pedal is released, there is a reduction of hydraulic pressure. On disc brakes, the caliper piston seals flex and retract the piston slightly. The rotating **brake disc** also helps to push the pads away from the disc surface. On drum brakes, springs on the brake shoes contract and pull the shoes away from the braking surface of the **brake drum.** The movement of the wheel cylinder pistons forces fluid back to the master cylinder to replenish the reservoir.

Power-Assisted Brakes

The operating principle of the power-assisted system is the same as for a non-power system. However, braking force is increased through the use of a *vacuum booster unit*, a chamber that stores vacuum supplied by the engine.

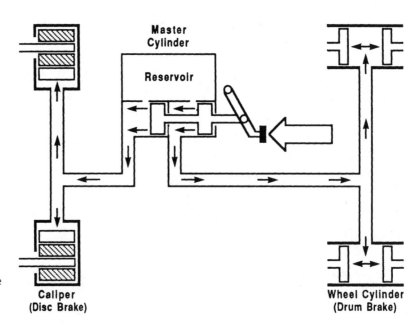

Figure 8–8: For safety purposes, separate master cylinder plungers and lines are used to operate each half of the non-power brake system.

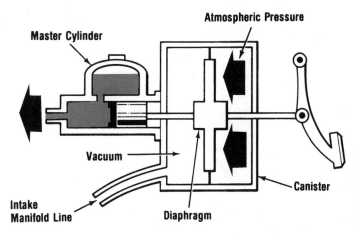

Figure 8–9: A power brake booster uses atmospheric pressure to apply force and reduce the physical effort required to operate the brakes.

Atmospheric pressure helps to push the brake pedal, as shown in Figure 8–9. Thus, less muscle effort is required.

The vacuum booster unit typically is mounted at the rear of the master cylinder. In the middle of the vacuum booster unit is a flexible rubber wall called the *diaphragm*. The diaphragm separates the booster into two chambers. Both chambers are connected to the intake manifold of the engine to create a vacuum. Atmospheric pressure may be admitted to the rear chamber on demand.

When the brakes are not in use, vacuum from the engine's intake manifold is supplied to both chambers and holds the diaphragm in a neutral position. As the brake pedal is applied, a valve opens to allow atmospheric pressure to enter the rear chamber and apply a force of 14.7 pounds per square inch on the diaphragm. As the diaphragm moves, it helps to push the rod that connects the brake pedal to the master cylinder. This action assists the driver so that brake pedal effort is minimized.

Another, less common way to increase the force applied to the master cylinder plunger is to use a belt-driven hydraulic pump. When the brake pedal is depressed, fluid pressure from the pump is applied to the rear of the master cylinder plunger through a series of tubes and hoses. This system may be found on many diesel engines, as these engines do not develop vacuum.

Thus, with power-assisted brakes you can travel at highway speeds in a vehicle that weighs several thousand pounds and yet bring it to a stop quickly and easily with only light foot pressure on the brake pedal.

Master Cylinder

In the brake system the key to reliability is the fluid level. The master cylinder incorporates a reservoir for the brake fluid. Hydraulic pressure can't be created unless there is enough fluid in this reservoir.

Master cylinders must supply pressure independently to each half of the brake system. The lines may be split to operate both front brakes together and both rear brakes together. A diagonally split system may be used to operate two sets of brakes. One set is composed of the front left and the right rear brakes and the other set is composed of the right front and the left rear brakes. This arrangement is a safety feature. Should one half of the brake system fail, the other half can still operate, although stopping distances will be increased.

The positions of the fluid reservoirs in the master cylinder do not necessarily coincide with the position of the brakes on the car. In other words, fluid for the front brakes may be contained in the rear compartment, and vice versa. Typically, the larger of the containers holds fluid for the front disc brake calipers, which require more fluid to operate. Cars with four-wheel disc brakes usually have the same size reservoirs.

As you apply the brakes, two pistons are pushed through the master cylinder. Each piston forces fluid ahead of it through a check valve and into the brake lines. A predetermined amount of fluid goes toward the front brakes and a predetermined amount is pushed toward the rear brakes. This movement is accomplished by the *proportioning valve*, which is a small hydraulic device. When the pedal is released, brake fluid is forced backwards through the line of the master cylinder by the action of piston seals on disc brakes and by springs on drum brakes.

Brake Linings

For many years, brake linings have been made of a combination of asbestos fibers, metal particles, and a binder material. As the brake linings wear due to friction, brake dust containing asbestos is created. Asbestos dust is a severe health hazard that has been linked conclusively with lung cancer in humans. Thus, caution must be used when you inspect brake units.

In particular, you should avoid breathing brake dust from wheel brake units. Ordinary paint and dust masks will not protect you from breathing asbestos dust particles. Special procedures and equipment are used by brake shops to contain and remove brake dust during service.

Nonasbestos brake linings are manufactured by major brake lining suppliers. These linings contain organic, synthetic, and metallic fibers that do not pose the serious health hazard of asbestos-based linings. However, brake dust is still considered unsafe.

Disc Brakes

As explained in Chapter 1, disc brakes operate on the same principles as hand brakes on a bicycle. Two friction material pads are held in a caliper on either side of a disc that is attached to the wheel hub. The disc, or *rotor*, turns when the wheel rotates. When the brakes are operated, frictional, or rubbing, force is applied to the disc to slow its rotation. Thus, the wheel attached to the disc slows and eventually stops.

The disc brake is simpler in design and easier to service than drum brakes. Because of the disc brake's design, in which the largest part of the rotor is exposed to the air, the heat produced by braking can be dissipated more easily. Thus, *brake fade*, caused by extreme heat on the braking surfaces, is reduced in disc brakes. Brake fade can occur when you stop quickly from a high speed or when the brakes are overused. The cause of brake fade is a layer of gas that builds up between the surface of the brake lining and the metal frictional surface of a brake unit. This layer of gas keeps the two surfaces from touching and reduces friction.

Because of the better heat dissipation qualities of disc brakes, they can be made more powerful and more effective than drum brakes of similar size. Also, the wiping action of the disc brake pads as the disc rotates between them helps to maintain the brake's effectiveness in wet weather.

Three basic components make up the disc brake: the rotor, the caliper, and the disc pads. The rotor is a flat steel or cast-iron disc that is mounted to the wheel hub and rotates with the wheel. The caliper is a stationary, C-shaped unit that surrounds part of the rotor. Disc pads provide friction to both sides of the rotor when the brakes are applied. A splash shield protects the unit from dirt and debris.

On the majority of disc brakes, the action of the caliper is similar to that of a C-clamp. When you step on the brake pedal, hydraulic pressure forces a piston against one pad inside the caliper. This piston pushes out a brake pad that makes contact with the rotor. This action pinches the rotor between both brake pads within the caliper to slow the vehicle. As the brake pedal is released, the caliper piston seals cause the piston to retract slightly, and brake fluid is returned to the master cylinder.

To operate, disc brakes require more hydraulic pressure than drum brakes. For this reason, a braking system that includes disc brakes is typically equipped with a power-assist device.

Disc brakes are self-adjusting. As the frictional material of the disc brake pad wears down, the caliper piston moves closer to the disc. Fluid from the master cylinder reservoir fills the additional space behind the caliper piston. The result is that the brake pad surface always remains next to the disc, lightly brushing it, and ready to react as soon as the brake pedal is pushed. Thus, as the disc brake pads wear down, the fluid level in the master cylinder reservoir drops.

The disc brake pad friction material is mounted on a steel plate. If the lining becomes completely worn down, the steel plate will rub against the brake rotor and ruin its surface. Some disc brake pads have a warning device built in (Figure 8–10). A metal tab is provided to contact the surface of the rotor before the lining is completely worn down. The metal tab produces a high-pitched screeching or rubbing sound when it contacts the disc during brake application. This noise alerts the driver that brake service is needed.

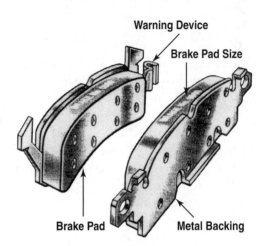

Warning Device

Brake Pad Size

Brake Pad

Metal Backing

Figure 8–10: Some disc brake pads include a metal tab that contacts the disc when the friction linings are worn excessively.

During hard braking, the front end of the car dives, or moves downward as most of the weight of the vehicle shifts forward. Thus, the front brakes must provide 75 to 80% of the total braking force required to stop a car. Because disc brakes can dissipate heat from hard braking better and provide better wet-weather operation, they are used on the front of all vehicles and the rear of many vehicles.

There are two major variations of the basic disc brake:

1. Sliding or floating caliper
2. Fixed caliper

Sliding- or Floating-Caliper Disc Brakes In sliding- or floating-caliper disc brakes, only one side of the caliper has a piston. When the brakes are applied, the activating piston pushes out on one pad and forces the opposite pad against the other side of the rotor (Figure 8–11). The caliper slides or pivots to clamp the rotor between the pads.

Fixed-Caliper Disc Brakes A fixed-caliper disc brake includes a rigidly mounted caliper with pistons mounted on both sides of the rotor. When you apply the brakes, the caliper does not move. Instead, pistons on each side push out pads that clamp against the rotor's frictional surface (Figure 8–12).

Drum Brakes

A drum brake applies force to move two semicircular brake shoes against the inside of a hollow brake drum. The parts of a drum brake include a drum, a backing plate, brake shoes, a hydraulic wheel cylinder, and return springs. One type of drum brake mechanism is shown in Figure 8–13.

The backing plate is stationary. It holds the brake shoes, hydraulic wheel cylinder, and return springs. It also helps keep dirt and debris out of the brake assembly.

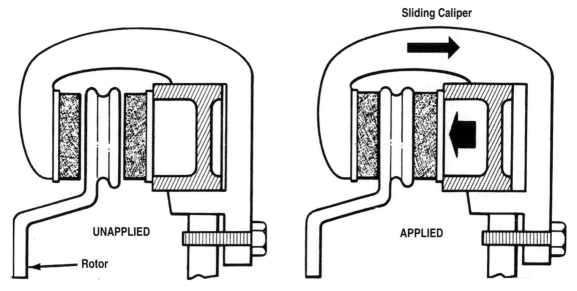

Figure 8–11: When the master cylinder applies pressure to a disc brake caliper, the rotor is pinched between the friction pads.

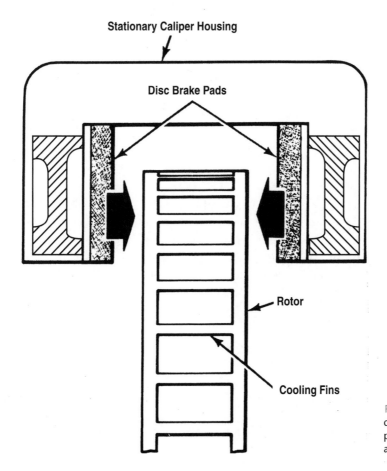

Figure 8–12: A fixed-caliper disc brake includes two caliper pistons to clamp the pads against the brake rotor.

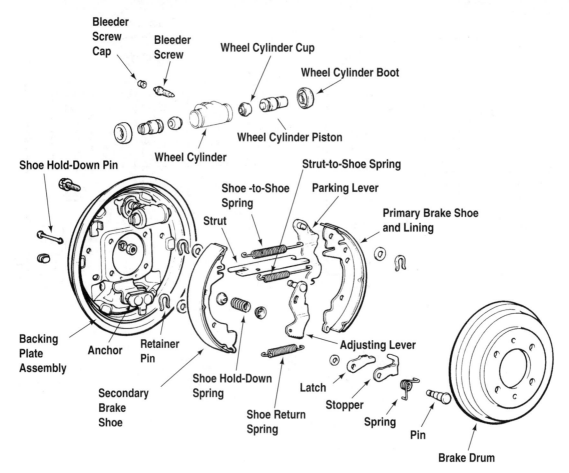

Figure 8–13: A leading/trailing brake, a typical drum brake.

The drum rotates with the wheel. When the brake pedal is depressed, the master cylinder creates fluid pressure that operates the wheel cylinder. The wheel cylinder forces the brake shoes outward against the inside circumference of the drum. The friction slows the turning of the drum. When the drum is slowed, the wheel is slowed.

When the brakes are released, the return springs pull the shoes away from the drum surface, the wheel cylinders contract, and the brake fluid is returned to the master cylinder.

Drum brakes require less force to operate than disc brakes. Therefore, it is easier to design a mechanically operated emergency brake mechanism. The mechanical leverage that can be provided by a system of cables and levers will lock rear drum brakes more securely than similar arrangements used to lock and hold disk brake units. Some vehicles with rear disc brakes make use of a small drum brake for the parking brake.

Self-Energizing Drum Brake   As the brake pedal is applied, the primary—or front—shoe makes contact first. On initial contact with the rotating drum, the primary shoe tries to wrap around, or rotate, with the drum. The lower end of the primary shoe pushes against the adjuster mechanism, which pushes against the lower end of the secondary—or rear—shoe. The secondary shoe also begins to wrap around the drum. As it does, it is stopped by the anchor pin at the top and begins to pivot outward. Thus, the secondary shoe is pressed toward the drum with great force. After the brake shoes have wrapped into a locked position against the drum, hydraulic pressure from the master cylinder holds the lining securely in position.

In a self-energizing brake, the secondary shoe does most of the braking, so it has a larger lining than that of the primary shoe.

Leading/Trailing Drum Brake   The other variation of the drum brake is the leading/trailing brake. (Refer to Figure 8–13 as you read this explanation.) It is similar to the self-energizing brake. In this system both the front and the rear shoes have the same size lining. As the brake pedal is applied, the shoes begin to move outward. The shoes are anchored at the bottom so each shoe acts independently of the other. The leading shoe contacts the drum at the front and provides friction. The trailing shoe contacts the drum at the rear and also provides friction. Thus, each shoe is essentially a separate brake on the same drum, but both operate at the same time.

Leading/trailing brakes are the most common drum brakes in use, but drum brakes are being replaced with disc brakes.

Drum Brake Adjustment   Unlike disc brakes, drum brakes require adjustment as the linings wear down. If the adjustment is neglected, the space between the shoe lining surface and the drum increases. To operate the brakes, the brake pedal must travel farther and farther down before the drum brake shoes make contact with the drum. Thus, if the adjustment is not automatic, periodic adjustment of the rear drum brake units is required.

Self-Adjusting Drum Brake   Self-adjusting drum brakes include a mechanism to adjust the clearance between the surface of the brake shoes and the inner surface of the drum.

Self-adjustment typically occurs in one of two ways. One type functions when you back up and stop. A mechanism is activated by the self-energizing action of the shoes wrapping around the drum as the car moves in reverse. If there is excess clearance between the shoe and the drum, an adjuster will turn that spreads the brake shoes slightly and holds them in a slightly expanded position.

For drum brakes that are not self-energizing, such as a hand-operated emergency, or parking, brake, the second type of self-adjustment is used. If the brake shoes move outward far enough, a ratchet device locks and holds them in a slightly expanded position.

Disc–Drum Combinations

In a combined system, the disc brakes are always on the front and the drum brakes are always on the rear. To coordinate disc and drum brakes, two valves are incorporated into the system: a proportioning valve and a metering valve. The proportioning valve regulates hydraulic pressure as the fluid travels to the rear brakes, providing less hydraulic force. The metering valve delays pressure to the front disc brakes for a split second. This action balances the action of the front and rear brakes to prevent rear brake lockup.

ANTILOCK BRAKE SYSTEM

The function of an antilock, or antiskid, braking system is to prevent the wheels from locking under hard braking. Maximum braking force is obtained just before the wheels lock and skid.

When the wheels begin to skid, the rubber of the tires heats, burns, and liquefies. If you have ever had the decidedly unpleasant experience of violent braking, you may remember the odor of smoke from liquefied, burning rubber. The liquefied rubber acts as a lubricant between the tire and the road surface. Thus, the car does not slow down as fast as it would before the wheels lock. In addition, once the wheels lock, steering control is lost. The car may swerve sideways or spin.

Antilock braking systems, commonly abbreviated as ABS, help to prevent wheel lockup. As shown in Figure 8–14, an ABS typically includes:

- A master cylinder (power assisted)
- Speed sensors at each wheel
- A computer-controlled hydraulic unit
- An electronic control unit (computer module)

The brake lines are routed through the hydraulic control unit (HCU). When the brakes are applied, the computer determines whether the hydraulic unit should be activated to modify brake pressure.

During normal braking, the ABS and non-ABS brake pedal feel will be the same. During ABS operation, a pulsation can be felt in the brake pedal, accompanied by a fall and then rise in brake pedal height and a clicking sound.

The speed sensors provide information on the speed of each wheel to the electronic control unit (ECU). The ECU compares and averages all the input signals to determine the vehicle's road speed. If the antilock brake control module senses a wheel is about to lock, based on wheel speed, it

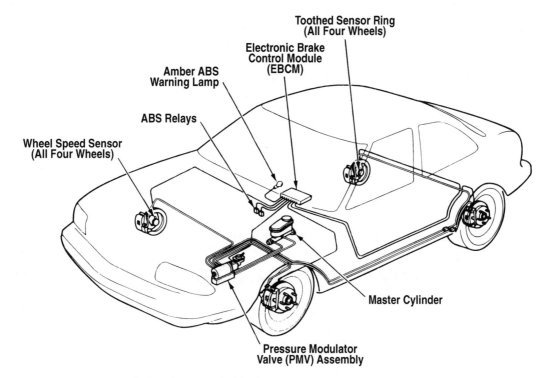

Figure 8–14: Components typically found in an antilock brake system.

closes the normally open solenoid valve for that circuit. This action prevents any more fluid from entering that circuit. The antilock brake control module then looks at the wheel speed signal from the affected wheel again. If that wheel is still decelerating, it opens the solenoid valve for that circuit. Once the affected wheel comes back up to speed, the antilock brake control module returns the solenoid valves to their normal condition, allowing fluid flow to the affected brake.

The hydraulic unit contains motors and solenoid valves similar to the valves in an automatic transmission valve body. The valves open or close to modify the fluid pressure flowing from the master cylinder to the wheel calipers. The hydraulic unit can decrease, maintain, or increase hydraulic pressure to either of the front two brakes independently or to the rear brakes either independently or together.

This entire course of events occurs in a fraction of a second. The ABS monitors and adjusts the brake pressure many times a second, from 10 to as many as 15 times every second. The result is that the wheels do not lock or skid. Braking distances may be slightly longer, but directional control can be maintained, even on an icy curve.

Antilock brakes cannot change the laws of physics. They will stop the vehicle as quickly as possible given the road and tire conditions. Contrary to popular belief antilock brakes will not stop your car faster. The main idea behind the ABS is to maintain control of your vehicle by avoiding a dangerous condition when the wheels lock up. When your wheels lock up, you have no steering control, so turning the steering wheel to avoid a collision will do you little good.

When driving on slippery roads, you need to allow for increased braking distance since the wheels will lock up much easier and the ABS will cycle much faster. Speed is a factor. If you're going too fast, even the control the ABS gives you will not be enough to overcome inertia. You may turn the wheel to the left or right, but inertia will keep you going forward.

If any part of the system should fail to work, the system goes into a failsafe mode: The brakes operate normally, as they would on a car that is not equipped with ABS.

ABS is a feature that is now standard on many cars.

TRACTION CONTROL

Traction control is yet another safety feature developed for today's automobiles. Traction control systems improve vehicle stability by controlling the amount the drive wheels can slip when you apply excess power. There are several variations, but most of these systems are more similar than you might think. One system works by limiting engine torque. Other traction control systems may use both the brakes and reduced torque to improve traction.

Traction control works just the opposite of ABS—dealing with acceleration rather than deceleration. Since many of the same principles apply to both systems, it is easy to visualize it as sort of ABS in reverse. The system automatically adjusts the engine power output and, in some systems, applies braking force to selected wheels during acceleration. When the wheel starts to spin, the ABS wheel speed sensors provide that data to the computer, which then adjusts or limits engine torque. If the wheels continue to spin, the computer can command the ABS to apply braking force to slow the wheels. The wheels are thus prevented from spinning and the car maintains maximum traction. Traction control is mainly found on vehicles with four-wheel antilock brake systems.

PARKING BRAKE

On most cars the parking brake uses a mechanical linkage to tighten and lock the car's rear brakes only. This brake is applied by a hand lever between the front seats or by a foot pedal under the dashboard. A cable runs from the driver's compartment to the rear of the car to operate this brake.

As its name implies, this brake should always be used when the car is parked. It can be used in case of an emergency, should your normal braking system fail. However, don't expect it to stop the car as well as the normal braking system would. Because the emergency brake system activates the rear brakes only, stopping distances typically will be much longer than when all brake units are operating.

SAFETY PRECAUTIONS FOR BRAKE SYSTEM MAINTENANCE

When performing maintenance on the brake system, be sure to follow the general safety precautions presented in Chapter 2. In addition, avoid using compressed air or any method to remove brake dust that will cause it to become airborne and easily breathed in. Special safety enclosures and vacuum cleaners are used by professional brake technicians to remove brake dust. A brake spray cleaner may be used as it washes the dust down without it becoming airborne.

CAUTION

Avoid breathing brake dust from worn brakes. If the brake dust contains asbestos, it may lead to lung cancer. Avoid using compressed air or any other method to remove brake dust that causes the dust to become airborne.

ROUTINE BRAKE SYSTEM MAINTENANCE

To assure the proper functioning of the brake system, the most important safety system in your car, regular checks and maintenance are required. Just one failure could result in serious injury or even a fatality. Use the following information as a general guide to perform brake system maintenance.

Check Fluid Level

Brake fluid levels are checked at the master cylinder reservoir, which may be a single container or two separate containers. Cars typically have screw-on or push-on caps on plastic reservoirs (Figure 8–15). Molded marks on the outside of plastic reservoirs indicate safe levels. If no level is indicated, fill the container or containers until the fluid is ¼ inch below the top edge of the reservoir.

As brake linings wear on both disc and drum brake units, fluid from the reservoir is transferred to the expanded wheel brake cylinders. Thus, the level slowly drops in the reservoir as the linings wear. If the reservoir

Figure 8–15: **A typical master cylinder.**

fluid level is noticeably low, refill the reservoir; then check brake linings on all four wheels for wear. If the linings appear to be in good condition, check for leaks from the calipers and wheel cylinders.

If there are no visible leaks at the brake units, fluid could be leaking from the rear of the master cylinder. On a non-power brake system, check at the rubber boot under the dash into which the actuating rod from the pedal fits. Spread the boot gently with a small screwdriver to check for brake fluid.

On a power-assisted brake system, it is usually possible to unbolt the master cylinder from the power booster. Pull the master cylinder forward gently about half an inch and check for brake fluid leakage at the rear of the master cylinder.

If any brake fluid is found at the rear of the master cylinder, it must be replaced. For power-assisted brakes, gently push the master cylinder back into place and tighten the fastening bolts after you have made your inspection. Replacing a leaking master cylinder is a job for a certified brake technician.

Check Linings

Brake linings are made of materials that cause friction when pressed against rotating metal surfaces. If the linings are badly worn, they must be replaced. Further information on checking brake lining wear is presented in Chapter 11.

BRAKE SYSTEM PROBLEMS

At some time, you may notice one or more of the following conditions related to brake problems. The following list is intended to help you identify brake problems correctly. However, the list is not exhaustive and does not cover every possible brake problem. Have brake problems diagnosed and fixed by certified brake technicians.

Noisy Brakes

Brakes may squeal when they are applied. If the squeal goes away after a few applications of the brakes, it is "cold brake squeal," an annoying but common problem, especially on cars with disc brakes.

If the brakes squeal even after they have been used and warmed up, the linings may have overheated and become glazed, or excessively smooth. Badly glazed brake linings should be replaced.

If you hear a metal-to-metal contact—a loud, high-pitched squeal or a dull, grinding noise—the brake lining is worn excessively. Metal parts of the pads or shoes are contacting the rotors or brake drums and can cause severe and expensive damage.

On some newer automobiles, the brakes may chirp. A warning device built into the linings causes this noise. It indicates that it's time to have the linings replaced.

No Brake Pedal or Low Brake Pedal

After starting and before you drive away, depress the brake pedal; it should move only an inch or two and then remain firm. If the pedal goes all the way to the floor, the brake system is not holding pressure. Don't drive the car—have it towed to a service center.

If the brakes work normally, but the pedal must be depressed farther than normal, the problem may simply be improperly adjusted rear drum brakes. A low brake pedal condition can also be caused by a malfunction in a power booster unit.

Spongy Pedal

If the brake pedal is not hard when you depress it, and it feels mushy, you have a spongy pedal. Although you can stop the automobile, you will notice it does not stop as well as it should.

Try the following test: Press hard against the pedal and hold it down for 30 seconds. If the pedal slowly sinks to the floor, the brake system is not holding pressure. Try pumping the pedal a few times and holding it down on the last pump. If the brake pedal stops at a higher level, or "pumps up," there are three possibilities. First, the master cylinder reservoir is empty of brake fluid, and air may have entered the brake lines. Try refilling the reservoir with brake fluid, and then repeat the test several times. If the problem still exists, don't drive the car. Have it towed to a service center.

A second possibility is that the plunger seals in the master cylinder are defective. Press and hold the pedal down. If the pedal sinks slowly to the floor, you know the master cylinder seals are defective. The master cylinder will need to be replaced by a technician.

Third, there is a leak in the brake lines or wheel cylinders. Look under the car for evidence of a brake fluid leak on the ground or inside the wheels. Check for master cylinder leaks at the rear of the master cylinder. On cars without power brakes, check under the dash where the brake pushrod passes through a protective rubber boot. If any of these conditions exists, have the car towed to a service center. Don't risk driving it.

Brake Pull

If your car begins to swerve, or pull, to the right or left when you apply the brakes, there may be brake problems. One or more of the brake units may not be working properly.

Any of a number of problems can be responsible for brake pull. There may be grease on the linings. A wheel cylinder may be leaking or may have become stuck. A stuck piston in a disc brake caliper will cause a pull on the opposite side of the car. Or a drum brake shoe may be hanging up on a backing plate.

Loose or defective steering or suspension parts also can cause pulling when the brakes are applied.

Just as with the human body, automotive parts and systems are interrelated. A problem with one part or system can affect others. For example, an unbalanced tire can bounce excessively and cause rapid and uneven tread wear. The excessive bouncing also can cause a shock absorber to wear out and fail. In turn, a worn-out shock absorber can cause damage to steering and suspension parts.

Systems and parts are often designed to help other systems operate more efficiently. For example, the suspension and wheels are designed to help keep the car stable during braking and turns. The next chapter discusses the steering and suspension systems.

1. What components make up a wheel?
2. What are the three types of tire construction?
3. Explain the metric tire sizing system.
4. Explain the UTQG and the information it provides.
5. What are the advantages and disadvantages of run-flat tires?
6. Where might the tire pressure placard be found?
7. Why is tire pressure important?
8. What is minimum tread wear, and how is it checked?
9. What are the two kinds of wheel balancing? Which is best?
10. What problems can an out-of-balance tire cause?
11. What powers the "power brakes"?
12. What is brake fade and what causes it?
13. What are the two kinds of brake mechanisms used?
14. List the two types of disc brake calipers.
15. What are disc brake pads made from?
16. Explain how a caliper works.
17. List two types of drum brakes.
18. Explain how a drum brake works.
19. What are the main components of ABS?
20. How does ABS work?
21. How may a driver tell if the ABS is working?
22. What is traction control?
23. How does traction control work?
24. What are some maintenance items for the brake system?
25. List some common brake system problems.

SUSPENSION AND STEERING

After reading this chapter, the reader should be able to:

- List the systems that support the car's weight and control its direction.
- List the types of springs used in suspension systems.
- Explain the operation of the shock absorber.
- List the different types of front suspension systems.
- List the different types of rear suspension systems.
- List the types of steering systems.
- Explain how power steering operates.
- List the different alignment angles for wheel alignment.
- Explain the importance of proper wheel alignment.
- List the maintenance items for the suspension and steering systems.
- List some common suspension and steering system problems.

As mentioned in Chapter 8, there are systems in the vehicle that support your car's weight, control its direction, and bring it to a stop. These parts and systems include:

- Tires
- Brake system
- Suspension system
- Steering system

This chapter focuses on the suspension and steering systems.

THE SUSPENSION SYSTEM

The suspension system supports the weight of the chassis, powertrain, and body as your car travels down the road over bumps and around curves. The suspension system and tires are responsible for the stability and ride of your car.

The suspension system design is based on the way that the weight of your car is distributed over its length. For example, the engine and transmission are typically mounted near the front of the vehicle. Thus, the front portion of a vehicle is heavier than the rear, and the suspension must be designed to account for this difference.

There are two extremes of suspension design. One, for high-performance cars, is known as a "hard" suspension and sacrifices ride comfort for maximum high-speed cornering power. The other extreme is a "soft" suspension, typically found on luxury cars. This suspension sacrifices cornering performance for ride comfort. Most suspension systems are a compromise between these two. They combine an acceptable level of performance and a comfortable ride.

Today, some cars feature electronically controlled suspension systems that can provide soft, intermediate, and hard suspension on a single vehicle. These systems can operate automatically or can be adjusted manually by the driver to suit driving conditions and individual preferences.

Springs

All suspension systems require some sort of springing device to support the vehicle's weight. After being compressed, the springing device springs back, or rebounds. After being stretched, or extended, the springing device contracts. The common types of springing mechanisms include:

- Coil springs
- Leaf springs
- Torsion bars
- Air springs

A **coil spring** is a specially tempered piece of steel wound around a cylindrical form. It is the most common type of springing device. A **leaf spring** consists of one or more flat pieces of tempered steel or may be made from a single leaf composed of composite materials. A **torsion bar** is a steel rod that can twist and untwist to provide springing action. An **air spring** is a rubber cylinder filled with air much like a tire.

Shock Absorbers

The purpose of the **shock absorber,** or the shock, is to control the action of the springs. In countries other than the United States, the term *shock absorber* is seldom used; *damper* is used because the device dampens spring oscillations.

If you took a spring, put tension on it, and then let it go, the spring would bounce around (oscillate) numerous times. This same effect would cause your car to bounce constantly and would cause an uncomfortable and unstable ride.

A shock absorber includes a tubular cylinder filled with fluid and a piston attached to a metal rod. The fluid is a special oil designed for this application. In addition, many shocks have gas pressure, usually nitrogen gas, applied to the fluid within the shock absorber to minimize foaming of the oil in the shock. These are called gas shocks (Figure 9–1). The piston is not solid, but contains small orifices, or holes, through which the shock absorber fluid can travel as the piston moves up or down.

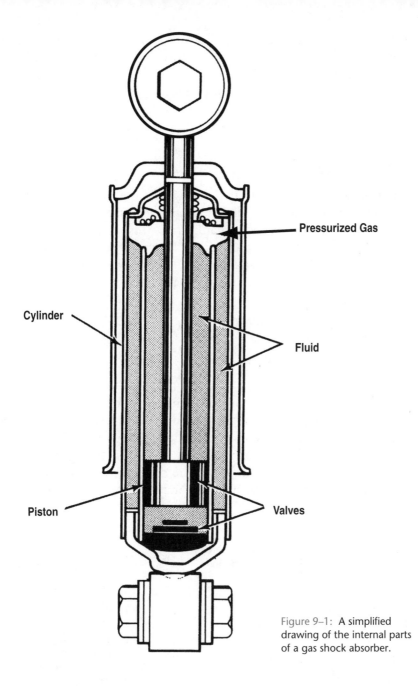

Pressurized Gas

Cylinder

Fluid

Piston

Valves

Figure 9–1: A simplified drawing of the internal parts of a gas shock absorber.

When the wheel hits a bump or drops into a hole, the shock absorber piston moves through the fluid. The difficulty of forcing the piston to move through the fluid creates the damping action that helps to stop the springs from bouncing.

Typically, one shock absorber is used for each wheel. Off-road vehicles may use several supplemental shock absorbers per wheel. Shock absorbers may be mounted through the center of a coil spring, beside the spring, or in any number of locations. Shock absorber placement and method of attachment depend on suspension design.

In an electronically controlled suspension system, solenoids can position metering needles within shock absorber passages to allow more or less fluid to pass through the plunger or other passages.

Front Suspension

The front suspension holds up the weight of the front part of the vehicle, including the powertrain. The action of the suspension system stabilizes the vehicle over bumps and around turns. Most front suspensions are **independent suspensions,** which allow either of the two wheels to react to bumps individually, without affecting the other wheel. Some basic types of front suspension in common use today are:

- MacPherson strut
- Short and long arm (SLA)
- Double wishbone
- Longitudinal torsion bar
- Leaf spring

MacPherson Strut   The MacPherson strut is an independent front suspension design consisting of a lower control arm mounted to the chassis and a strut assembly that connects to the body (Figure 9–2). The strut assembly includes a large shock absorber and a coil spring. The bottom of the shock absorber is connected to the **steering knuckle.** The steering knuckle is a pivoting metal part that includes the **spindle** on which the front wheel mounts.

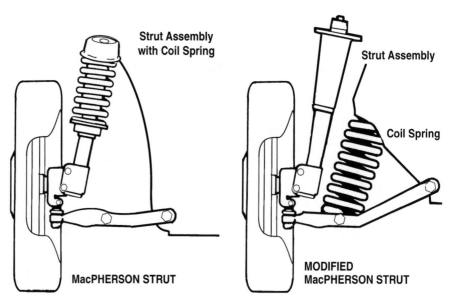

Figure 9–2: MacPherson struts are used on many front suspensions.

At the top of the strut assembly, the coil spring is connected to a tower in the body sheet metal. The tower is part of a reinforced inner fender. The top of the strut assembly rotates in the tower when the steering system operates to turn the wheels.

At the base of the strut is a **ball joint.** A ball joint is similar to the hip joint in your body. A round ball is positioned within a hollow socket. The ball can rotate and swivel to provide three-dimensional movement. The ball joint at the base of the MacPherson strut acts as a pivot point for the steering knuckle arm to turn the front wheels.

When the wheel goes over bumps and dips in the road, the coil spring and shock absorber are compressed or extended. The shock absorber helps to damp the oscillations, or bouncing movements, of the spring.

Short and Long Arm **Short- and long-arm (SLA)** suspension is a type of independent front suspension used on high-performance cars, larger vehicles, and some trucks. Its name derives from the size of the two control arms on each side of the vehicle that connect the steering knuckle to the chassis (Figure 9–3). This system may also be known as a **double A-arm suspension** or an A-frame **wishbone suspension.**

On the inner ends, the control arms pivot on shafts attached to the car's chassis. At the outer ends, the control arms are connected to the steering knuckles by ball joints. Between the upper and lower A-arms, a coil spring and a shock absorber are mounted.

When the wheel hits a bump, both control arms pivot upward to compress the spring and shock absorber. As the control arms move downward, the spring and shock absorber expand.

Double Wishbone The double wishbone (Figure 9–4) is similar to the short- and long-arm suspension system except that the lower control arm is narrower and shaped like an I rather than an A. The upper arm is also

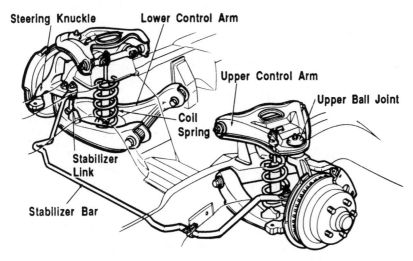

Figure 9–3: A short- and long-arm, or a double A-arm, suspension system.

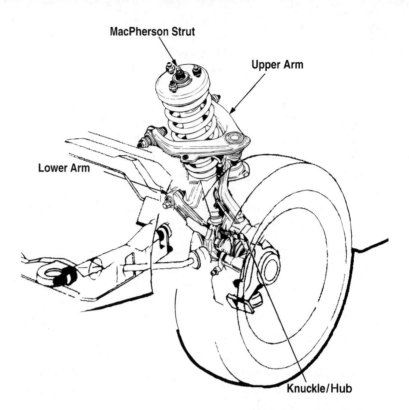

MacPherson Strut

Upper Arm

Lower Arm

Knuckle/Hub

Figure 9–4: A double wishbone suspension for a FWD car.

similar except it looks more like Y with a short tail. Both the upper and lower arm assemblies each look like a wishbone. This system borrows its design from Formula One racing cars. It provides excellent cornering, braking, and acceleration characteristics, while providing a firm but not harsh ride. There are variations on this theme, using different lengths, numbers, and combinations of arms, but they all work in the same way. Other suspension systems that are nearly identical to double wishbone are single lower control arm and narrow lower control arm.

Longitudinal Torsion Bar The **longitudinal** torsion bar independent front suspension design also uses pivoting control arms that mount to the vehicle's chassis (Figure 9–5). This design is often found on light trucks and 4WD vehicles, as well as some cars. Torsion bars, specially tempered steel rods, are used instead of coil springs to provide springing action.

In a typical case, each torsion bar is connected to a lower control arm. The torsion bar extends longitudinally, or lengthwise, along the chassis of the vehicle and its end is mounted to the chassis. When the wheel moves up or down, the control arm twists the torsion bar. The torsion bar resists this twisting motion and returns the control arm to its normal position.

Leaf Spring Leaf springs are used mainly on 4WD vehicles and trucks. In most leaf-spring suspensions, the leaf springs are mounted longitudinally, as with torsion bars. As a wheel moves upward, pressure is exerted on

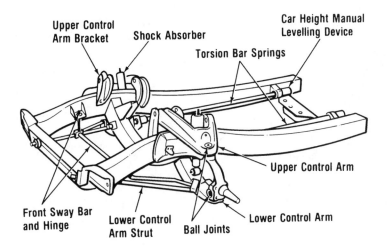

Upper Control Arm Bracket

Shock Absorber

Car Height Manual Levelling Device

Torsion Bar Springs

Upper Control Arm

Lower Control Arm

Front Sway Bar and Hinge

Lower Control Arm Strut

Ball Joints

Figure 9–5: Torsion-bar front suspension is commonly used on small RWD and 4WD trucks.

the leaf spring. The leaf spring flattens. Then as the wheel moves downward, the spring expands. Shock absorbers control the oscillations of the spring.

In most applications, a leaf-spring suspension is *not* an independent suspension. A beam axle or differential forms a solid connection between the driven wheels. Thus, as one wheel moves, the other is forced to move. The result is that if one wheel hits a bump, both sides of the vehicle react.

Leaf springs are used to provide a rugged, heavy-duty suspension system. In 4WD and RWD vehicles, leaf springs help transfer torque from the differential housing to the chassis to move the vehicle.

Rear Suspension

The purpose of the rear suspension is to support the weight of the rear of the vehicle. As with the front suspension, this system contributes to the stability and ride of the vehicle. Independent suspension systems can be used at the rear of the vehicle as well as at the front. However, most rear-end suspensions don't have to cope with steering, so they are simpler in design. Three rear-end suspensions are discussed below:

- Live axle suspension
- Semi-independent suspension
- Independent suspension

Live Axle Suspension   In a RWD suspension system, leaf springs or coil springs support the differential and help to transfer torque to move the vehicle (Figure 9–6). Leaf springs are held to the rear axle by U-bolts and aligned with a center bolt. Brackets on the vehicle frame hold the ends of the springs. The leaf springs also serve to locate the axles so control arms are not required. If coil springs are used, control arms must also be used to locate and hold the axle in position. The live axle is a simple system and is

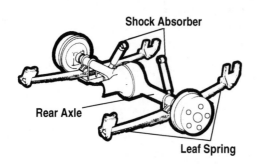

Figure 9–6: Leaf-spring rear suspension systems help to position the differential securely and to transfer torque to the driving wheels.

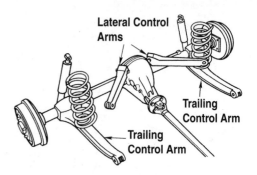

Figure 9–7: Typical live axle suspension with coil springs.

still found on some RWD cars and pickups (Figure 9–7). This system has a lot of unsprung weight, meaning weight that must move up and down when hitting a bump in the road, which creates a harsher and rougher ride. It also has some instability when hitting a bump. If a wheel on one side hits a bump and moves up or down, then the other wheel will tip in and out at the top.

Semi-Independent Suspension A semi-independent suspension system commonly uses coil or torsion springs, but leaf and air springs can also be used. The function of the rear springs is the same as it is on the front-end suspension. This system uses a cross-member or an axle beam to connect the two sides. Figure 9–8 illustrates a rear suspension using a cross-member, coil springs, and a separate shock absorber. The wheels are located behind the torsion bar on separate members called *trailing arms*. Figure 9–9 shows a rear suspension system using an axle beam and a

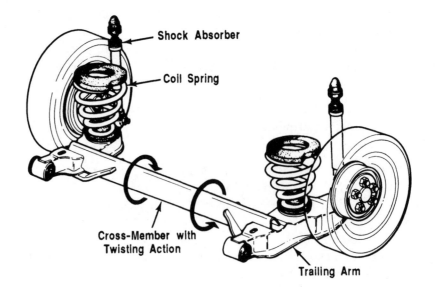

Figure 9–8: Coil-spring rear suspension is used on most late-model passenger cars.

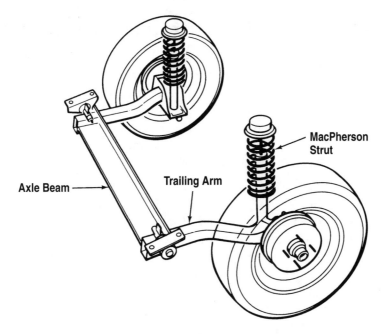

Figure 9–9: A beam-shaped solid rear axle can act as a torsion bar to help control leaning on turns.

MacPherson strut. Both the cross-member and axle beam act as a torsion bar, or antiroll bar, to help prevent excess body roll during turns.

Independent Suspension   The introduction of the independent rear suspension was prompted by the same concerns for improved traction, handling, and safety as for front suspensions. Independent rear suspensions may have several variations. One uses trailing arms and connects to the vehicle frame at the front and to the wheel at the rear. Springs and shocks or MacPherson struts may be used. Figure 9–10 shows a trailing arm using coil springs and shocks.

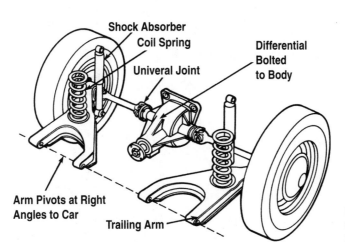

Figure 9–10: Trailing arms are often used with independent rear suspensions.

Electronically Controlled Suspension

The suspension system may also include variable, electronically controlled shock absorbers or struts that can produce a soft, medium, or hard ride. Some of these systems may adjust vehicle ride height. Some may do both.

The ride and handling of the car can be changed for different road conditions or to suit individual driver preferences. Such suspension systems may also lower the vehicle automatically at highway speeds to decrease wind resistance and improve gas mileage. As the car slows down, the vehicle rises to normal height.

Major parts of an electronically controlled suspension may include:

- Adjustable strut/shock absorber units (actuators)
- Front and rear height sensors
- Vehicle speed sensor
- Throttle-position sensor
- Steering wheel angular velocity sensor
- Information/control panel
- Electronic control unit (ECU)
- Air compressor

The ECU, or computer module, receives electrical input signals from sensors on the vehicle. Input signals indicate vehicle speed, vehicle height, the throttle position (how far the driver has depressed the accelerator), and steering wheel angular velocity (how fast the steering wheel is being

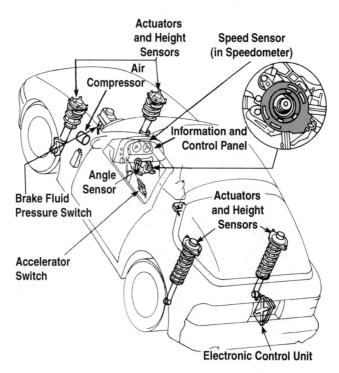

Figure 9–11: The major parts of one type of electronically controlled suspension system.

turned). Switches may be available to the driver to select a high, hard, or soft position, or to set the system for automatic operation. Some systems are fully automatic. At times, the ECU may override the manual settings to improve safety, for example, if the driver pushes the SOFT button at high speeds.

Based on a program in the computer and driver's requests, the ECU determines whether to make the shock absorbers dampen spring oscillations more firmly or more softly. Valves within the shock absorbers can be positioned to make damping more firm, medium, or soft.

Ride height may also be adjusted by some systems. These actions are accomplished by actuating the air compressor or valves in the shock absorbers. Air can be added or bled off to raise or lower the car, respectively. In some of the most recently designed systems, the height of the front and rear of the car can be varied separately to angle the front of the car downward for decreased wind resistance at high speeds. When the headlights are switched on, the rear end lowers to correct headlight aiming.

ROUTINE SUSPENSION MAINTENANCE

Suspension and steering maintenance procedures typically are performed at the same time. Therefore, routine suspension checks and maintenance are covered along with steering checks and maintenance in a single section. This maintenance section follows the discussion of steering systems.

STEERING SYSTEMS

The function of the steering system is to provide the driver with a means of controlling the direction of the vehicle as it moves. On today's cars, two types of steering systems are commonly used to provide steering control:

1. Rack and pinion
2. Recirculating ball.

Either of these two types of steering mechanisms may be a fully mechanical system or a power-assisted system.

Rack and Pinion

Most cars use rack-and-pinion steering, a simple design that provides quick and positive response (Figure 9–12). A pinion gear is connected to the end of the steering column shaft. This pinion gear meshes with teeth machined into a steel bar, or *rack*, inside a housing. The rack housing is mounted parallel to the front axle.

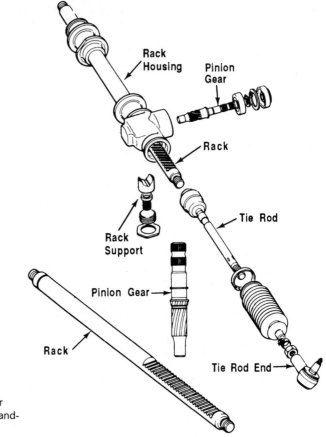

Figure 9–12: The major internal parts of a rack-and-pinion steering system.

As the steering wheel turns, it turns the pinion gear. This action causes the rack to slide within the housing. The ends of the rack are connected to both front-wheel steering knuckles through **tie rods,** or linking bars. As the rack slides, it moves the front wheels and causes the vehicle to turn. As shown in Figure 9–12, minimal **steering linkage** is required to connect the steering mechanism to the wheels.

Recirculating Ball

Larger automobiles and trucks may use recirculating-ball steering gearboxes and a more elaborate system of steering linkage. The recirculating-ball steering gearbox (Figure 9–13) contains steel ball bearings to reduce friction between the internal parts and reduce steering effort.

As the steering wheel is turned, it rotates a worm gear in the steering gearbox. The worm gear has a rounded spiral groove along its length. A ball nut with an internal rounded spiral groove fits over the worm gear. Ball bearings are positioned in the space between the worm gear and the ball nut. As the steering wheel turns, the wormshaft turns and forces the balls to move the ball nut. As the ball nut moves, it rotates a sector shaft. The sec-

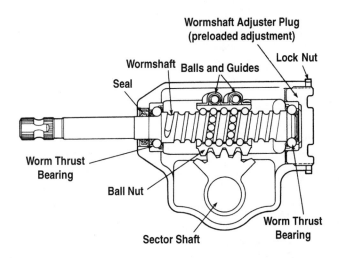

Figure 9–13: The internal parts of a recirculating-ball steering gearbox.

tor shaft is connected through steering linkage to the front wheels. To transfer the motion of the sector shaft to the front wheels, a series of connecting parts is required. A **pitman arm** is connected to the sector shaft and swings in an arc as the sector shaft turns. These parts are the steering linkage. The pitman arm then moves the tie rods and steers the wheels.

Perhaps the most common type of steering system used with the recirculating-ball steering box is the parallelogram steering system. Figure 9–14 shows a typical parallelogram steering system and its common components.

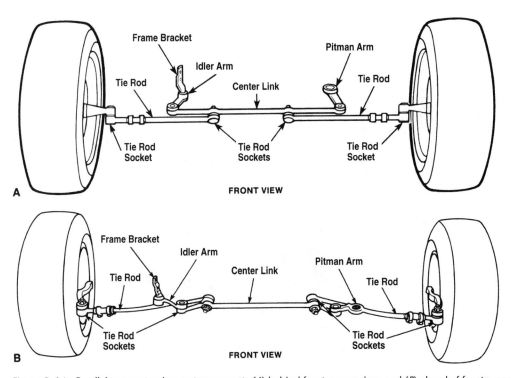

Figure 9–14: Parallelogram steering system mounts (A) behind front suspension, and (B) ahead of front suspension.

On some vehicles, a small steering damper is connected to the linkage. This damper helps to reduce the tendency of the linkage to transmit force backward to the steering wheel when the vehicle hits a bump.

POWER STEERING SYSTEM

The purpose of a power steering system is to help reduce steering effort. An engine-driven pump is used to create hydraulic force. The hydraulic force is applied to the steering system to help move the steering linkage. Either rack-and-pinion or recirculating-ball steering systems can be power assisted. Figure 9–15 shows a typical power rack-and-pinion steering system.

The hydraulic pump, driven by a belt on the engine, provides hydraulic pressure to operate the steering mechanism. A series of valves and passages in the steering mechanism directs power steering fluid (an oil) to coincide with the movement of the steering wheel.

A fluid reservoir stores power steering fluid for use by the pump. The reservoir may be located either on the pump itself or in a separate location.

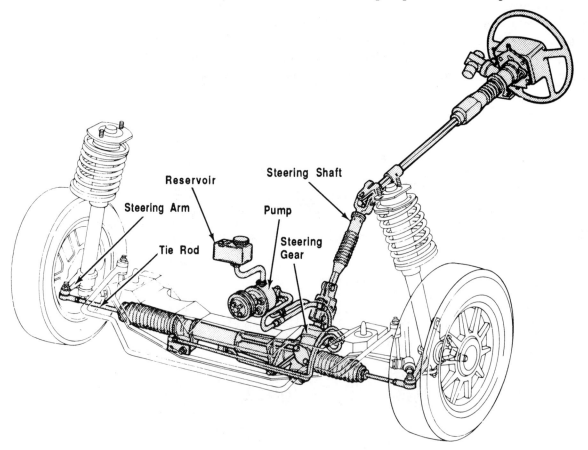

Figure 9–15: A typical power rack-and-pinion steering system includes a belt-driven pump that provides hydraulic pressure to help move the steering linkage.

Connecting hoses are used to provide fluid to the pump and to return fluid to the reservoir.

As the belt turns the power steering pump, fluid is forced through the system to pistons within the steering mechanism. The pistons move and apply force to the steering linkage.

Two types of power steering systems are available:

1. Variable assist
2. Variable ratio

Variable Assist

The variable-assist approach to power steering allows the driver to have a feel of the road. The amount of hydraulic pressure that acts on the pistons in the system is varied according to driving conditions. For example, when you turn the wheels sharply for parking or going around a tight corner, the system provides maximum assist. When you steer around curves or change lanes, the power steering system provides a small degree of assist. When you are driving down a straight road, the system provides little or no assist. This allows for minor steering corrections and provides a good feel of what the front wheels are doing.

Variable Ratio

Variable-ratio steering has two purposes. First, it decreases the number of turns of the steering wheel that must be made during parking and slow-speed operation. Second, variable-ratio power steering makes the steering less sensitive to steering wheel input when driving in a straight line. For example, on an interstate highway a slower steering gear ratio is preferred so that corrections can be made gradually. However, for rounding sharp corners or parking, you want the steering ratio to be faster so that fewer rotations of the steering wheel are required to turn the front wheels sharply.

Variable ratio steering is accomplished primarily through the design of gear teeth in the steering gear box. Variable-assist and variable-ratio power steering can be combined in a single unit.

WHEEL ALIGNMENT

To ensure that you can steer safely and easily, and to minimize tire wear, a number of adjustments may be made to the front wheels. These adjustments, known as **wheel alignment,** affect the angles at which your front wheels contact the road.

Periodically, a wheel alignment may be necessary because of wear on steering or suspension parts. Wheel alignment also can go out of adjustment when you jump a curb or hit a hard bump solidly. Newer cars with

MacPherson strut front suspension systems typically have fewer adjustments than double A-arm suspension systems. A wheel alignment includes checks of the following:

- Vehicle ride height
- Toe-in and toe-out
- Camber
- Caster
- Turning angle

Refer to Figure 9–16 as you follow the descriptions of wheel alignment factors.

Vehicle Ride Height

Every vehicle is designed to ride at a specific height known as *curb height* or *ride height*. A vehicle that is not at the proper height will cause havoc with the wheel alignment of the vehicle. Over time the springs will begin to sag and not support the vehicle as well as they once did. A spring may become damaged or cracked, but this is not too common. Overloading the vehicle may also cause ride height problems. Problems in ride height may be noticed by an unusual vehicle lean when parked, problems in handling, or with excess tire wear.

Ride height is usually checked during a wheel alignment. The vehicle should normally have half a tank of fuel when ride height is checked.

Toe-In and Toe-Out

When the wheels roll, they tend to angle themselves outward in the direction of movement. This angle is known as **toe-out.** Any looseness in the steering system tends to increase this tendency. To counteract this force and keep the wheels parallel and pointing straight ahead, the wheels can be adjusted to angle inward, or **toe-in,** slightly. The proper amount of toe-in on cars is very small, ⅛ inch or less. Incorrect toe adjustment will result in a feathered edge on the tread surface of the tires. In addition, steering stability is directly affected by toe adjustment. On a properly aligned car, the wheels should be pointing straight ahead when the car is moving forward in a straight line.

Camber

To assure good traction and steering control, the tread surface of the front tires should rest flat against the ground when you are driving. Ideally, the tire should be standing vertically, at a right angle to the road.

However, as you load more people and luggage into a car, the extra weight tends to tilt the tires from vertical. Also, most highways have a

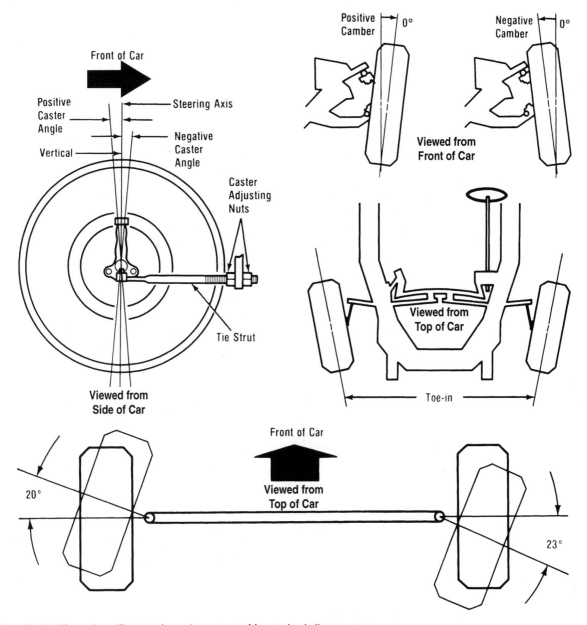

Figure 9–16: These views illustrate the major aspects of front-wheel alignment.

curved sloped surface from side to side, called *road crown*, to allow the rain to drain off to the sides. To compensate for these problems, the tires are purposely tilted from the vertical. The angle at which the top of the tire leans inward or outward is known, respectively, as negative and positive **camber.** As weight is added, the tires tend to straighten themselves to the vertical position.

The front tires on most passenger vehicles have a slight positive camber. Proper camber settings assure effective tire contact with the road and minimize tread wear.

Caster

The **caster** angle of your car's front wheels helps to maintain directional stability and control. In addition, caster helps to reduce steering effort and to return the steering wheel to the center position after a turn.

To help understand caster on your car's front wheels, imagine a furniture caster. A furniture caster is the small wheel assembly used to make it possible to roll furniture from one location to another. The center of the caster wheel is not positioned in a straight line with the shaft that attaches the caster to the furniture. As you move the furniture, the caster wheel will pivot automatically so that it is pointed in the direction you are pushing. Similarly, the caster of your car's wheels causes them to seek the most stable position, which is pointing straight ahead.

The lean, or angle, of the steering knuckle to which the front wheel is attached determines caster. If it leans to the rear, the wheel has positive caster. If it leans to the front, the wheel has negative caster. Today, many cars have positive caster because it offers greater directional stability. An incorrect caster setting will not cause abnormal tire wear. Hitting a deep pothole or other impact may alter the caster and other alignment angles.

Turning Angle

During a turn the inner wheel must follow a smaller diameter circle than the outer wheel. Thus, the **turning angle** of each of the front wheels must be slightly different. The inner wheel might turn at an angle of –5 degrees, and the outer wheel at an angle of 18 degrees. This relationship is also known as *turning radius* and *toe-out on turns*. If the inner wheel didn't turn at a sharper angle, the inner tire would be dragged around the turn and would wear excessively.

Four-Wheel Alignment

Older cars only required alignment of the front end as the rear was generally not adjustable. The toe-in, caster, and camber were set using the centerline of the vehicle as a reference. This process is a *two-wheel alignment* or a centerline wheel alignment. However, due to slight variations in the alignment of the rear of the vehicle, it became necessary to align the front wheels based on the toe-in or toe-out of the rear wheels. The rear wheels have just as much influence over directional stability as those at the front, and that's why many vehicles need to have all four wheels aligned. The total of the toe-in or toe-out of each rear wheel is known as the *thrust line* of the vehicle. This makes for a more accurate alignment and is usually performed on vehicles with nonadjustable rear ends. This process is a *four-wheel alignment* or a thrust line alignment. Newer cars are lighter and tend to be driven farther distances and at faster average speeds. Tires are also lasting much longer than those on older cars. These factors combine to make today's cars more sensitive to misalignment.

On many cars, wheel alignment is necessary for the rear wheels as well as the front wheels. Typically, cars with independent rear suspensions may require rear-wheel alignment for the best steering stability, directional control, and tire wear on all of the tires. This type of alignment is also a four-wheel alignment, or some call it a *full* four wheel alignment. In general, only camber and toe are adjustable on independent rear suspensions. Thus, the preceding descriptions of camber and toe also apply to the rear wheels of some cars.

Some angles, especially on smaller cars, are not factory adjustable and may require an aftermarket modification to bring the vehicle into specification.

Wheel alignment must be performed according to the manufacturer's recommendations. A certified technician who has the necessary equipment and knowledge will best perform this work (Figure 9–17).

Figure 9–17: An automotive technician performs a wheel alignment using computerized wheel-alignment equipment.

ROUTINE SUSPENSION AND STEERING MAINTENANCE

Since suspension and steering components interact with each other constantly, wear is inevitable. However, wear can be minimized with a regular maintenance program.

There are two ways to detect problems with wheel alignment. The first is by the feel of your car's steering. Does it feel loose or have excess play, or do you notice problems when driving? For example, does your car pull or wander to the left or right when driving on a flat and level road or when braking? Does it seem to take more effort to turn the steering wheel, or do you have to help return the steering wheel after a turn? Steering problems such as these can indicate the need for a wheel alignment or other steering system service.

The second way to detect wheel alignment problems is by feeling the wear on your front tires with your fingertips. Camber and toe problems produce characteristic wear patterns. If you notice that your tires are not wearing evenly across the tread surface, it is a good indication that wheel alignment is necessary.

Symptoms of suspension or steering problems include:

- Steering wheel not centered when car is moving straight ahead
- Steering wheel pulls to the left or right
- Feeling of looseness or wandering
- Excessive or uneven tire wear
- Steering wheel vibration or shimmy

Check Steering Wheel Play

Roll down the driver's side window and insert the key in the ignition to unlock the wheel. If your car is equipped with power steering, make this check with the engine running in PARK (automatic transmissions) or NEUTRAL (manual transmissions) and make sure that the parking brake is set securely. Stand outside the car and grasp the steering wheel rim. Lightly move the steering wheel back and forth and watch the front tire closely while noting what you feel. Notice how much you can move the steering wheel before the tire just begins to move or jiggle. Ideally, there should be less than 2 inches of play, or looseness, measured at the steering wheel rim. Use one of the spokes of the steering wheel to judge this distance. If you hear any clunking sound when moving the wheel, or if there is excessive play, perform the activity described next.

Check for Loose Steering or Suspension Parts

Grasp the top of a front wheel, near the inside edge of the tire. Attempt to pull it outward and push it inward in a shaking motion. Excessive play could indicate a loose or defective wheel bearing or ball joint. Replacing a ball joint is best left to a service garage.

The next procedure is to crawl underneath the car. Don't jack up the car; leave it on the ground. Have someone get inside and move the steering wheel back and forth. Watch the movements of the steering linkage. If any part shows play, or looseness, it must be corrected. After a loose or worn steering part has been replaced, a wheel alignment must be performed. Thus, servicing steering systems is best left to shops that can perform wheel alignment.

Check Tire Wear

Uneven tire wear can indicate tire inflation, suspension, and steering system problems. Assuming that you have maintained proper tire pressures, there are definite signs that indicate steering system problems (refer to Figure 9–18).

If no irregular wear is visible, the next step is to close your eyes and lightly rub your fingertips back and forth across the surface of the tire tread. In most cases, you can feel potential problems before they result in a visibly worn-out tire. Closing your eyes helps to heighten your sensitivity to touch.

For example, if there is more wear on either the inside or outside shoulder of the tread, the camber angle is incorrectly set. Wear on the outside shoulder means too much positive camber. Wear on the inside shoulder of the tread means too much negative camber.

If you feel sharp ridges pointing toward the inside of the tread, it indicates a toe-in problem. If you feel sharp ridges pointing toward the outside of the tread, it means a toe-out problem.

Next, run your fingertips around the tire on the tread area. You may notice low spots or wavy ridges around the circumference of the tire tread. These defects usually indicate unbalanced or out-of-round tires, worn shock absorbers, or worn steering/suspension parts.

Don't Worry about Lubrication

Manufacturers now install permanently lubricated ball joints that have no grease fittings and cannot be lubricated. Rubber or polyurethane bushings that cannot be greased are currently used in suspension systems.

SUSPENSION AND STEERING PROBLEMS

The following descriptions of common steering and suspension conditions are intended to help you diagnose simple problems. In most cases, the best advice is to consult a certified professional technician for any problem that affects the tires, brakes, suspension, or steering systems. These systems directly affect the safety of riding in your car.

Uneven Tire Wear

Your tires should wear flat and level across the tread surface. If they are wearing unevenly, you could have problems with inflation pressures, suspension or steering parts, alignment, or possibly poor driving habits. Figure 9–18 shows common wear problems, their causes, and corrective actions to be taken. Adjustments to camber and toe are performed during a wheel alignment.

Bouncing Ride or Wheel Tramp

Wheel tramp is a rapid, up-and-down movement of a wheel. You may notice it as a steering wheel vibration or as excessive noise as you drive. It typically occurs within a specific speed range—say, between 40 and 45 mph. At other speeds, the vibration disappears.

The most common cause of wheel tramp is an out-of-balance condition in the front tires. The tires should be inspected and balanced. It also can indicate that a tire is out of round or that a shock absorber is worn or defective.

Shimmy

A shimmy is a rapid, side-to-side movement of the front wheels. This movement is transmitted through your steering wheel, causing it to move the steering wheel quickly back and forth.

Unlike wheel tramp, a shimmy can be felt most of the time as you drive. You can have either a low-speed shimmy or a high-speed shimmy. At intermediate speeds, this harsh movement lessens and is sometimes imperceptible.

Shimmy can be caused by looseness or a bent part in the steering linkage or steering mechanism. An unbalanced or bent wheel, loose wheel bearing, or loose lugs on a wheel can also create this condition.

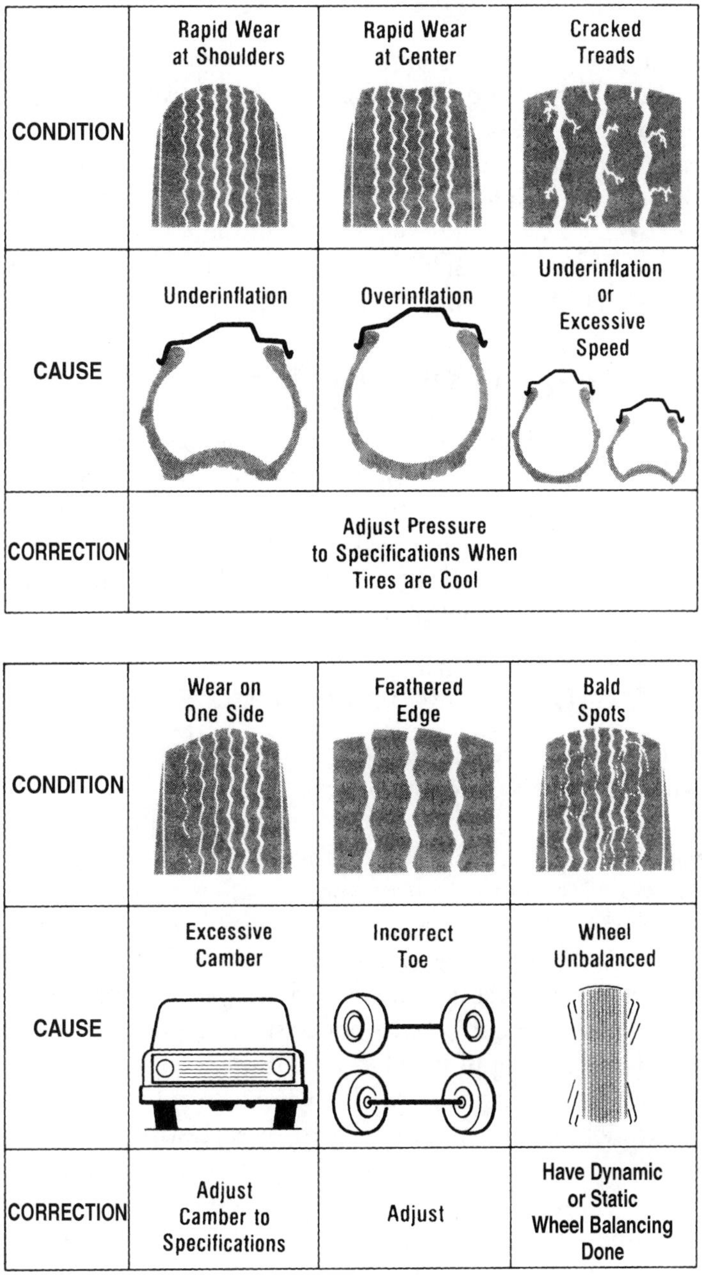

	Rapid Wear at Shoulders	Rapid Wear at Center	Cracked Treads
CONDITION			
CAUSE	Underinflation	Overinflation	Underinflation or Excessive Speed
CORRECTION	Adjust Pressure to Specifications When Tires are Cool		

	Wear on One Side	Feathered Edge	Bald Spots
CONDITION			
CAUSE	Excessive Camber	Incorrect Toe	Wheel Unbalanced
CORRECTION	Adjust Camber to Specifications	Adjust	Have Dynamic or Static Wheel Balancing Done

Figure 9–18: Chart of unusual-tire wear patterns, their causes, and corrective measures to be taken.

Bouncing and Floating

After going over a bump, your car should steady itself. It should not bounce, or rebound, more than one or two times. If your car does bounce more than once or twice, you can double-check the rebound when the car is standing still. Use your body weight to push down and get one corner of the car bouncing rapidly. Release pressure at the bottom of a pushing stroke. After you have stopped pushing, the car should come to a normal stop after one or two rebounds. If it continues to bounce, you need new shock absorbers.

Bottoming Out on Bumps

A hard, loud thump on the underside of the car when you hit a bump can indicate that the suspension is bottoming, that is, the suspension members move so far upward that they hit chassis parts. Bottoming usually indicates that the springs or torsion bars have lost their resiliency. They are too weak to cushion the chassis from road shocks. Bottoming may also be caused by faulty shock absorbers.

Whining in Power Steering

If the power steering pump makes excessive or unusual whining noises when you turn the steering wheel, there may be a problem. If this noise occurs just after the car has been started, it could be caused by cold power steering fluid that is thick and not flowing properly through the system. However, whining also may indicate a low fluid level or defective power-steering pump.

Check the power steering reservoir. If the fluid level is low, there is a leak. The reservoir must be refilled, and the source of the leak must be repaired. If the oil level is correct, the power steering pump itself may be defective.

A loose and slipping belt on the power steering pump can cause a high-pitched screeching sound. The belt needs to be tightened or replaced.

Do not confuse a high-pitched screech with the low-pitched moaning or groaning noise that occurs if you turn the wheels all the way to the right or left. The low-pitched groaning sound typically occurs when you park in a tight spot. This is normal and does not indicate a problem.

Hard Steering

A number of conditions can make it difficult to turn the steering wheel. Front tire pressures may be so low that there is more friction when you turn the wheels. If you have power steering, the power-assist unit may not be working. Your wheels may be out of alignment or there could be a problem in the steering unit.

Poor Return of Steering Wheel During Turns

After you round a corner, the steering wheel should return to center, or a straight-ahead position, by itself. You should not have to steer the car to straighten the wheels. If you car's steering does not return to center, the most probable cause is incorrect caster, which can be corrected during a wheel alignment.

Pulling or Wandering

If you are continually making steering corrections as you drive, there may be looseness in the steering linkage. To check, drive on a flat and level road. Hold the steering wheel lightly, with your fingertips. If the car drifts by itself in one direction or the other, there is a problem. Your tires could have manufacturing defects, or they may be unevenly inflated. You may have a worn steering mechanism or improper wheel alignment, or there may be a brake problem. You will have to check and repair the problem.

Sway Around Corners

If you car leans excessively when you turn corners, suspect the sway bar, or antiroll bar. It reduces body roll, or sway, for better control in turns. If your car sways more than normal, the sway bar may be loose or broken, or it may have worn bushings.

Your tires, brakes, suspension, and steering system are responsible for your safety. In addition, they help to produce a comfortable ride. Chapter 10 discusses accessories that are designed specifically to contribute to your comfort and ease of driving: heating, air conditioning, and other optional equipment.

1. List the different suspension designs.
2. List the types of springs found in suspension systems.
3. What is the purpose of a shock absorber?
4. How is a shock absorber made?
5. What is a gas shock?
6. List the types of front suspension.
7. Which types of front suspension are most used today?
8. List the types of rear suspension.
9. Which types of rear suspension are most used today?
10. What are the advantages of electronically controlled suspension?
11. What are the components of the electronically controlled suspension?
12. List two steering system types.
13. Which steering system is used more on cars?
14. List the items checked during a wheel alignment.
15. Compare a two-wheel alignment and a four-wheel alignment.
16. What are some symptoms of a problem in the steering or suspension system?
17. What does a ragged, feathered edge on the tire tread indicate?
18. You examine a tire and find that there is more wear on one side of the tire compared to the other. What is the likely cause?
19. Both shoulders of the tire are worn more than the center. What is wrong?
20. The center of the tire is worn more than the sides. What is wrong?
21. What are the causes of wheel tramp?
22. What are the causes of wheel shimmy?

HEATING, AIR CONDITIONING, AND OPTIONAL EQUIPMENT

After reading this chapter, the reader should be able to:

- List some optional equipment that can make driving a more pleasant experience.
- Explain how the heating system operates.
- List the maintenance for the heating system
- List some common problems in the heating system.
- Explain how the air conditioning system operates.
- List the maintenance for the air conditioning system
- List some common problems in the air conditioning system.
- List the main components of the cruise control system.
- Explain how these systems improve safety and enhance the driving experience.

This chapter focuses on the heating and air conditioning systems. Information is also provided on optional equipment and devices such as the following that can make driving a more pleasant experience:

- Automatic temperature control system
- Audio systems
- Cruise control
- Power seats
- Power windows and mirrors
- Power door locks
- Power adjustable pedals
- Sunroof
- Rear-window defogger
- Antitheft system

Not one of these devices is required for an automobile to operate properly. As late as the 1950s, a car equipped with a radio and heater was considered to be "loaded" with options. The term "loaded" was used to indicate that the manufacturer would load the vehicle with most, if not all, of the options available. Today, heaters are standard equipment on all passenger cars and on most trucks. Thus, this chapter begins with a description of the heating system.

THE HEATING SYSTEM

Your car's heating system uses heat given off by the engine to provide passenger comfort. The operation of this system has many similarities with that of the cooling system. In fact, the heating system dissipates engine heat in exactly the same manner as the radiator.

Hot coolant is drawn from the cylinder head and passed through an inlet **heater hose** to the heater assembly (Figure 10–1). A valve controls the flow of coolant into the heater core, which is actually a small radiator, or heat exchanger.

Air is directed through ducts to pass through the heater core. The heater core transfers heat from the coolant to the air passing through the fins of the core. Typically, a blower, or fan, pushes the warmed air through a series of ducts and into the passenger compartment.

The lower-temperature coolant that has passed through the heater core is then returned to the engine block through a return heater hose. Descriptions of heating system parts in greater detail follow.

Heater Hoses

Heater hoses are smaller in diameter than radiator hoses, as shown in Figure 10–1. Clamps at each end of the hoses prevent the hose ends from leaking. Heater hose is made from natural or synthetic rubber that is reinforced with fabric. Standard or heavy-duty hoses can be used.

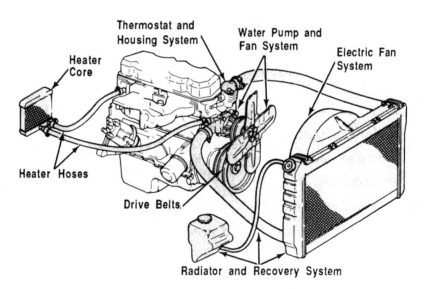

Figure 10–1: Heater hoses on most cars run from the engine's cooling system through the firewall to a heater core.

Heater Temperature Control

The temperature output to the passenger compartment by the heater can be controlled one of two ways: by a heater control valve or by a heater blend door.

The heater control valve opens and closes to allow heated coolant to flow to the heater core. The valve typically is located in the inlet heater hose. As with a water faucet, the flow of coolant can be controlled to produce different levels of heating. Heater control valves can be operated by metal cables or by vacuum tubing.

Most cars today use a heater blend door, which will blend cooler air with the heated air from the heater core. This door may be manually controlled with a lever-operated mechanism, or typically controlled with an electric servomotor. The motor will move the door back and forth to blend more or less cool air with the heated air from the heater core.

Some heating systems are integrated with air-conditioning systems to provide automatic temperature control, covered later in this chapter. These systems control the temperature of air from the heater core and air conditioning automatically, in response to control panel settings.

Heater Core

When hot coolant flows through the heater core, heat is transferred from the coolant to the metal tubes and fins (Figure 10–2). The heated metal of the core transfers heat to the air passing through the unit. Thus, air passing through the heater core from a venting system is warmed as it enters the heater chamber.

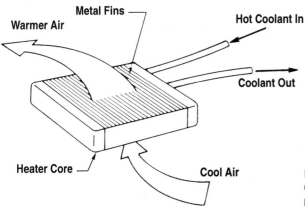

Figure 10–2: A heater core uses hot coolant to heat the passenger compartment.

Blower

An electric blower, or fan, can be used to provide airflow through the heater core. The fan is operated by a control switch inside the car. On vehicles equipped with air conditioning (AC), either fresh outside air or recirculated air from the passenger compartment can be passed through the heater core. Vehicles without AC receive outside air for the heater core.

On automatic temperature control systems, a computer, in response to a temperature setting, regulates the blower motor automatically.

Outside Air Intake

Ducts are located either behind the grille or in front of the windshield to increase airflow and replenish the fresh-air supply.

Distribution Plenums

A **plenum** is a chamber inside which pressure is greater than the surrounding atmosphere. As air is forced into the duct network, it becomes slightly pressurized. Doors in the plenum chambers open and close to control airflow over the heater core. Ducts connect the chambers to heater and defroster outlets. In a manual system, metal cables or vacuum motors and hoses are used to move the air distribution doors. In an automatic temperature control system, electric motors can be used to move the doors.

By varying the controls, you can redirect airflow to the defrosters, floor, or occupants' feet or faces. Combinations of airflow direction are also available to the driver as well. Figure 10–3 shows the components involved in airflow direction. Varying the plenum controls has the same effect as operating heater vents in your home.

ROUTINE HEATING SYSTEM MAINTENANCE

The heating system in your automobile is reasonably trouble-free because its operation is relatively simple. However, heater assembly parts can become blocked, corroded or stuck, or develop other defects. Heater problems can also affect the operation of your engine. There are certain checks you should make to prevent the unexpected.

Check Heater Functioning

Checking the heater functioning is a simple test. Just start the engine and let it warm up. Then set the heater controls for maximum heat and turn on the blower. You should feel a voluminous flow of heated air.

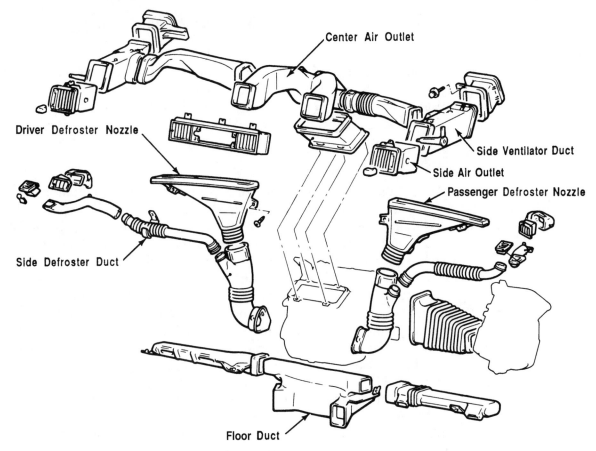

Figure 10–3: Plenum chambers, doors, and ducts are used to control and direct the flow of air through a heating/air conditioning/ventilation system.

Check Hoses for Leaks

Open the hood and locate the heater hoses (refer to Figure 10–1). The heater hoses run from the firewall to the engine. Check along the length of each hose and at the ends for any signs of leaking.

If no leaks are found, feel each hose. It should be firm but pliable. If the hose feels spongy or hard—signs of aging—it should be replaced. The constant heating and cooling of these hoses promotes deterioration.

Also look for bubbles or cracks. Enlarged areas often appear near the clamps at the engine block and firewall. Any of these signs indicate that the hoses should be replaced.

You may find a crack or leak at the end of a hose. If the hose is long enough, cutting off just the damaged end of the hose and replacing it on the connection can often remedy this problem. However, if the hose end shows signs of crumbling or rotting, replace the entire hose.

HEATING SYSTEM PROBLEMS

The heating system is reasonably reliable. However, to assure continued comfort, you should be able to recognize symptoms that indicate heater problems. These symptoms generally require corrective measures that are usually simple and relatively inexpensive. Don't allow problems to continue so long that heater components are destroyed.

Cold Air

Any number of conditions can cause the heating system to blow only cold air. First, check the control settings. Move the controls back and forth a few times to make sure that you have set the temperature control properly.

If a heater control valve hasn't been used recently, it can become stuck. When the valve is stuck in either the open or the closed position, it may not operate correctly when you set the dash controls. In many cases, you can fix a stuck control valve by moving the working parts with your fingers or with a pair of pliers. For maximum reliability, it is recommended that you operate the heater as described for a few minutes each month, even in hot weather. This procedure will prevent the heater valve from becoming stuck.

Check the heater hoses to see if they are carrying hot coolant. If both hoses are hot, it indicates the heat is not being radiated in the heater and is returning to the engine hot. Check the fuses to ensure that the blend doors are receiving power. If both hoses are hot, it could indicate a heater core blocked with debris or a blend door stuck or not functioning. If one hose is noticeably cooler than the other, heat is being radiated in the heater. If heat is being radiated there should be hot air inside the car. Usually, cold air indicates that the flow of coolant is not sufficient to adequately supply heat to the inside of the car.

If the controls are properly adjusted, a control lever, cable, or vacuum hose connection may be at fault. A cable may be broken, bent, or disconnected. A vacuum hose may be cracked, kinked, or disconnected. On automatic temperature control systems, a blown fuse may prevent the heater valve or plenum (air flow control) doors from operating. If so, replace the fuse.

Other conditions that can produce cold air input include problems with the fresh-air ducting, low engine coolant, kinked heater hoses, a clogged heater core, or a defective thermostat. Flushing the cooling system properly and refilling it with a fresh coolant mixture can cure many problems. The procedure for flushing the cooling system is described in Chapter 11.

Insufficient Heat

Small, insufficient flows of lukewarm air typically are caused by blockage in the heater hoses, control valve, or heater core. Another common problem that leads to this condition is an inoperative or removed cooling system thermostat.

Coolant Dripping into Passenger Compartment

If you find coolant dripping into the passenger compartment, a hose or clamp near the firewall may be leaking. Otherwise, the heater core has a hole in it and must be replaced. As a temporary measure it might be possible to bypass the heater to prevent coolant loss and loss of pressure in the cooling system, which can lead to overheating. Locate the heater core inlet and outlet hoses near the firewall (Figure 10–1). Trace the hoses back to the engine. Disconnect one of the hoses from the engine. Disconnect the other hose from the heater core and reconnect it to the engine where the other hose had been connected. This arrangement essentially removes the heater core from the coolant flow.

Controls Inoperative

If the controls don't work on automatic temperature control systems, check all the fuses. If the system is completely inoperative while the engine is running, take your car to a technician.

Intermittent Blower Operation

If the blower operates at times and stops unexpectedly, the most common causes include a poor electrical connection at the blower switch, a loose ground connection on the blower motor, or a defective blower motor.

Blower Not Functioning

If the blower isn't working, first check the blower fuse in the fuse box. If the fuse isn't blown, check the electrical connections between the blower motor and the switch. Any of a number of relays or switches may be at fault. A ground connection or resistor block may be defective. In some cases the blower may need to be replaced.

If your car is equipped with an automatic temperature control system, have the system diagnosed by a competent technician who is familiar with your system.

Windshield Doesn't Defrost

If you have plenty of heat, but the defroster isn't working, the heated air is not being directed properly. This problem can be caused by defective cable, vacuum, or electrical motor controls that operate the plenum doors. The plenum doors may be jammed or ducting may be disconnected or broken.

Look for anything that may be preventing the air from being directed to the defroster duct. The outer wire sheath of cable controls must be firmly and properly positioned. Vacuum lines can crack, break, kink, or become disconnected. If you are handy with tools and motivated, you may be able to fix these problems yourself. However, an automatic temperature control system with electric motors to operate the plenum doors should be checked only by trained technicians.

AIR CONDITIONING SYSTEM

The purpose of the air conditioner is to cool the air in the passenger compartment. As the air is cooled, moisture condenses and is removed. Thus, cool, dehumidified air is circulated in the passenger compartment for the occupants' comfort.

The air conditioner operates on the principle that a liquid absorbs heat when it evaporates. The air conditioning system uses a special gas, known as **refrigerant,** sealed in a closed system. (The same type of gas is found in the refrigerator in your home.) When the refrigerant is under high pressure, it changes from a gas into a liquid. If the pressure is relieved, the refrigerant will evaporate rapidly and turn into a gas again. As the refrigerant evaporates, it absorbs heat energy.

The evaporation of the refrigerant produces a rapid cooling effect on an **evaporator,** or heat exchanger, which resembles a radiator or heater core. Air is forced through the cold evaporator by a blower. The air becomes cooled and is directed into the passenger compartment.

Two basic types of air conditioning systems have been used in cars: the clutch cycling orifice tube (CCOT) and the thermostatic expansion valve (TXV). The CCOT system cycles the compressor on and off based on a desired temperature. The TXV system varies the refrigerant flow to the evaporator while the compressor runs continuously. There are numerous variations on these two systems. The two have nearly the same components. Most vehicles use the CCOT system, so this is the system we will look at.

An automotive CCOT air conditioning system (Figure 10–3) includes:

- Compressor
- Air conditioning hoses
- Condenser
- Expansion device
- Evaporator
- Accumulator
- Refrigerant (and oil)
- Cycling thermostat or pressure switch

- Service valves
- Controls and plenum doors

Refer to Figure 10–4 and Figure 10–5 as you read the following descriptions of these parts.

Radiator
Radiator Cap
Condenser (in front of radiator)
Compressor
Fan
Blower
Accumulator
Expansion Device
Passenger Compartment Unit
Evaporator

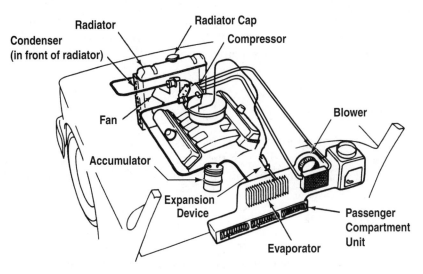

Figure 10–4: The basic components of an automotive air conditioning system.

BASIC HEAT EXCHANGER
Evaporator
Hot Air from Car
Condenser
Heat Flow
Outside Air
Cool Air to Car
Outside Air (forced in by vehicle motion)
Absorbs Heat from Passenger Compartment
Releases Heat to Outside Air

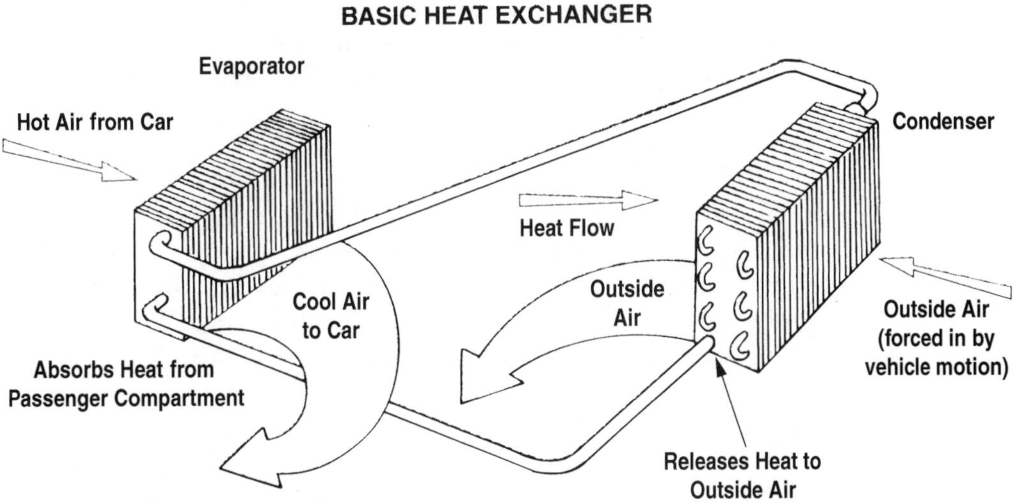

Figure 10–5: Heat transfer in an air conditioning system is done by heat exchangers. The evaporator absorbs heat into a liquid refrigerant. The condenser gives off heat from the refrigerant.

Compressor

The compressor is a belt-driven pump that circulates the refrigerant through the air conditioning system. It has a suction inlet through which low-pressure refrigerant gas is drawn from the evaporator. High-pressure refrigerant is discharged through an outlet into the condenser.

The operation of the compressor is controlled by an electrical clutch mechanism that engages and disengages to operate the compressor mechanism. This clutch has a coil of wire that operates the clutch by magnetism. When a magnetic field is present, the clutch engages, and the belt turns the compressor. The computer uses a relay to control current flow, which turns the electrical clutch on and off.

Air Conditioning Hoses

Hoses direct refrigerant throughout the system and keep the gas from escaping as it circulates through the system. These hoses are specially made to withstand high pressures and prevent the refrigerant from leaking out.

Condenser

The purpose of the condenser (Figure 10–5) is to cool the heated refrigerant gas from the evaporator. The gas from the evaporator has absorbed heat from the air passing into the passenger compartment. The compressor raises the pressure and temperature of the vapor refrigerant. Like the cooling system radiator and the heater core, the condenser is a heat exchanger. As the refrigerant reaches the condenser, cool air passing through the condenser removes the heat from the refrigerant and returns it to a liquid state. The condenser typically is mounted near or in front of the radiator. The cooled refrigerant condenses into a high-pressure liquid that travels to the expansion device that is placed in the line leading to the evaporator.

Expansion Device

The expansion device contains a small orifice. The orifice, or small opening, creates a restriction in the line to the evaporator. Thus, high-pressure liquid refrigerant slowly enters the low-pressure area of the evaporator core. The change in pressure causes the refrigerant to evaporate rapidly. This rapid evaporation cools the evaporator core.

Evaporator

Fresh air or recirculated air from the passenger compartment is forced around and through the evaporator fins and tubes (Figure 10–5). The air becomes cooled, and moisture condenses on the evaporator, much as water beads form on a cold glass of liquid. This cooled, dehumidified air is directed into the passenger compartment.

As the refrigerant evaporates and takes on heat from the air passing over the evaporator core, it turns into a low-pressure gas. This low-pressure gas is drawn toward the suction side of the compressor pump.

The evaporator is located within the blower housing and ductwork that also houses the heater core.

Accumulator

The function of the accumulator, or receiver/dryer, is to store excess liquid refrigerant and to remove water vapor from the refrigerant. Any liquid refrigerant that did not evaporate will be caught and stored in this container. Liquid refrigerant would seriously damage the compressor. Excess water vapor in the system will combine with the refrigerant and can cause corrosion and system component damage.

Refrigerant (and Oil)

R134a is the only safe and environmentally friendly gas recommended for car air conditioning. R134a requires the use of synthetic polyalkylene glycol oil (PAG). Your car's air conditioning self-oils as it runs, transporting oil around with the refrigerant to the spots where it is needed most. If your air conditioning runs out of gas, there is a leak in the system. A leak may show up as an oily spot on a connection or hose. Using your air conditioning regularly will help maintain it in good working order.

If your air conditioning loses efficiency, get it checked immediately. If the system has lost gas, it has also lost oil and the compressor may overheat, causing damage and possible failure.

Cycling Thermostat or Pressure Switch

Two devices may be used to control temperature. A temperature-sensing device or thermostat signals the computer that a desired evaporator temperature has been reached. The computer then cycles the clutch on the compressor. A pressure switch may be used instead of a thermostat. Since there is a direct relationship between refrigerant temperature and refrigerant pressure, many manufacturers choose to use a pressure switch. When the pressure reaches a certain point, it opens a circuit to the computer. The computer then cycles the clutch on the compressor. The switch has one more advantage: If the system is low or out of refrigerant, the system will not operate, thus preventing damage to the compressor.

Since the cool surfaces of the evaporator will condense water, this temperature will not be below 35 degrees Fahrenheit (2 degrees Centigrade). Should the temperature drop below 32 degrees Fahrenheit (0 degrees

Centigrade), ice would form on the evaporator. If ice forms on the evaporator, the flow of air will be blocked. The thermostat or pressure switch and the computer prevent this problem. By regulating the system in this way, the surface temperature of the evaporator can be kept above the freezing temperature of water. The moisture that does condense on the evaporator drains through an opening in the bottom of the evaporator housing onto the road.

Service Valves

Two small valves are located within the system. These are used to charge, discharge, or evacuate (clean out) the system. They also are used to check pressures.

CAUTION

The air conditioning system should be checked and serviced by trained and licensed technicians only. If a service valve is opened, refrigerant will escape and evaporate rapidly. This evaporation can cause severe skin burns (frostbite) or permanent injury to eyes.

According to the law, a licensed technician must use a recovery unit when servicing the air conditioning unit. The law was passed to help prevent the escape of refrigerant into the atmosphere. The average car owner usually lacks both the licensing and the equipment to service the air conditioning system (but you can maintain your air conditioning system; see later in this chapter).

Controls and Plenum Doors

Air conditioning controls are often incorporated with heater controls into a single system. The temperature control regulates how often the compressor cycles, or operates, or regulates the flow of low-pressure gas from the evaporator. Other controls permit plenum doors to be adjusted to direct the flow of air upward or downward for comfort. On many cars, the computer controls the operation of the air conditioning system.

The control to turn on the air conditioner is actually a switch that "requests" that the computer turn on the system. The computer may ignore the request and may sometimes turn the system off. If a fault is detected, the computer will not turn the system on. Some faults might include low or no refrigerant, excess pressure in the system, or an overheating engine. The computer may turn off the system if the driver floors the accelerator pedal, perhaps when passing another car. Turning off the AC takes some of the load off the engine and makes a little more power available for acceleration.

In addition, on many integrated heater/air conditioner systems, the air conditioner can be operated at the same time as the defroster to remove moisture from the air and clear mist from the inside window surfaces on damp, cool days.

AUTOMATIC TEMPERATURE CONTROL SYSTEMS

The purpose of an automatic temperature control system is to maintain a desired temperature in the passenger compartment. Typically, the driver selects a desired interior temperature by adjusting a single thermostatic control. The driver also may select an AUTO mode, in which the system operates all controls automatically. As an alternative, the driver may select blower speeds, airflow direction, fresh or recirculated air, and so on.

Multiple temperature sensors monitor temperature at different locations in the passenger compartment, outside the car, and in the ducts. Some systems use sensors that determine the amount of heat entering the car as solar radiation through the window glass. Signals from the sensors are sent to an electronic control unit (ECU), or computer module. The ECU, based on a program, controls actuators (motors, relays, and solenoid vacuum valves) that operate heating, air conditioning, and plenum door systems. Some automatic temperature control systems include self-diagnostic capabilities that can produce trouble codes to help technicians locate problems.

In most automatic temperature control systems, the heater core and air conditioning evaporator are located near each other in the same plenum housing.

ROUTINE AIR CONDITIONING SYSTEM MAINTENANCE

Servicing an air conditioning system properly requires a high level of technical skill. Leave repairs and refrigerant recharging to qualified service personnel. However, there are certain checks you can perform to help assure continuous, reliable service from the air conditioner.

Check Belt Tension

As with all belts, the compressor belt can stretch and wear. Check the belt frequently to make sure the compressor is at peak efficiency. Press down firmly on the belt. There should be no more than a half-inch of movement, or play. If it is loose, it must be tightened. If there are any cracks or glazing (a smooth, slippery surface), the belt must be replaced.

Inspect Hoses for Leakage

Unlike heater hoses, air conditioning system hoses should feel very hard and solid. Check all hose connections and the hoses themselves for cracks. An oily deposit—especially at metal hose connections—indicates a refrigerant leak. Special lubricant circulates with the refrigerant through the system to lubricate the compressor seals. If you find a leak, have the hose replaced by a professional technician. Professional air conditioning service personnel also use electronic and other types of detectors to find small leaks.

> ⚠ **CAUTION**
>
> Never disconnect an air conditioning system hose. If the pressure on the refrigerant is released, it will evaporate instantly and produce tremendous cold that can freeze your skin severely or do permanent damage to your eyes.

As mentioned earlier, moisture that condenses on the evaporator core drains through an opening in the plenum housing. Thus, you may find a puddle of clear water on the ground under your car after using the air conditioner. This water condensation is normal. The evaporator drain typically is located in the engine compartment, just in front of the passenger's seat foot well.

Check for Noises

The two most common noises from the air conditioning system come from the compressor. The first noise may be a squealing sound caused by a glazed belt. If the belt is glazed, it must be replaced.

With the hood open, have someone start the car and switch the air conditioning controls on and off a few times. Watch the magnetic clutch at the front of the compressor. If you notice that the center portion of the compressor does not turn when the air conditioning is switched on, the clutch mechanism is not operating.

If the clutch engages, listen for a rumbling, grinding, or clashing sound when the air conditioner is switched on.

These noises indicate that the bearings inside the clutch or internal compressor parts are defective. If the noise continues, have it checked by a professional automotive air conditioning and heating technician.

Check Condenser

A blocked condenser will result in poor cooling inside of the car. Because the condenser is located in front of the radiator, just behind the grille, the condenser fins can become blocked with bugs or debris. Use a paintbrush to gently brush the debris from the condenser fins.

AIR CONDITIONING SYSTEM PROBLEMS

Air conditioning service requires a high level of technical expertise. However, you can diagnose some of the more common air conditioning system problems with the following descriptions.

No Cooling

Lack of cooling is usually an electrical problem. The most common causes include:

- Blown fuses in the blower or compressor clutch circuits. If a replacement fuse blows, there may be a short circuit or an accidental ground connection in the affected circuit.
- Loose wiring to the compressor clutch making an intermittent connection.
- A defective air conditioning on/off switch.
- A loose or disconnected ground, an accidental ground, or a short in the electromagnet coils in the magnetic clutch on the compressor.
- Defective vacuum lines or electrical connections to motors that operate plenum doors.

In addition, the system may be low or out of refrigerant and so will not operate. If the refrigerant has leaked out, the system will need to be evacuated so the leak can be repaired, and then recharged with the proper amount of refrigerant.

Inadequate Cooling

If cooling is inadequate, first check if the condenser fins behind the grille are blocked. With the engine turned off, remove any debris from the condenser fins. Use a paintbrush or flush debris out with a garden hose.

Start the engine and move the car to an outdoor position where air can flow easily over the radiator. Also, make sure that the radiator is not clogged with bugs or other debris. Set the air conditioning controls for maximum cooling and open the car doors and windows. The compressor clutch should engage and operate until the air becomes cool. If the clutch engages and disengages many times during a one-minute period, an action called "short cycling," the system is likely low on refrigerant. It will need to be taken to a technician for service.

> ⚠️ **CAUTION**
>
> Keep your hands, arms, and other body parts away from the fan, belts, and pulleys at the front of the engine. Severe injuries can result from becoming caught in moving machinery. Also avoid touching hot engine parts. Severe burns can result.

Next, put one hand on the suction hose that enters the accumulator and determine if the line is cold. *This hose or line will be cold in a normally operating air conditioning system.* Then place your hand on the hose or line that exits the compressor and leads to the condenser. *This hose or line will be very hot in a normally operating air conditioning system.* (Be careful. A severe burn can result.) If the hose on the outlet side of the compressor is hot and the accumulator line is cold, the system is functioning normally.

Odor

If the evaporator drain is blocked, mold and bacteria can grow in the moisture at the bottom of the evaporator housing. Dirt and debris may enter through the fresh air inlet. Mold and bacteria may grow on this debris in the wet environment of the evaporator. If you can locate the evaporator drain, probe the rubber drain hose gently with a small wood dowel or cotton-tipped swab. Avoid using metal tools that can puncture the evaporator core tubes.

Lack of Air Circulation

In some cases, only a weak flow of cooling air comes out of the ducts, even when the blower control is set on HIGH. A nonfunctioning blower motor can cause this problem. Check the blower motor fuse. If the fuse is not blown, the blower motor switch may be defective. A plenum door may be stuck or a duct may be broken or disconnected. The vacuum or electric lines that operate plenum door motors may be defective.

Automobile air conditioning is one of those areas where your car only requires limited maintenance. You can help ensure proper operation through checks such as those just described and by checking belts and hoses routinely. However, a certified technician or air conditioning specialist should perform servicing and repairs.

OPTIONAL EQUIPMENT

Most of the optional equipment described in the following sections is operated electrically. In most cases, a car owner is limited to checking or replacing fuses. Service can require special tools, electrical test equipment, and considerable experience in diagnosing electrical problems.

Audio Systems

Radios and cassette tape or compact disc (CD) players are popular options. For all practical purposes, there are only a few things the typical car owner can do to repair them if they fail to function. First, check for blown fuses and loose or unplugged electrical connectors. Replace blown fuses. Check to see if the antenna cable has become disconnected. Disconnect and reconnect electrical connectors and antenna connectors several times to make good electrical contact. If the unit has a separate ground wire, check for a solid, tight electrical connection.

If these steps don't correct the problem, take the unit to a qualified electronic repair service.

Cruise Control

Cruise control is designed to maintain a selected, steady speed for cruising on highways and turnpikes where the traffic is light. Most cruise systems operate the same. Typically, a cruise control system (Figure 10–6) includes:

- Vehicle speed sensor
- Control switches (main switch, set/resume switches)
- Brake switch
- Clutch switch (manual transmission)
- Electronic control unit (ECU)
- Actuator unit

When the main switch is turned ON, the speed sensor input signals are directed to the ECU. When the SET switch is depressed, the computer operates a servomechanism, or output actuator, that holds the throttle linkage at the position set by the driver's foot on the accelerator pedal. The driver then can release his or her foot from the accelerator, and the cruise control system will maintain the set speed.

If the vehicle begins to climb a hill, speed may fall below the setting. The computer output signal operates the servomechanism to pull the throttle linkage open slightly more. When speed sensor input signals indicate that the set speed has been reached again, the computer operates the servomechanism to hold the throttle position steady.

If the vehicle descends a hill, its speed may increase due to gravity. The ECU eases off on the throttle linkage and tries to maintain the set speed.

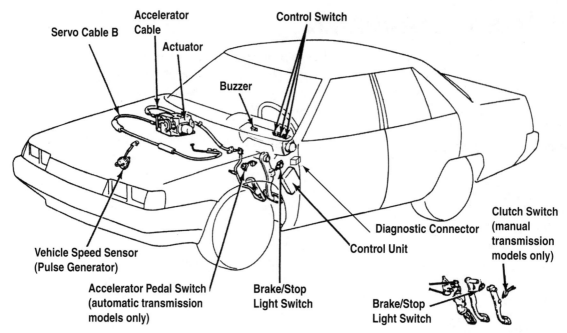

Figure 10–6: The major parts of a cruise-control system.

Some systems, on vehicles with automatic transmissions, may downshift to a lower gear to increase engine braking further and reduce vehicle speed.

If the driver depresses the brake pedal to slow the vehicle, the ECU releases control of the throttle linkage. Also, on vehicles equipped with manual transmissions, the ECU will release control of the throttle mechanism if the driver depresses the clutch pedal. To accelerate to the previously set speed after braking or pressing the clutch, the driver presses the RESUME switch. The ECU again receives vehicle speed input data and operates the throttle to accelerate to the previously set speed.

The throttle servomechanism and speed sensor are mechanical devices. In some cases, a vacuum unit is used to provide force to pull the throttle linkage. In other systems, an electrical solenoid or stepper motor is used to hold, release, and press the throttle mechanism.

Speed sensors can be located in the transmission, axles, or wheels.

Servicing of cruise control units is best left to qualified professional technicians. Some units include self-diagnosis capabilities that allow technicians to call up trouble codes on special diagnostic equipment.

Here is one useful tip: If the cruise control function does not release the accelerator when the brake is depressed, immediately shut off the main switch to cancel the cruise control function. Then check the brake fuse, brake light switch, or stoplight bulbs. Typically, the ECU receives a signal from the brake light switch to release control of the system. If one or more of these devices are defective, the cruise control may fail to release the accelerator. Do not use the cruise control system if it continues to malfunction.

You should not operate the cruise control unit when the temperature drops below 38 degrees Fahrenheit (4 degrees Centigrade) or when it is raining. Should the wheels experience a reduction or loss in traction, the wheels could spin and vehicle control would be lost.

Power Seats

Power seat mechanisms consist of electric motors operated by switches. The motors operate gear mechanisms that reposition seats (Figure 10–7).

Some cars include power seats with a memory function. An electronic control unit stores two or more individual driver settings. The seat can be reset automatically to a particular driver setting by pressing a control button.

Power Windows

Power windows are essentially similar to power seats. A switch and relay are used to operate an electric motor. The motor turns a gear to raise or lower the window through a mechanical lever arrangement (Figure 10–8). To gain access to the switch or motor, the door panel must be removed.

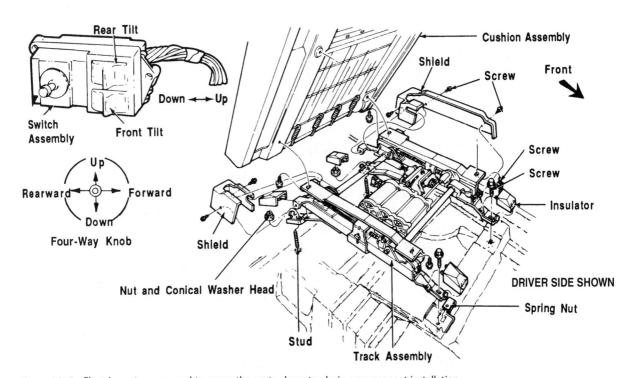

Figure 10–7: Electric motors are used to move the seats along tracks in a power-seat installation.

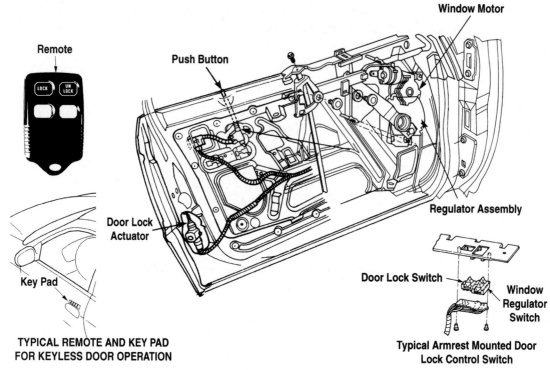

Remote

Push Button

Window Motor

Regulator Assembly

Door Lock Actuator

Door Lock Switch

Window Regulator Switch

Key Pad

TYPICAL REMOTE AND KEY PAD FOR KEYLESS DOOR OPERATION

Typical Armrest Mounted Door Lock Control Switch

Figure 10–8: A door with a power window unit and a power door lock.

Power Mirrors

Power mirrors are essentially similar to power seats. A switch is used to operate electric motors behind the mirror. The motor turns a gear to raise or lower the mirror or to turn it left or right. If the vehicle has memory seats, often the mirror settings are stored as well.

Power Door Locks

Power door locks operate when the key is turned in the driver's door lock, when operating a remote or keyless entry, or when the driver's doorknob is opened. In general, small electrical actuators, or solenoids, are used to move the door lock linkages (Figure 10–8).

A switch in the driver's door linkage energizes a relay that provides power to the other doors' actuators. When the driver's door is operated, the other actuators move. Thus, opening or locking the driver's door typically causes the other locks to operate in the same manner.

Power Adjustable Pedals

Power adjustable pedals are a relatively new feature seen on a few cars. This feature allows the driver to bring the pedals closer to the seat rather than bringing the seat to the pedals, allowing the driver to be at a comfortable distance from the steering wheel and other controls. A motor turns a gear to move the pedals closer or farther for the driver. This feature may also be a memory function.

Sunroof

A sunroof consists of both body parts and an electric motor drive mechanism. A switch controls the action of the motor. In most cases, if the motor fails, there is a way to close the sunroof mechanically by inserting a special crank and turning the drive mechanism by hand.

If the power seats, windows, door locks, or an electrically operated sunroof do not function, check the fuse box. If a fuse is blown, replace it. If the replacement fuse blows or there is another problem with the system, consult a qualified technician.

Rear-Window Defogger

Rear-window defoggers operate by heating a grid of electrically conductive material applied to the rear window. Electrical power is supplied to the lines in the grid by a relay that is operated by a switch on the dash.

If the rear-window defogger does not operate, first check and replace any blown fuses. If the unit still does not operate, inspect the conductive lines in the grid carefully. A small break in any one of the lines can render the grid nonconductive.

To repair a broken conductive line, small bottles of conductive paint are available through car dealers, auto parts stores, and at electronics supply shops. Typically, the glass must be cleaned thoroughly before application of the conductive paint. An artist's brush is used to apply several coats of the material at 15-minute intervals to allow thorough drying (Figure 10–9). Full instructions are included on the containers of conductive paint.

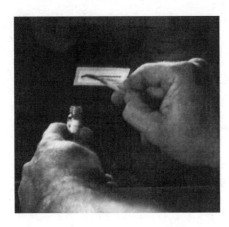

Figure 10–9: Application of conductive paint to repair a rear window defroster.

If the unit still doesn't work, it may be necessary to replace the power relay, the entire grid, or the rear window to affect a repair. Leave such servicing to a professional body shop or dealership.

Antitheft System

Each year, tens of thousands of automobiles are stolen or broken into. In response to this problem, car manufacturers have offered optional or standard-equipment antitheft systems.

The components needed to operate an antitheft system (Figure 10–10) include:

- Electronic control unit
- Front and rear door switches
- Trunk key cylinder unlock switch
- Hood switch
- Starter inhibitor relay
- Horn relay
- Horn
- Special ignition keys

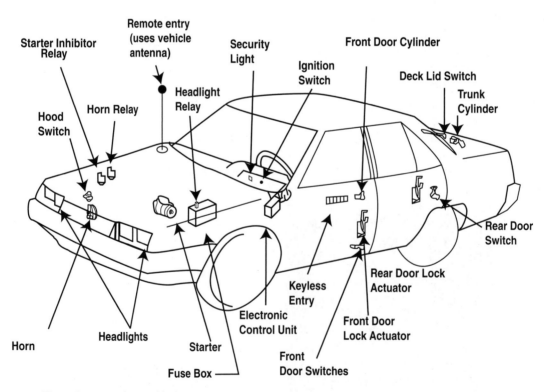

Figure 10–10: The major parts of an antitheft system.

Antitheft systems typically disrupt the ignition or starter systems, sound the horn at intervals, and flash headlights to alert the car owner and attract attention to the thief. In addition, the car will not start, even if the ignition key switch is broken out and turned with a screwdriver. A trend on the rise is the use of special ignition keys. While the systems vary, the concept is the same. The key may contain special electronic codes that the vehicle computer must read and confirm before allowing the car to start.

Consult the owner's manual for the specific procedure for your car. To arm most systems, the driver's door is locked with the key or the driver's doorknob is depressed and the handle is lifted when the door is closed. A security light on the dash typically illuminates for 30 seconds to signal that the system will function. When the light goes out, the alarm and other functions are armed.

If a door, hood, or trunk is forced open, the alarm is triggered. The system will continue to sound and flash a warning for two to three minutes. Then, if the door, hood, or trunk has been closed, the system will shut itself off and reset the alarm automatically.

The trunk can be opened with the key to load packages or cargo. The alarm will not be triggered, and the alarm will reset automatically 30 seconds after the trunk lid is closed. To disarm the system completely, the driver's door must be opened with the key.

Refer servicing of antitheft systems to qualified service personnel.

Other Systems

There are many other optional systems that may be found on a car. Many of these systems are available on select models of cars. While these systems undoubtedly improve driver safety, they also provide driver comfort. Just a few of the possibilities are listed here, followed by a brief description.

Speed Sensitive Wipers and Stereos   As the vehicle speeds up, the rate at which the wipers operate will increase. The volume of the stereo will also increase and help overcome increased wind noise.

Satellite Navigation System   Navigation systems are available that use satellites to plot your current position. Some systems can actually guide you to your destination. One manufacturer is offering a plethora of services with its system, which includes air bag notification and emergency services, remote door unlock, cellular phone communications, vehicle diagnostics while you drive, roadside assistance, and other optional services. Other manufacturers have their own offerings.

Heads-Up-Display (HUD)   Much like fighter aircraft, visual images and information may be projected onto the windshield in a heads-up-display so the driver can watch the road ahead. Information such as vehicle speed, turn signals, warnings, and headlight high beam indicators may be displayed.

Night Vision   Night vision technology has it origins in the military and has been adapted to the automobile. It allows the driver to see objects that are not visible under adverse lighting conditions. The system works by detecting differences in the heat levels of different objects. With this system it is possible to see through fog, rain, smoke, and at night. It cannot see through concrete, glass, metal, or other materials. The image from the sensor is projected using the HUD.

Collision Avoidance   Automotive researchers are working on various types of collision-avoidance systems. One is a side-mounted sensor that warns you if there's a car in your blind spot when you attempt to change lanes. Another is used to warn of objects or people in the way while backing up.

Satellite Radio   Satellite radio is available today and comes on some new cars. It receives a signal from an overhead satellite and the sound quality is comparable to a CD. You can listen to the same radio station from coast to coast. There are a wide variety of channels to choose from and some of the channels are commercial-free.

This chapter completes your introduction to the parts and systems found on today's vehicles. Chapter 11 focuses on practical, vitally important maintenance jobs that you can perform on your own car.

1. List some optional accessories.
2. What is the purpose of the heater?
3. List the common components of the heater system.
4. Describe the construction of the heater core.
5. Explain where the heat from the heater comes from.
6. Describe the air distribution for the passenger compartment.
7. List common heater system problems.
8. List the components of the air conditioning system.
9. What are the two types of air conditioning systems?
10. Explain how the air conditioning system cools.
11. What is the purpose of the accumulator?
12. What is a sign of a refrigerant leak?
13. What are the maintenance items for the air conditioning system?
14. What are some common air conditioning system problems?
15. What can cause insufficient cooling?
16. What can cause no cooling?
17. What could cause an odor in the air conditioning system?
18. Can a rear window defroster be repaired? How?
19. How does the antitheft system work?
20. Discuss some other optional accessories for the automobile.

YOUR MAINTENANCE PROGRAM

After reading this chapter, the reader should be able to:

- Explain the benefits of maintenance.
- Explain the importance of keeping a maintenance logbook.
- List and explain the additional maintenance concerns for cold weather.
- List and explain the additional maintenance concerns for hot weather.
- List environmental concerns affecting maintenance.
- List the types of driving and how they affect the maintenance needs of the vehicle.
- List maintenance items and regular maintenance intervals.
- List the steps for changing oil.
- List the steps for changing the air filter.
- Describe the maintenance needed for tires.
- Describe how to check brakes.P
- Explain how to flush the cooling system.
- Describe the maintenance needed for the ignition system.
- Describe how to change the automatic transmission fluid.
- List the steps for interior and exterior care.

THE IMPORTANCE OF MAINTENANCE

The responsibility for preventive maintenance is yours (Figure 11–1). If you want your car to perform safely, reliably, and economically, it is in your own best interest to take an active role in performing or scheduling regular maintenance. Surveys reveal that an owner who takes an interest in the car's maintenance has a much higher level of satisfaction with the vehicle's performance and reliability.

Most of the major parts and systems of the chassis, body, and powertrain have predictable use and wear patterns. These parts and systems require regular attention and preventive maintenance. Regular, scheduled maintenance can prevent most common problems and assure that your car will be as safe and reliable as possible. The sooner you recognize the telltale signs of wear or potential defects, the sooner you can take corrective action. Preventive maintenance reduces the chance of breakdowns and lowers your overall costs.

Maintenance Logbook

It is relatively easy to get 100,000 miles out of your car, and you could get 300,000 with the proper maintenance. You don't need to know how the car works, but you do need to maintain your vehicle according to a schedule and keep a maintenance logbook.

Keep an up-to-date logbook with the date, type of car repair or maintenance done, and mileage at the time of the service. A logbook will make you aware of abnormal fuel, oil, or other fluid consumption. Abnormal or unusual fluid usage could be the clue to a leak or other potential problem. A properly kept logbook can also serve as a reminder that the vehicle is overdue for an oil and filter change, transmission service, or other important maintenance. Having a record of gas purchases versus distance driven can point out a decrease in gas mileage, warn of a malfunction, or indicate the need for tune-up.

Drivers whose vehicle expenses are tax deductible know the importance of maintaining good records. In addition to the benefits noted, when it's time to sell or trade, detailed documentation can make a difference in the value of the vehicle. A used-car buyer likes to know what maintenance has been done and when.

Keeping a good record isn't difficult. Some people use a computer spreadsheet or dedicated maintenance software to track maintenance tasks and their costs. These tools make it very easy to print reports for different categories of maintenance. However, these are not very effective if you don't get the information from the car to the computer. It's a matter of keeping a pencil and notepad handy to record anything of importance about the maintenance and operation of the car. Just make entries into your logbook when they occur. Essentially there are several types of entries you may want to make.

Figure 11–1: In a self-service marketplace, motorists assume responsibility for maintaining their own cars.

- Gasoline purchases and mileage
- Addition of fluids such as engine oil, coolant, and so on, and at what mileage
- Maintenance services performed: when, what, by whom, and cost
- Repair services performed: when, what, by whom, and cost
- Reminders for service such as changing oil, rotating tires, and other maintenance intervals
- Renewal reminders for insurance, license, registration, and tags.

Other entries in the logbook may also include basic part numbers or specifications that may not be readily found in the owner's manual (belts, hoses, filters, lamps, spark plugs, etc.).

Having a logbook may be of assistance in case of car trouble. It may be useful to report to a technician exactly what work has been done and may offer clues to the root cause of a problem. Your logbook can offer an "automotive medical history" of your car should a repair be needed.

Having a dependable vehicle is an integral part of life for many of us today for both work and leisure. Keeping your car in good operating condition is made easier by keeping track of maintenance tasks. These records are important to the reliability, longevity, safety, and value of your car.

INDIVIDUALIZING YOUR MAINTENANCE PROGRAM

Preventive maintenance works for all cars, in all locations and climates, and under all driving conditions. However, there is no such thing as a universal maintenance program that applies to all automobiles, all of the time. Thus, your goal is to individualize your car's maintenance program.

Take a few minutes to read your owner's manual from cover to cover. You will notice that there are recommendations as to when maintenance jobs should be done based on time or miles driven, whichever occurs first. There are footnotes and other explanations that stress that these maintenance intervals will vary according to conditions and usage. For example, maintenance must be performed more often in extremes of heat and cold, and in dusty conditions. Also, cars that are used for short trips, continuous stop-and-go driving, or pulling trailers must be serviced more frequently.

With knowledge of your own driving habits and the conditions under which you drive, you are the best judge of your car's maintenance needs. Some of the most common maintenance-requirement factors are reviewed in this section.

Cold Climate

Extreme cold, snow, and ice affect virtually every part of your car's performance. Systems that should be checked and serviced before cold weather sets in include the cooling system and the electrical system. The engine oil should be changed and replaced with the recommended grade and proper viscosity oil for your climate. Steps also should be taken to prevent fuel-line blockage due to freezing. Proper servicing of these systems can prevent breakdowns in cold weather.

Fuel System   In the fuel tank, excessive amounts of condensed water can freeze and block fuel lines. Even small amounts of moisture may damage fuel-injection systems severely. The occasional or regular use of super-unleaded fuel will help keep moisture out of the fuel system. The ethanol in super-unleaded can absorb the moisture. Additives, or so-called gas dryers, are available to prevent moisture buildup. However, there is a simple precaution that can help to reduce moisture in the fuel system: Keep your fuel tank filled. A full tank has less room for air, which contains moisture.

Electrical System   The battery loses power in cold weather. The chemical action that produces the electrical power slows down at lower temperatures. At 0 degrees Fahrenheit (–17.8 degrees Centigrade), a typical battery will have lost nearly 40% of its power, as compared with its performance at 80 degrees Fahrenheit (26.7 degrees Centigrade). In addition, at 0 degrees Fahrenheit, the engine requires more than twice the normal amount of power to crank and start. Thus, the battery should be in top condition for cold weather use.

Electrical wires and cables that are corroded or damaged will not carry sufficient electrical current during cold weather. Check the battery cables and any exposed wires; replace any that appear to be damaged or corroded. Also, be sure all electrical connections are tight. If moisture gets into a connection and freezes, it may damage the wire or cause a short circuit.

Lubricating System   Lubricating oil increases in viscosity, or thickness, as temperatures drop. Multiviscosity oils, such as SAE 5W30 and SAE 10W30, are designed to flow and lubricate engine parts in both cold and hot weather. Check your owner's manual to determine the recommended engine oil viscosity for your climate.

Also, moisture condenses in your car's oil pan more readily in cold weather. Moisture in the oil causes acids and sludge to form. In addition, short trips in cold weather cause the fuel system to provide a rich mixture that can wash lubricant from cylinder walls and dilute the oil supply. To prevent engine damage, oil changes should be more frequent in cold weather.

Cooling System   The entire cooling system should be flushed out and refilled with the proper mixture of antifreeze and water to prevent freezing in cold weather. Also remember that a fresh coolant mixture prevents rust, corrosion, and engine damage caused by electrolysis.

The freezing point of coolant is checked with an instrument called a coolant hydrometer, shown in Figure 11–2. If the coolant freezing point is not low enough for expected temperatures in your area, more antifreeze should be added to the coolant, according to directions on the coolant container. Typically, you'll be instructed to drain part of the liquid from the system and refill with straight antifreeze.

The coolant hydrometer does not measure the antirust and anticorrosion properties of the coolant. The antifreeze chemicals that provide these protections lose their potency over time. Thus, flushing and refilling the system with a fresh coolant mixture regularly is a wise preventive maintenance step.

All cooling system hoses, including heater hoses, must be checked. Figure 11–3 shows typical trouble spots. If any hose shows signs of deterioration, replace it.

Engine-Block Heaters   If temperatures in your area are extremely low, an engine-block heater might be a good investment. These heaters are available

Figure 11–2: A coolant hydrometer can be used to determine the freezing point and relative concentration of the coolant in your car.

as options on most new cars sold in cold climates. Block heaters can also be purchased at auto parts stores. The engine-block heater plugs into your household current and keeps the engine coolant, block, and cylinder head warm. It is much easier to start a warm engine than one that is dead cold.

Snow and Ice Problems   Snow and ice may collect and build up on the body of the car, particularly inside fenders and under the car. These deposits may weigh hundreds of pounds and help to trap rust-causing moisture next to chassis and body parts. Snow and ice deposits affect safety,

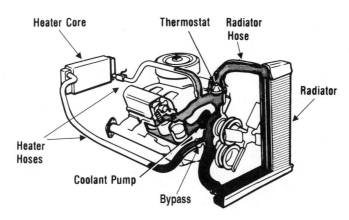

Figure 11–3: Check all the hoses in the cooling system for cracks, hardening, and leaks. Shown here are typical trouble spots.

performance, and fuel economy. Ice also can build up on various linkage components and make steering, shifting, or braking difficult.

To melt snow and ice and make driving safer, salt and other chemicals are often spread over roads in cold parts of the country. However, salt increases the corrosion of most metals and promotes rusting. The best way to attack the salt/chemical problem is to keep you car clean and well waxed. Wash your car frequently and hose off ice and snow deposits. It is especially important to wash away deposits inside fender wells and around drivetrain, suspension, and steering system parts. There are car washes available that offer underbody washing.

Frozen Door Locks   Moisture may condense and freeze in a door lock. You may not be able to insert or turn the key. To open the door, you can heat up the key with a match and then insert it into the lock, but just heating the key doesn't always work. As a preventive measure, keep a can of de-icer spray handy outside of the vehicle. These sprays are available at auto parts stores and other retail outlets.

Hot Climate

In very hot climates, such as desert regions, heat and dirt are your car's major enemies. Hot, dusty weather has its greatest effect on cooling, fuel, and lubricating systems.

In hot and dry climates, your car's fuel system filters can become clogged with dust, dirt, and sand. Fine dirt particles float freely on hot, dry air, especially when the wind blows. Dirt can enter the fuel tank and lines, and clog fuel-injection systems. Dirt particles that enter the air that passes through the intake manifold can cause severe damage to internal engine parts. Thus, air and fuel filters should be replaced frequently in extremely dusty conditions.

Lubrication System   In hot, dusty climates, oil and oil filter changes should also be made frequently to prevent engine damage. Dust and dirt can enter the lubricating oil and act as abrasive agents between moving parts.

Proper oil viscosity is also important in hot climates. High-viscosity (thick) engine oils are recommended because heat tends to thin oil and reduce the thickness of the oil film between metal parts. Check your owner's manual for the recommended oil viscosity for hot climates.

Cooling System   The coolant in the cooling system should contain the proper proportions of water and antifreeze for expected temperatures in your area. The antifreeze/water mixture helps to prevent boiling, as discussed in Chapter 6.

In hot weather, special attention should be given to all cooling system components. Hot weather and high engine temperatures dry and help deteriorate rubber parts, including hoses and drive belts. Gaskets throughout the engine and transmission also dry and become brittle in high-heat conditions.

Additional cooling devices may be a good investment in areas that experience extremely high summer temperatures. These devices include engine oil coolers and additional automatic transmission fluid coolers.

Sun Damage   It is wise to keep your car out of the hot sun whenever possible. The sun causes your tires and windshield wiper blades to dry and crack. Sunlight also fades body paint and deteriorates the interior of the car. On a 95 degree Fahrenheit (35 degree Centigrade) day, the air trapped inside a vehicle can heat up to a scorching 140 degrees Fahrenheit (60 degrees Centigrade), the steering wheel can reach 150 degrees Fahrenheit (65 degrees Centigrade), and the dashboard can reach up to 180 degrees Fahrenheit (83 degrees Centigrade). One of the most vulnerable parts of the interior is the dashboard, which suffers direct exposure to the sun. If not protected adequately, the vinyl can crack and split. Typical car windows offer some degree of sun protection. Many people will use a sunshade to help prevent sun damage to the interior which also helps reduce interior temperatures. Sunshades come in many sizes and styles. If you decide to use one, choose one that is convenient to use. You can help protect your car by keeping it waxed and keeping the interior conditioned with a good preservative. If your car has a vinyl roof or convertible top, keep that well conditioned, too.

Environment

Dust, wind, moisture, and altitude are environmental conditions that require special attention. Some environmental conditions are seasonal, while others remain constant. As these conditions change, be aware that your car may need more than just routine maintenance.

Moisture   Any moisture in the air will create condensation in the fuel and lubrication systems. Fuel-injection systems are affected by even small amounts of water. Excessive moisture can cause the engine to run poorly. Some fuel tanks may rust internally if enough water is present. Rust can clog fuel lines and damage fuel system and engine parts.

The fuel filter must be replaced more frequently in moist climates than in dry areas. In extreme conditions or on vehicles with fuel injection (gasoline or diesel), it may be necessary to use a special water-separating filter system.

If significant amounts of water have collected in the fuel tank, it may be necessary to drain the tank completely. The job of draining the fuel tank is best left to a certified technician.

As discussed earlier, condensation also forms in the crankcase and combines with sulfur to produce sulfuric acid in the oil. Therefore, the crankcase oil and oil filter should be changed frequently in humid or wet weather.

High Altitude   As the oxygen content of the air decreases with altitude, the car will produce a little less power. Electronic fuel injection includes sensors for barometric pressure or changes in altitude that will direct the computer to correct the air/fuel to maintain the best power and economy.

Type of Driving

There are many types or styles of driving, which can generally be organized into categories. Some of the categories are listed, and the three most common are discussed.

- Short trips driving less than 10 miles
- Stop and go
- Highway driving
- Extensive idling
- Trailer towing
- Sustained high-speed operation
- Off-road driving

Short Trip Problems   Many people think the ideal used car is one owned by a person who drove the car only a few miles a week, to the market and on Sundays. In reality, an engine subjected to this kind of driving would become worn excessively at very low mileage. Short-trip driving causes many serious problems.

Cars that are driven only on short trips rarely have an opportunity to heat up to normal operating temperatures. Average trip-lengths of less than 10 miles are considered short trips and call for following the severe service recommendations for your car. Many cars fall into this category.

As discussed earlier, your engine operates best when it is warmed up. If the main use your car gets is a short drive to work or to neighborhood markets and so on, extra maintenance is necessary. Under normal driving, oil should be changed every three months or 3,000 miles, whichever occurs first. For short-trip driving, the oil should be changed more frequently.

In addition, regular inspections should be made with a more critical eye. For example, approximately 100 miles of driving per week minimum is required to keep a battery fully charged. Partial charging can produce a battery condition called sulfation, in which a battery will not accept or hold a charge and thus needs to be replaced.

Stop-and-Go Driving   Stop-and-go driving on city streets and clogged freeways is hard on all of your car's systems. You may not have a choice about the traffic conditions in which you drive, but you should be aware of its effects on your car. Overheating problems are common, brakes wear more quickly, and the oil and filter need changing more frequently. In addition, automatic transmission lubricant and filter service should be performed once a year or every 12,000 miles, whichever occurs first.

Highway Driving The best way to get the most use from a car is to drive it at steady, moderate speeds over long distances. If you have ever taken a long trip in a car, you may have noticed an increase in your gas mileage. Thus, this type of driving often is known as "good mileage driving." When the engine is heated to normal operating temperature and all parts are moving and being lubricated, everything functions the way it was designed to work. If you do mainly good-mileage driving, your major engine components will have a long life.

Generally, you have little or no choice about the type of driving you must do. However, with the information presented in the remainder of this chapter, you will have a basis for creating an individualized maintenance program. Consult the owners' manual for the recommended service intervals for your type of driving. Consider your own driving habits and the conditions under which you drive.

Realize that extra maintenance, over and above the minimum required, can never hurt your car. Extra maintenance can only help to keep parts and systems in top working order.

As a car owner, part of your challenge is to balance the benefits of doing extra maintenance against the cost of the services. If you learn to do simple maintenance jobs yourself, you can save the typical $55 to $85 per hour of labor costs charged by service garages and dealership service departments.

Also consider long-term economy, not just short-term expenses. For example, in the short run, it is cheaper not to do oil and oil filter changes at all. In the long run, the cost of an engine overhaul ($4,000 or more) outweighs the small amounts saved by not doing oil changes. With proper maintenance, the life of the engine and all other parts and systems on your car can be extended considerably.

DAILY MAINTENANCE

In the process of using your car every day, you can make some simple maintenance checks. These checks can help in early diagnosis of problems that can cause breakdowns.

Visual Checks

Whenever you approach your car, take a minute or two to examine its appearance critically.

- Check your tires for low air pressure, nails, or other evidence of defects.
- Check to see if the car leans or tilts in any direction from side to side or from front to back.
- Look for any puddles of fluid underneath the car.

Make these visual checks a part of your normal routine. If you spot a problem, you can have it fixed before it causes a breakdown.

Driving Checks

Try to be aware of the way your car operates. Notice how quickly it starts, the sounds that it makes, how long it normally takes to warm up, and so on. Permanent changes from normal operating conditions usually indicate a problem.

Brake Pedal   Before you turn on the ignition, step on the brake pedal firmly. The pedal should be high and should feel firm. If the pedal is low, check the brake fluid. If the fluid level is okay, a brake system problem is indicated. If you pump the pedal and it rises to normal height, you may have a loss of pressure in the brake system.

Warning Lights and Gauges   Turn the ignition switch to the ON position, but don't start the car. The warning lights should work. If any light does not work, a bulb may be burned out or a switch may be defective. (On many General Motors cars, the key must be turned slightly past the ON position to operate the TEMP warning light.) Electrical gauges should activate the register correctly.

Power Brake Booster   Press lightly on the brake pedal and start the car. If the pedal drops slightly, approximately one-half to three-quarters of an inch, the power-assist system is operating correctly. Within 10 to 15 seconds, all warning lights should go out. If the parking brake is set, the light for this system will not go out until the parking brake is released. Gauges should indicate normal operating conditions. If a warning light remains on or a gauge indicates a problem, shut off the engine and check the system indicated.

A brake warning light that remains on after the parking brake lever is released may indicate any of the following:

- The parking brake light warning switch is misadjusted, stuck, or shorted. This switch is located on or near the parking brake linkage.
- The brake fluid in one or more master cylinder reservoirs is low. Check the reservoirs and add fluid if necessary.
- There is a loss of pressure in the brake system. The warning light operates when the brake pedal is depressed to alert the driver of a potentially dangerous situation.

If you have any doubt about the proper operation of your brake system, do not drive the car. Have the car checked by a qualified technician.

Exterior Lights and Turn Signals   As a final check before driving off, try the turn signals and brake lights. Look for reflections on surrounding cars or buildings to determine whether these lights are functioning. At night, use the same techniques to check your headlights, parking lights, side marker lights, and taillights. Watch for the reflection of your lights on cars ahead of and behind you, or in store windows when you are stopped at a traffic light.

WEEKLY MAINTENANCE

Gauges and warning lights on the dashboard won't indicate that something is wrong until a problem arises. Rather than risk a breakdown, take a few minutes each week to check under the hood.

The best time to make these checks is in the morning, before the car has been started, when the engine and fluids are cool. For all of the following checks except automatic transmission fluid level, the engine should be *off*.

Procedures for these simple, quick checks are discussed in the maintenance sections of the appropriate chapters, as indicated.

- Check engine oil level (Chapter 6).
- Check coolant level in the recovery tank and the radiator (Chapter 6).
- Check brake fluid level (Chapter 8).
- Check belts for defects and proper tightness (about ½-inch deflection, as described in Chapter 5).
- Check hoses (Chapter 4).
- Check for coolant leaks from hoses or oil leaks from gaskets (Chapter 6).
- Check for fuel leaks or a strong smell of gasoline (Chapter 4).
- Check battery electrolyte level and cables (Chapter 5).
- Check automatic transmission fluid (Chapter 7).
- Check tire pressure (Chapter 8).

If you would rather leave these routine matters to someone else, have a local service center or a qualified technician make these checks on a regular basis.

3,000–7,500-MILE MAINTENANCE

For maximum reliability, preventive maintenance should be performed on a regular basis. The intervals at which maintenance should be performed are specified in your owner's manual. However, consider your driving habits as well as the climatic conditions you have encountered. If you drive short distances, or drive in extremely cold, wet, dry, or hot weather, preventive maintenance should be done more frequently than if you drive moderate distances in temperate climates. Maintenance to be performed at approximately 3,000 to 6,000 miles, or three to six months, whichever occurs first, includes the following:

- Change engine oil and oil filter.
- Check fluids.

- Inspect hoses and belts.
- Inspect or change air filter.
- Examine driveline and CV boots.
- Inspect brakes.
- Inspect tire wear and pressure (including the spare).
- Check interior and exterior lights, turn signals, and horn.
- Inspect wiper blades and washer operation.
- Lubricate hinges and latches.
- Inspect air filter.
- Check exhaust system for leaks, damage, and loose parts.

Change Engine Oil and Oil Filter

To drain the old oil from your car, you may need to raise the car and support it safely with ramps or jack stands (refer to Chapter 2). If there is enough room to work under the car, you may be able to perform an oil change without raising it. For ease and comfort, you may need a creeper to lie on and roll into position. To allow the oil to flow out quickly and remove the maximum amount of contaminants, the engine should be warmed up (but *not* running) when you drain the oil. It is a good idea to do an oil change after a 20-minute drive. You should not drain the oil if it is too hot, as you do not want to burn yourself. Let it cool down for a few minutes. Hot oil drains faster and carries more contaminants out of the engine. But again, having it hot enough to burn you is not necessary.

To do the oil change, you will need the following tools and supplies:

- An oil drain pan
- A wrench or socket to loosen the drain plug
- An oil filter wrench to loosen the oil filter
- Rags or paper towels
- New oil and filter

Figure 11–4 through Figure 11–10 illustrate the steps required to perform an oil change. Refer to them as you read the following procedures.

Step 1. Locate the oil drain plug on the oil pan at the bottom of the engine. The plug may be at the front, rear, bottom, or on either side of the oil pan. Position the drain pan under the plug or nearby. Use a wrench or socket to loosen the plug (Figure 11–4).

After loosening the plug, put the drain pan under it and then use a rag or paper towel to grasp the plug with your fingers. Hold the plug to prevent it from dropping out as you unscrew it with your fingers. When it's ready to come out, move it out of the way quickly. Remember, the plug and oil are hot.

Step 2. Allow the oil to drain for at least five minutes (Figure 11–5). The oil will continue to drip for several hours, but five minutes should be enough time to drain almost all of the oil from the pan. While the oil is

Figure 11–4: Oil change step 1: After raising and supporting the car safely, blocking the wheels, and positioning a drain pan under the car, loosen the drain plug.

Figure 11–5: Oil change step 2: Allow the oil to drain for at least five minutes.

draining, use a rag to clean the drain plug and its threads thoroughly. Inspect the drain plug gasket. If it is cracked or damaged, get a replacement gasket or a new drain plug.

Step 3. After the oil has drained, wipe off the area around the drain plug on the oil pan. Screw the plug back into the hole with your fingers

Figure 11–6: Oil change step 3: Clean the drain plug and oil pan, then screw the plug in by hand and tighten it securely.

(Figure 11–6). It should thread all the way in with your fingers. After the plug has been seated, then tighten it firmly with a wrench or socket, but don't overtighten the drain plug because the threads are easily damaged. Next, reposition the drain pan carefully—it's full of oil now—under the oil filter.

Step 4. Loosen the oil filter by turning it counterclockwise with an oil filter wrench (Figure 11–7). When it is loose, use a rag to grasp the filter

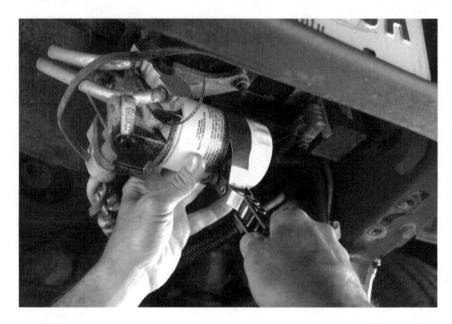

Figure 11–7: Oil change step 4: Move the drain pan underneath the oil filter, then use an oil filter wrench to loosen it. Unscrew and remove the filter by hand.

Figure 11–8: Oil change step 5: Rub a few drops of new oil on the new oil filter sealing gasket.

and unscrew it by hand. At this point, the filter may still be full of oil. Remember, the filter, too, may be hot, as may other parts of the engine. Position the filter carefully and pour the oil into the drain pan. When the filter is empty, dispose of it properly. Inspect the filter and make sure the old rubber gasket came off with it; if not, look at the mounting surface and make sure it's not still stuck on there.

Step 5. Before installing a new filter, clean the filter mount on the engine with a rag or paper towel. Then take a few drops of clean oil on your finger to lubricate the rubber gasket on the filter base (Figure 11–8). If the gasket is dry, it may start to grab and pull out of position as the filter is tightened and cause a leak.

Step 6. Screw the new filter onto the mounting threads by hand until the filter base begins to tighten against the filter-mounting surface on the engine (Figure 11–9). After the filter contacts the engine, tighten the filter an additional one-half to three-quarters of a turn. Follow the manufacturer's instructions printed on the oil filter. Never use a filter wrench to tighten a filter. Overtightening a filter will make it extremely difficult to remove during the next oil and filter change.

Figure 11–9: Oil change step 6: Screw the new filter on by hand. After the filter contacts the engine, tighten it one-half to three-quarters of a turn more.

Step 7. After the oil drain plug and new filter have been installed, add the new oil to the engine (Figure 11–10). Usually 4–5 quarts of oil are required when an oil and filter change are done. Refer to your owner's manual for specific refill capacities. Choose the recommended API Service oil and the recommended viscosity for temperatures where you drive.

Step 8. After the specified amount of oil has been added and the oil filler cap has been replaced, start the engine and let it idle. Watch the OIL

Figure 11–10: Oil change step 7: When the drain plug and filter are tightened securely, add the recommended amount of new oil.

warning light or gauge. When the light goes out or the gauge reaches the normal level, check for oil leaks. Inspect the oil drain plug and the base of the oil filter. If no leaks are evident, the job is complete. Tighten dripping oil drain plugs and oil filters as necessary to stop leaks.

Used engine oil may be considered toxic waste because it contains sulfuric acid and metals from the engine. Some service stations and garages will allow you to dispose of used oil in their facilities. In some localities, authorized recycling centers have been established. Dispose of the oil in accordance with all federal and local regulations. Recycling information may be found at http://www.recycle.org and at http://www.earth911.org.

Driveline Inspection

On many late-model cars, lubrication points have been eliminated. On most cars, permanently lubricated, sealed joints are installed when the car is built. When these joints wear out, they are replaced.

Rear suspension lubrication is usually minimal. The main job in rear-end lubrication involves checking the differential.

Look carefully at steering linkage and all suspension components. If anything appears loose or looks bent, broken, or otherwise damaged, have the part checked by a certified technician.

Constant-velocity (CV) joints are permanently lubricated, so they need no service unless their boots are torn or leaking.

Manual Transmission and Transaxle Lubrication

To check lubricant in a RWD car's manual transmission or a FWD car's manual transaxle, the vehicle should be level. Thus, both ends of a RWD vehicle must be raised and supported with jack stands so that all four wheels are off the ground. To check a FWD transaxle, it may be more convenient to have all four wheels on the ground. Some FWD cars with manual transaxles include a dipstick that can be reached through the engine compartment.

To check a manual transmission or transaxle without a dipstick, remove the plug on the side of the case and put the tip of your finger in the hole (Figure 11–11). The lubricant should be at the lower edge of the hole. If you can feel the lubricant with your fingertip, the lubricant is at the proper level.

If the lubricant level is low, you can pour the recommended manual transaxle lubricant in with a funnel that includes a rubber or plastic tube. For some manual transmissions, a pump unit or suction gun is required to force liquid into the transmission case.

Transmission

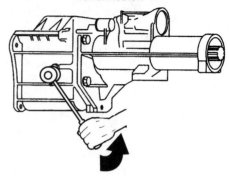

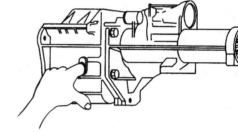

Figure 11–11: Remove the transmission's plug to inspect the fluid level and condition. Differentials have similar plugs.

After checking or refilling the transmission or transaxle, replace the plug and tighten it. Also inspect the front of the transmission or transaxle for evidence of lubricant leakage. If there is a leak, a seal is defective and should be replaced.

4WD Transfer Case Lubrication

Checking and refilling a transfer case is similar to the procedure for checking a RWD manual transmission. Refer to your owner's manual for the proper lubricant to be used.

Differential Lubrication

On FWD cars, the lubricant supply for the differential is included in the transaxle lubricant. Thus, no further checking is needed for a FWD vehicle.

For RWD and 4WD vehicles, locate the check plug on the front or rear of each separate differential. Follow the procedure previously outlined for checking and refilling a RWD manual transmission. Refer to the owner's manual for the recommended lubricant.

Also inspect differentials at the front and at the point where the axle housings meet the backing plates in the center of the drive wheels. Leakage indicates a defective seal or gasket.

When you have finished underneath, raise the car, remove the jack stands, and lower the vehicle gently onto the ground. To complete a lube job, use a little graphite powder to lubricate all door locks to prevent the lock cylinders from sticking and oil the door hinges and hood hinges. This simple maintenance will keep these parts from sticking and squeaking.

Air Filter

Check the condition of the air filter each time you change the engine oil and filter. Change the filter if it is coated noticeably with dust and debris or if the paper element is soaked with oil. Also remember that if you do a lot of driving under dusty conditions, the useful life of the air filter is reduced.

Compare the inner, or clean, side of the filter with the outer side to determine how much debris has accumulated. A clogged or oiled air filter reduces airflow to the fuel delivery system. Engine performance and fuel economy suffer. (Oil on the air filter indicates a clogged PCV or a major engine problem.)

Figure 11–12 through Figure 11–16 illustrate the steps in checking and replacing an air filter. Refer to them as you read the following procedures.

Step 1. To remove the air filter, unclip the housing on the top of the air filter housing. Lift off the top and remove the pleated paper element (Figure 11–12; Figure 11–13 shows another style of element). On some cars, it may be necessary to remove several screws or bolts to remove the air filter housing. It may also be necessary to remove some ductwork.

Figure 11–12: Air filter replacement step 1: Remove the cover of the air filter housing and the air filter element.

Figure 11–13: Another style of air filter element.

Step 2. To check the filter, hold it up to the light or place a trouble light near the inside surface of the filter (Figure 11–14). If you can see light through the filter, it is okay. If any holes are noticed the filter must be replaced. If there is any doubt about whether the filter is dirty enough to require replacement, it is better to install a new filter.

Step 3. Before you put in a new filter, the inside of the filter housing must be cleaned. Use a slightly oiled rag or paper towel to wipe and remove any debris inside the housing (Figure 11–15). Be careful not to sweep any debris into the air intake or carburetor throat.

Step 4. Put in a new filter, replace the top of the housing, and fasten the clips, screws, or bolts (Figure 11–16). Inspect the ductwork to make sure it is properly fastened, and the job is done.

Figure 11–14: Air filter replacement step 2: Hold a light behind the filter to check for any holes and to determine if it is dirty.

Figure 11–15: Air filter replacement step 3: Clean the inside of the filter housing, including the bottom of the cover.

Figure 11–16: Air filter replacement step 4: Install the filter and replace the housing cover and fasteners securely.

Check Spare Tire Air Pressure

Most people don't think about their spare tire until it is needed. In many cases, the spare is underinflated and would not support the care safely if a flat tire must be changed. Get in the habit of checking the inflation of the spare tire every time you check your other tire pressures. Also remember that many newer cars are equipped with temporary spare tires that must be inflated to 60 psi to support the weight of the car. Check the tire pressure placard or the owner's manual for tire pressure information.

Rotate Tires

In most cars, the weight of the engine rests on the front wheels, which must turn for steering. Therefore, the front tire tends to wear out sooner than the rear tires. The front tires of FWD cars also must supply the torque to move the vehicle forward, so they can quickly become worn.

Tire rotation, changing the positions of the tires on the car periodically, helps to even out the wear of the front and rear tires. Tire rotation is recommended at six-month or 7,500-mile intervals. Consult the tire manufacturer's recommendations, as some recommend the first rotation for radial tires at the 7,500-mile interval and then every 15,000 miles thereafter. The same 7,500-mile interval is recommended for checking the brakes (described in the next section). With a little planning, you can do both a tire rotation and a brake check at the same time. If you want to do both procedures, read through each of the two descriptions before you start.

Refer to the tire rotation patterns in Figure 11–17. Decide which tire rotation pattern is suitable for your vehicle.

To rotate the tires, you will need two jack stands in addition to the jack. A jack stand is used to support the car where you have removed a wheel while you use your jack to raise another corner of the car. Whenever you prepare to jack up a car, use wheel blocks on both sides of the wheels that will remain on the ground. When you raise the rear end of a RWD car, neither the parking brake nor the transmission will prevent the front wheels from turning. Always set the parking brake securely and place the transmission in PARK (automatic transmissions) or in gear (manual transmissions).

First remove any wheel covers, lug nut covers, or hubcaps from the wheels. Pry them off either with the tapered end of the lug wrench or with a large screwdriver. Work your way around the edge slowly until the cover comes off.

Start at any position on the tire rotation diagram that you will use. Use the lug wrench to loosen the wheel lugs about one or two turns before you jack up the car. Then refer to your owner's manual to locate the correct position for the jack. Use the jack as directed in your owner's manual to raise the first wheel.

When the first tire is about an inch or more off the ground, position a jack stand under the frame or other supporting member of your car's chassis. Refer to your owner's manual or a manufacturer's service manual to locate recommended support members. Adjust the jack stand until it fits under the supporting member. Then lower the jack so that the car is supported securely and safely by the jack stand. Remove the wheel lugs by hand and remove the tire from the car.

After you remove the tire, check the tread all the way around the tire. Look for nails, glass, or rocks imbedded in the tread. If you find a small

Rear and four wheel drive vehicles

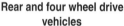

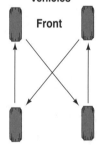

Front wheel drive vehicles

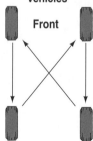

Figure 11–17: Tire rotation patterns for FWD and RWD vehicles.

chunk of rock or other debris, pry it out gently with a small screwdriver. Rocks or debris left in the tread can damage the tire. If you find a nail in the tire, take the tire to a service station or tire center that can fix it before you remove the nail.

Also inspect the inner and outer sidewalls of the tire. If you find any deep cracks or gouges, have the tire checked by a tire store or certified technician. A tire with a severely damaged sidewall must be replaced.

Check the tire treads for evidence of unusual wear, as described in Chapter 8. If you notice any unusual wear patterns, make a note of where the tire was on the car before the rotation to help in the diagnosis of problems. If there are no problems with the tire, roll it over to its new position in the tire rotation pattern.

Move the jack to the next position and repeat the process. Remove the wheel lugs and the second tire. Then install the first tire at the second position. Secure the wheel lugs on by hand, then tighten them with the lug wrench. Lower the jack until the car rests on the tire. Then use the lug wrench to tighten the lugs securely. Inspect the second tire as described for the first tire. If there are no problems, roll it over to the next position in the rotation pattern.

Repeat this procedure for the other wheels. After you are done, check the lugs on all four wheels one final time to be sure that they are tightened securely. A torque wrench may be used to precisely tighten the lugs. A click-type torque wrench is the best and easiest to use (Figure 11–18).

Figure 11–18: Use a click-type torque wrench to tighten wheel lugs precisely.

Torque specifications for the lug nuts can be found in the owner's manual. Then put your wheel covers back on and replace the jack and spare in the trunk.

As a final step, check and change the pressures on all tires to the manufacturer's recommendations as indicated on the tire placard or in the owner's manual.

Check Brakes

Brake checking should be a regular procedure at about six-month or 6,000-mile intervals—more frequently if you do a lot of hard driving. Stop-and-go driving is harder on the brakes than highway driving. Though there are several brake combinations, the most common has disc brakes on the front and on the rear. Figure 11–19 shows a typical four-disc system.

Normally, a certified technician should service the brakes. However, you can learn to check for brake problems yourself. The initial steps are similar to the procedure for tire rotation. Block the wheels that will remain on the ground before you raise the car. Set the parking brake securely and place automatic transmissions in PARK (manual transmissions in gear).

Start with the front wheels. Use the lug wrench to loosen the wheel lugs about one or two turns before you jack up the car. Jack up the front of the vehicle to a good working height. Use two jack stands for support, positioned on each side of the car under the frame or other support member of the chassis, according to the manufacturer's service manual. Lower the car until it is supported securely and safely on the jack stands. (Never work under your vehicle unless it is supported safely by jack stands or by a commercial lift. Working under a car supported only by a jack can lead to severe physical injury.) When the car is supported properly on the jack stands, remove the lugs from the wheels and then remove the wheels.

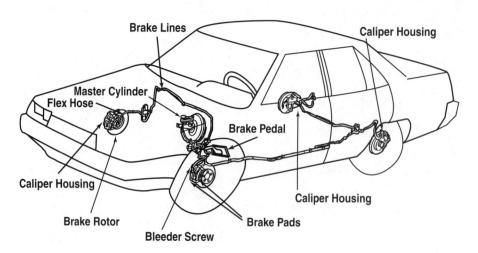

Figure 11–19: Many cars use disc brakes at all four wheels.

Checking Disc Brakes   You will be able to see the brake discs and the caliper assemblies, including the brake pads (Figure 11–20). Check both the inner and outer sides of each brake disc, or rotor, for grooving. The surface of the rotor should be very smooth. If there are deep grooves in the rotor or the surface is rough, it must be *turned*. Turning is a process in which a layer of metal is removed to make the rotor surface smooth again. A rotor should be turned only if a problem is noted in its surface. If you want to remove the rotors yourself, consult a service manual for the proper procedures for your car.

Inspect the brake hose connected to each front caliper and check for any cracks or swelling. If the brake hose has become cracked, brittle, or swollen, it should be replaced. Leave this work to a certified technician.

Look closely at the caliper for brake fluid leakage. A leaking caliper must be replaced. If brake fluid leaks on the linings, the brakes may lose their stopping power. If the calipers show any sign of leakage, have them replaced by a certified brake technician.

Avoid breathing the dark brown or blackish dust on disc brake assemblies. The material is brake dust, which could contain asbestos. Ordinary paint or dust masks will not protect you from asbestos fibers should they be present. Avoid blowing or brushing the dust so that it becomes airborne.

Figure 11–20: **Check the thickness of the brake pad, both inner and outer pads (A). Also check the surface of the rotor (B), which should be smooth with no gouges or grooves.**

Use a brake spray cleaner to clean brake dust; catch the dust in a disposable container and dispose of it properly.

Next, check the thickness of the disc brake pad linings (you may need a screwdriver). A disc brake pad consists of brake lining material riveted or bonded (glued) to a flat metal plate that fits inside the disc brake caliper. Most disc brakes use a bonded brake material. The metal plate is approximately $\frac{1}{8}$-inch to $\frac{1}{4}$-inch thick. Take this component into consideration in judging how much lining material is left as you view the edge of the pad.

If all the lining wears off a disc brake pad, the metal backing will rub against the disc and ruin it. A disc rotor costs $50 to $150 or more. Disc brake pads for both front wheels cost about $10 to $30. It is less expensive to replace worn brake pads than to replace ruined rotors.

There are two disc brake pads in a disc brake unit. One pad presses against the outer surface of the disc, and the other presses against the inner surface. Because they may wear unevenly, you must check the lining thickness on both pads.

Many calipers have an opening through which you can see the edge of the disc and the edges of one or both disc brake pads. On some cars you may have to use a screwdriver to pry off, gently, a protective sheet-metal cover over this opening. Then you can look inside the opening to judge how much lining is left on the disc brake pads. If either of the pads has $\frac{1}{8}$ inch or less of lining material remaining, all of the pads on both front wheels must be replaced.

Remember, the linings are fixed to metal plates. Don't include the metal plate in the thickness of the brake material. You will need to measure the brake material attached to the plate, as shown in Figure 11–21.

Replacing the disc brake pads is not too difficult a job. If you want to replace the pads yourself, refer to the procedure in the manufacturer's service manual.

Assuming you have completed your inspection, you can go ahead and replace the wheels and tighten the wheel lugs. Then jack or raise the car slightly, remove the jack stands, and lower the car. When the wheels are resting on the ground, tighten the wheel lugs securely again.

Now block the front wheels and raise the rear of the car. Place the jack stands solidly under the frame or other supporting member of the chassis according to the manufacturer's service manual. Then lower the car so that it is supported securely and safely by the jack stands. Now remove the rear wheel lugs and the wheels and tires. Inspection of rear disc brakes is the same as for the front.

Checking Drum Brakes   Some vehicles have drum brakes on the rear or a combination of a disc brake with a drum parking brake. The rotor may need to be removed to inspect the drum-brake portion of the brake system. Consult the manufacturer's service manual should you wish to do this procedure. Normally, you would have a certified technician check the rear drum brakes in this system.

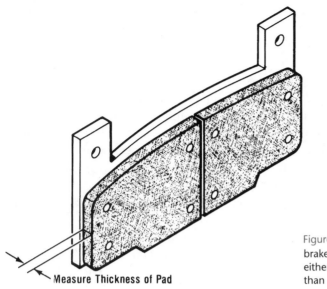

Measure Thickness of Pad

Figure 11–21: For safety, brake lining thickness on either pad should be no less than ⅛ inch.

To remove the rear drums on a FWD car, it may be necessary to remove a dust cover and disassemble the rear wheel bearings before you can remove the rear drums. If you have a FWD car, consult a service manual for the proper procedure to remove the rear drums.

On RWD cars, the drums may be held in place with two small bolts or screws, or locking clips on the studs. If small bolts or screws hold the drum in place, use a wrench or screwdriver to loosen and remove them. If locking clips hold the drum in place, they can be snapped off with a screwdriver and pliers and then discarded. New locking clips are not usually required as they are used during vehicle assembly when the car is new.

Release the parking brake lever, and you should be able to pull off the drum. However, if the brake lining has worn a groove into a drum, you may not be able to pull the drum off. If you can't get the rear drums off easily, replace the wheels and tires and tighten the lugs securely. Then have a certified technician check the rear drum brakes.

Avoid breathing the brake dust that collects inside drum brakes. Any dust inside the drum, on the brake shoes, backing plate, or other parts must not be blown or brushed out. Brake dust could contain asbestos. Ordinary paint or dust masks will not protect you from asbestos fibers should they be present. Use a brake spray cleaner to clean brake dust; catch the dust in a disposable container and dispose of it properly.

Inspect the inner friction surface of the brake drums. The surfaces should have a smooth finish but not a mirrorlike shine. Neither should they have deep scratches or scorch (bluish discoloration) marks on them. If any defects are noted, the drum should be turned, as previously described, to remove surface scratches and to "true," or round, the friction surface smoothly. A special machine is needed to turn the drums. This work will need to be done by a certified technician or automotive machine shop. Drums should be turned whenever brake linings are replaced. If a drum or disc brake rotor is deeply grooved, or has cracks or other major defects, it must be replaced.

Look at the drum-brake linings. Brake linings may be riveted or bonded. If your car has riveted linings, check the distance from the top of the rivet in each brake lining hole to the edge of the lining. If it has bonded linings, measure the thickness of the lining from the metal shoe to the outer edge of the lining. In either case, you should have $\frac{1}{16}$ inch or more of lining at the thinnest point. Note that the linings do not always wear evenly. Judge the thickness of the lining at its thinnest point. If the remaining lining on any brake shoe is $\frac{1}{16}$ inch or less in thickness, all the brake linings on both rear wheels must be replaced.

If you have determined that the brake linings are okay, check the wheel cylinders. Use a screwdriver to push, gently, the rubber dust seals on each end of the wheel cylinder out and away from the cylinder. If brake fluid leaks or drips are noted, a certified brake technician must replace the cylinder.

Trace the metal brake line that connects to the back of the wheel cylinder on the other side of the backing plate. From there, trace the brake lines to each rear wheel. One or two rubber brake hoses will be used at the rear of the car. Flex each hose and check it as described in the procedure for checking disc brake hoses. A damaged hose must be replaced. Leave this job to a certified technician.

If everything has been checked on both rear drum brakes, replace the drums and any screws or bolts that held them in place. Replace the wheels and tighten the lugs with the lug wrench. Jack the car to raise it slightly; then remove the jack stands and lower the car. When the wheels are on the ground, tighten the lugs securely. Go back around the car to make sure you have tightened all the lugs on all four wheels securely. Then replace the wheel covers.

If you have noticed any problems in the brakes, take the car to a certified brake specialist. Good brakes are the most important safety equipment on your car.

And don't forget to repeat weekly checks.

12,000–15,000-MILE MAINTENANCE

After each 12,000 to 15,000 miles, or one year, additional preventive maintenance should be done. Remember to consider your own driving habits and driving conditions and to compensate for the type of driving that you do.

The first step in a 12,000- to 15,000-mile maintenance program is to repeat your weekly check and your regular maintenance program for the 3,000- to 7,500-mile period.

Air Filter Replacement

Check the condition of the air filter each time you change the engine oil and filter. The filter should be replaced every year or if it is dirty. Remember that if you do a lot of driving under dusty conditions, the useful life of the air filter is reduced. Air filter replacement is discussed earlier in this chapter.

30,000–60,000-MILE MAINTENANCE

Fuel Filter Replacement

The fuel filter removes impurities and small amounts of water from the fuel delivery system. Replace the fuel filter every two to four years or every 30,000 to 60,000 miles, whichever occurs first. Consult the owner's manual for the recommended service interval for your fuel filter.

Gasoline is extremely flammable and can cause serious burns. Keep a fully charged class B fire extinguisher handy whenever you work on the fuel system. When you replace a fuel filter, avoid spilling gasoline. Wrap a rag around the fuel line and place a metal container under the filter to catch any fuel that might spill. Store soaked rags in a tightly covered metal safety container.

A fuel filter can be mounted anywhere between the fuel tank and the fuel injectors. The filter typically is a metal container with a paper filter element inside. Check the owner's manual or a service manual for the location and replacement procedure for your fuel filter.

Fuel filters generally have markings to indicate which way the fuel flows to the fuel delivery system. Make sure the arrow points toward the fuel injectors.

Fuel injection system pumps often produce pressures of 40 psi or more. If the fuel filter is removed when it is pressurized, large amounts of fuel can be sprayed considerable distances. To change a fuel injection system filter, special procedures must be followed to depressurize the filter. Refer to a service manual to determine the proper procedure for replacing a fuel injection system filter, otherwise leave this job to a technician. Should you replace a fuel filter, make sure you check for any fuel leaks. Any leaking of any amount of fuel is unacceptable and poses a serious problem. As stated before, gasoline is extremely flammable.

Whenever you change a fuel filter, check all of the fuel lines and hoses. Any cracked or damaged hose should be replaced. Any fuel system service, besides changing the fuel filter, should be done by a technician.

By the time you have driven your car 30,000 to 60,000 miles, typically over the course of two to five years, wear has occurred to all moving parts.

To maintain reliability, several preventive maintenance tasks should be performed.

Before you start any of the following procedures, repeat all of the regular maintenance programs outlined earlier. These maintenance tasks include all of the services that were performed at the 12,000- to 15,000-mile interval.

For the following procedures, you will need a few more tools and supplies. However, the cost of these tools and supplies—as compared with the costs of major repairs—are relatively inexpensive. You may elect to have these maintenance jobs done for you by a professional technician. Even if you have the work done by someone else, you should understand the benefits of performing these preventive maintenance jobs.

PCV Valve

The PCV (for positive crankcase ventilation) valve is an important part of the engine even though it may not look it. The PCV valve was the first emission system component to be put into vehicles and dates back to the 1960s. Although the PCV valve is a simple device, it reduces the pressure inside the engine by directing the gases in the crankcase to the air intake system. Since the air is contaminated with oil, gasoline, and other engine residue, it will, in time, affect the operation of the PCV by clogging it.

It is good maintenance to check the PCV valve every 30,000 miles and even replace it every 60,000 miles. It is a simple and inexpensive maintenance task that will help prevent poor engine performance, rough idling, and stalling.

Most PCV valves are located on or near the rocker cover. Pull the PCV valve out of the valve cover and pull off the hose (Figure 11–22). Shake the valve and it should rattle. If it doesn't rattle, it needs to be replaced.

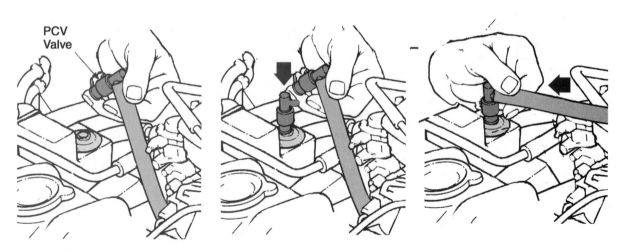

Figure 11–22: Removal and replacement of the PCV valve.

Another test is to pull the valve out of the valve cover, start the engine, and put your thumb over the hole in the end. If you hear a soft click, the PCV valve is functioning properly. If there is no vacuum at the hole, either the hose is bad or the valve is clogged. To see if the hose is bad, pull the PCV valve off and put your thumb over the end. If there is vacuum, replace the valve; otherwise, replace the hose. Be sure to check both the hose and the PCV valve. Be sure to purchase the proper PCV valve for your car. It is designed to allow a specific flow for a given engine size, and the wrong PCV valve could lead to drivability problems.

Flush Cooling System; Refill with Fresh Coolant

Vehicles using two-year antifreeze or green color antifreeze will need to have the cooling system drained, flushed, and refilled with fresh coolant every two years. However, with the increased use of the five-year antifreezes this service may be scheduled at the 100,000-mile interval or five years, whichever one comes first. Check the type of coolant used in your vehicle to determine the proper maintenance schedule.

Over time, the chemicals in antifreeze that prevent rust and corrosion become exhausted, or used up. To prevent damage to the cooling system and engine parts, flush the cooling system once a year and refill it with a fresh mixture of antifreeze and water. In cold climates, the best time to flush the cooling system is in the fall, before cold weather arrives. In hot climates, the best time to flush out and refill the system is in the spring, before the hot summer weather. Remember, a proper mixture of antifreeze and water also helps to prevent boiling and the severe engine damage that it can cause.

The amount of antifreeze to be mixed with water for your cooling system varies with the outside temperatures you expect to experience. For most climates in the United States, a 50/50 mixture—half antifreeze and half water—is sufficient. The freezing point of this solution is –34 degrees Fahrenheit (–36.7 degrees Centigrade).

If you expect colder temperatures, you add proportionately more antifreeze to lower the freezing point of the mixture. Charts are printed on antifreeze containers that indicate what proportions of antifreeze and water should be used for expected outside temperatures. Consult your owner's manual or a service manual for your car to find out the total capacity of the cooling system. Then you can buy the amount of antifreeze you need to make the correct mixture. The use of 100% antifreeze is not recommended for any climate, including Antarctica. In extremely cold temperatures, pure antifreeze can gel. In extremely hot temperatures, pure antifreeze can boil over and leave deposits in the engine.

If you have a garden hose, a knife, a screwdriver, and a pair of pliers, you can flush your own cooling system and save some money. You also will need sufficient antifreeze to make the correct coolant mixture for the temperatures you expect. For example, if you know that your car's cooling system capacity is 16 quarts (4 gallons), and you want to make a 50/50 solution, you need to buy 2 gallons of antifreeze.

The easiest way to flush the entire cooling system, including the radiator, engine, and heater system, is to use a flushing and filling kit, sold at auto parts stores. These kits typically cost less than $10.

The kit includes several plastic fittings of different sizes, one of which will be inserted permanently into a heater hose. A garden hose attaches to the flushing T connector with the use of a flushing adapter (Figure 11–23). The flushing and filling kit allows you to back flush the entire cooling system. Backflushing is a procedure in which the coolant flow is reversed from its normal direction. This reversal of flow helps to flush away contaminants that have built up in the cooling system passages. During backflushing, the old coolant and contaminants are forced upward and out through the radiator neck.

These inexpensive kits will enable you to do a complete backflushing of the heater core, engine block, cylinder head, and radiator. Some people try to flush the system by forcing water through the radiator neck with a gar-

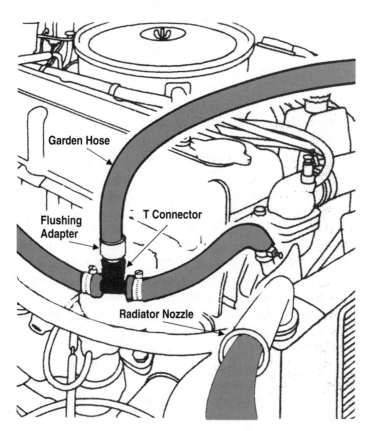

Figure 11–23: For backflushing, a garden hose is attached to a T-connector installed in the heater hose to force old coolant out of the radiator neck.

den hose. This practice doesn't flush any of the other parts of the system and doesn't even do a good job of flushing the radiator.

First, read the instructions on the flushing and filling kit package. For maximum safety, flush the cooling system when the engine is cold. Do not begin flushing on a hot engine. The cold water can do serious engine damage to a hot engine. Be very careful when using a garden hose to flush your cooling system. A typical home outdoor faucet will produce 40 psi of water pressure or more. Modern cooling systems are designed to operate at 15 psi. You do *not* want to turn the faucet on full blast. A steady, constant stream of water will do a satisfactory cleaning job, without the need to overpressurize the system. Doing so may damage the cooling system.

The procedure is fairly simple and easy to do. The flushing T-connector will usually be installed in the heater hose that leads to the cylinder head. You are going to cut it, preferably at a point slightly lower than the opening of the radiator neck. Cutting it there will help to get all the air bubbles out when you refill the system with your fresh coolant mixture. Air bubbles that remain in the cooling system can cause the engine to overheat and boil.

Attach the garden hose to the flushing T-connector by turning and threading the plastic ring of the flushing adapter onto the garden hose end and the flushing T-connector. Remove the radiator cap and push the plastic nozzle that came with the kit down into the radiator neck until it pops tightly into place. Then position the nozzle to direct the flow of liquid that will rise through the radiator neck. If you have a small radiator neck opening, the nozzle may not fit. If you have a pressurized expansion tank, remove the pressure cap. Continue flushing until the water that pours from the radiator is clean. When you are finished, disconnect the garden hose and install the cap. Drain the radiator once again. Do not run the vehicle while flushing the system. Even though running the engine would likely do a better job, the cold tap water running from your home may cause aluminum engine components to warp as the engine heats up. When only clear water flows from the nozzle, shut off the faucet and remove the garden hose and adapter from the flushing T-connector.

Typically, a drain is located on the outlet tank of the radiator, near the bottom. In some cases, you may have to get under the car to locate the drain. Open the drain by turning the handle counterclockwise. If it is extremely tight, you may need a pair of pliers to loosen it, gently. When it is loose, use your fingers to open the drain fully. Allow the water to drain until it stops.

Some cars don't have a radiator drain. If yours doesn't, loosen the metal clamp on your lower radiator hose at the radiator. Twist the hose to break it loose, and then pull the hose off the connector pipe to drain the radiator. After the liquid has finished draining, close the drain tightly—or replace the lower radiator hose and tighten the clamp. Be aware that draining the liquid this way does not remove all of the liquid from the engine. A considerable amount of clean water remains in the lower part of the engine block below the level of the coolant pump.

Add the proper amount of coolant. Check the owner's manual to determine the total capacity of the car's cooling system. Add the proper proportion of antifreeze to create a coolant mixture suitable for your climate. For example, if you are making a 50/50 mixture, you will add an amount of antifreeze that equals half the total capacity of the system. Add this amount of antifreeze to the cooling system and finish filling the system with water. After flushing the cooling system, some pockets of water will remain in the cooling system and this remaining water will dilute a 50/50 mix. Be sure to run the engine long enough to circulate and mix the coolant. Check the strength of the coolant to verify that you have a proper mix of antifreeze and water after the engine cools.

On some cars, the heater core is positioned higher in the cooling system than the radiator neck. Thus, when you refill the system up to the radiator neck, a large air pocket remains in the heater core. This air pocket can cause severe overheating and engine damage. While you are running the engine to mix the coolant, check the heater to see if it is delivering heat. If it is delivering heat, you have coolant flow through the heater core.

A number of cars use a bleeder screw to help eliminate air from the cooling system. Loosening the bleeder screw on top of the engine cooling system while the system is being filled will usually allow the air to escape. Close the screw when a steady stream of coolant runs from the bleeder screw.

To remove air pockets from the cooling systems, use ramps or a jack and jack stands to raise the front end of the car safely, as described previously in this chapter. Then, with the radiator cap off, start the engine and allow it to run for five minutes in PARK or NEUTRAL. Make sure the heater controls are set to provide full heat so that the heater control valve is open. If the heater is providing heat, coolant is flowing through, and all the air should be expelled from the heater core.

After five minutes, shut off the engine. Recheck the coolant level of the radiator. If a large air pocket has been expelled, you will need to add additional coolant to the system. Then replace the radiator cap securely.

If the engine overheats after you have flushed the cooling system, it is likely there is still an air pocket or bubble that is preventing coolant from flowing properly in the system. Recheck to make sure you have followed all the procedures for flushing and removing air in the cooling system.

To clean the coolant recovery container thoroughly, it must be removed from the engine compartment. Typically, bolts or screws hold the container securely in place. Remove the bolts or screws to allow the container to be lifted out. Remove the rubber or plastic tubing that connects to the radiator overflow pipe and remove the container from the engine compartment.

Scrub the inside of the container with a baby bottlebrush and household cleanser to remove rust and scum deposits. Thoroughly rinse the tank. After cleaning, replace the container in the engine compartment, tighten the screws or bolts, and reconnect the tubing to the radiator neck.

Pour in the same mixture of antifreeze and water that you used in the cooling system. In other words, if you used a 50/50 mixture in the cooling system, fill the container halfway with water and the rest of the way with antifreeze. Fill the container to the appropriate FULL mark.

Put the cap back on the recovery container and start the engine. Let it run for about 10 to 15 minutes to heat up the system thoroughly. Then shut the engine off and check for leaks. It may be necessary to tighten the plastic clamp on the flushing T-connector or the clamps on the heater hose. Also check the radiator drain or lower radiator hose connection for leaks. If there are any leaks, tighten the drain faucet or radiator hose clamp.

REPLACE SPARK PLUGS, INSPECT IGNITION SYSTEM

Considerable savings are available for car owners in jobs such as changing spark plugs and inspecting the ignition system. At one time, car owners performed these tasks as part of a complete engine tune-up.

Today, however, the complexity and reliability of electronically controlled fuel, ignition, and emissions control systems makes the traditional tune-up obsolete. Rather, regular maintenance and replacement of spark plugs and spark plug wires is important. Included with this maintenance is an inspection of the ignition system. Vehicles today include an on-board computer and employ a sophisticated method of determining faults in the vehicle engine, powertrain, and emission systems. A simple failure to properly tighten the gas cap can cause the computer to illuminate the malfunction indicator lamp (MIL), and other system problems can cause the MIL to illuminate. Only knowledgeable, certified technicians with approved test and diagnostic equipment should attempt to diagnose and repair problems detected by the computer system. An improperly running fuel-delivery or ignition system may result in excessive tailpipe emissions or severe engine damage.

However, the tasks discussed here are well within your capabilities. The completion of these tasks will improve the reliability and economy of your car, and should help to reduce harmful exhaust emissions. Step-by-step instructions follow.

Changing Spark Plugs

With the use of electronic ignition and fuel injection systems, spark plugs may continue to function acceptably for tens of thousands of miles. Recommended spark plug change intervals may range to as much as 35,000 miles for standard spark plugs and up to 100,000 miles with platinum plugs. However, it is strongly recommended that you replace spark plugs on a regular basis according to the mileage and spark plug type.

Spark plugs can tell a lot about how the engine is operating. When replacing plugs, look at the old ones, and they can tell a tale. Even if the plugs look normal, it is best to replace them regularly with new ones. The cost of new spark plugs is relatively low. If there is a problem with engine operation, such as an engine miss, cylinder misfire, or noticeable bucking or jerking while accelerating, replace the spark plugs and the spark plug wires.

The dimensions of spark plugs for different engines vary. The thread diameter and *reach*, or length of the threads, may be different. The correct spark plug for your engine must have the proper thread diameter and the proper extension into the combustion chamber.

In addition, the spark plug must be of the correct heat range. The heat range is an indication of the temperature of the tip of the spark plug during normal engine operation. If the tip temperature is too cold, deposits will form and possibly hinder the spark from jumping the gap. If the tip temperature is too hot, the tip will burn down quickly, and the spark plug will become prematurely worn. In addition, a tip temperature that is too high may result in detonation and engine damage. Spark plugs are available in different heat ranges for special driving conditions.

Spark plugs may also have platinum electrodes. These spark plugs have numerous advantages: reduced gap erosion, fewer misfires, and better performance. In addition, they are more durable.

In general, you will want to buy the recommended spark plug, one that is suitable for a wide range of driving conditions. Consult your owner's manual or a service manual for the proper spark plug for your engine. Auto parts stores also have charts that list the proper spark plug for different makes and models, as well as equivalent spark plugs in different brands.

Changing spark plugs is a fairly simple job. Don't be overwhelmed by the jungle of wires and tubes under the hood. Simply locate the cables that lead from each spark plug to the distributor. Each spark plug cable is connected from a specific spark plug location to a specific location on the distributor cap. If you attach the wires incorrectly, the engine may run poorly or not at all.

To avoid making the connections incorrectly, you can refer to a spark plug wiring diagram in a service manual. Or, you can devise your own identification system. Take some masking tape and number the wires in a logical manner. When you remove these wires from the spark plugs, you must be able to identify each wire and the spark plug position to which it attaches. The best procedure to avoid confusion is to remove and replace spark plugs one at a time.

To remove a spark plug cable from the spark plug, grasp the boot, or thick part near the spark plug. Rotate the boot slightly on the spark plug. Then, holding the cable by the boot, pull it straight off the end of the spark plug. Avoid pulling on the cable itself, because the inner conductor is fragile and can be broken easily.

If the vehicle is equipped with coil-on-plug, the coil will need to be removed to gain access to the spark plug. The coil is typically held in position by a small bolt.

Remove Spark Plugs   On engines with aluminum heads, be sure to remove spark plugs only when the engine is cool. After the cable or coil is disconnected, you will need tools to unscrew and remove the spark plug. Typical tools needed to remove and replace spark plugs include:

- Ratchet handle, $3/8$-inch drive
- Extension bars, $3/8$-inch drive, 3 inches and 6 inches long
- Swivel joint, $3/8$-inch drive
- Spark plug socket, $5/8$- or $13/16$-inch size.

On most four-cylinder engines, access to the spark plugs is easy. You may need only the ratchet handle and spark plug socket. However, on some V-6 and older V-8 engines, you may need all of the tools listed. In addition, sometimes you will have to get at the plugs from the top, and at other times you may have to get underneath the car to remove them. You may even have to loosen the alternator or air conditioning units to gain access to the spark plugs. If this is necessary, loosen the bolts around the alternator or air conditioner and pivot the unit out of the way.

Examine Firing Tip Color and Condition   As you remove each plug, check it against the chart in Figure 11–24. Inspect the firing tip—the center electrode and nose of the ceramic insulator—for proper colors. In general, medium colors—tan, gray, or brown—indicate proper running conditions in the cylinder from which the plug was removed. Excessively light colors—white, yellow, or pink—indicate that a cylinder is running excessively hot. Dark colors—sooty black or chocolate brown—indicate that a cylinder is running excessively cold.

If the firing tip is oily or contaminated with black, greasy deposits, engine problems are indicated. Oiled plugs are a symptom of worn piston rings, worn cylinders, or worn valve guides. If the outside of the spark plug is oily, you have an engine oil leak unrelated to the condition of the spark plug tip.

After you examine each spark plug, keep the plugs in order so that you can identify any cylinder in which a problem may be developing. Knowing which cylinder may have a problem will be helpful to you in diagnosing running problems.

Before the new spark plugs are put in, the gap must be checked and adjusted if necessary. You will need a plug-gapping tool. Figure 11–25 shows a typical gapping procedure.

Adjust New Spark Plug Gap   The proper gap for a spark plug to be used in your engine is stated in thousandths of an inch. For example, 0.035

GAP BRIDGED

IDENTIFIED BY DEPOSIT BUILDUP CLOSING GAP BETWEEN ELECTRODES.
CAUSED BY OIL OR CARBON FOULING. REPLACE PLUG, OR, IF DEPOSITS ARE NOT EXCESSIVE, THE PLUG CAN BE CLEANED.

OIL FOULED

IDENTIFIED BY WET BLACK DEPOSITS ON THE INSULATOR SHELL BORE ELECTRODES.
CAUSED BY EXCESSIVE OIL ENTERING COMBUSTION CHAMBER THROUGH WORN RINGS AND PISTONS, EXCESSIVE CLEARANCE BETWEEN VALVE GUIDES AND STEMS, OR WORN OR LOOSE BEARINGS. REPLACE THE PLUG.

CARBON FOULED

IDENTIFIED BY BLACK, DRY FLUFFY CARBON DEPOSITS ON INSULATOR TIPS, EXPOSED SHELL SURFACES, AND ELECTRODES.
CAUSED BY TOO COLD A PLUG, WEAK IGNITION, DIRTY AIR CLEANER, DEFECTIVE FUEL PUMP, TOO RICH A FUEL MIXTURE, IMPROPERLY OPERATING HEAT RISER, OR EXCESSIVE IDLING. CAN BE CLEANED.

NORMAL

IDENTIFIED BY LIGHT TAN OR GRAY DEPOSITS ON THE FIRING TIP.

PREIGNITION

IDENTIFIED BY MELTED ELECTRODES AND POSSIBLY BLISTERED INSULATOR. METALIC DEPOSITS ON INSULATOR INDICATE ENGINE DAMAGE.
CAUSED BY WRONG TYPE OF FUEL, INCORRECT IGNITION TIMING OR ADVANCE, TOO HOT A PLUG, BURNT VALVES, OR ENGINE OVERHEATING. REPLACE THE PLUG.

OVERHEATING

IDENTIFIED BY A WHITE OR LIGHT GRAY INSULATOR WITH SMALL BLACK OR GRAY BROWN SPOTS AND WITH BLUISH-BURNT APPEARANCE OF ELECTRODES.
CAUSED BY ENGINE OVERHEATING, WRONG TYPE OF FUEL, LOOSE SPARK PLUGS, TOO HOT A PLUG, LOW FUEL PUMP PRESSURE, OR INCORRECT IGNITION TIMING. REPLACE THE PLUG.

FUSED SPOT DEPOSIT

IDENTIFIED BY MELTED OR SPOTTY DEPOSITS RESEMBLING BUBBLES OR BLISTERS.
CAUSED BY SUDDEN ACCELERATION. CAN BE CLEANED IF NOT EXCESSIVE; OTHERWISE REPLACE PLUG.

Figure 11–24: Chart illustrating normal and abnormal spark plug tip conditions and recommendations for service.

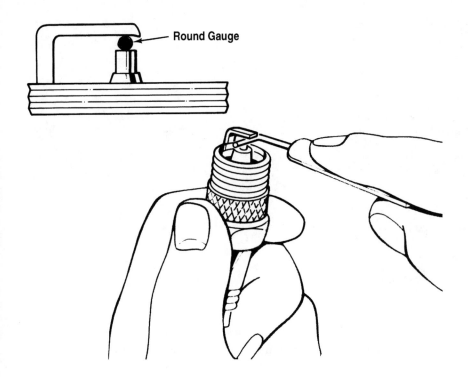

Figure 11–25: A round, or wire, gauge is used to gap spark plugs properly.

is 35 thousandths of an inch. Select a wire gauge of the recommended size for your spark plugs.

Gently attempt to slide the wire gauge between the center and side electrodes, as shown in Figure 11–25. If the gauge will not slide easily between the electrodes, the gap is too small. If the gauge can be moved up and down between the electrodes, the gap is too large. You should feel a light drag, or slight resistance, as you slide the gauge between the electrodes. If the gap needs adjustment, bend the side electrode upward or downward slightly. A tool on the gap gauge is used to bend the side electrode. It may take several tries until you feel the correct light drag on the gauge as you slide it gently between the electrodes.

Install New Spark Plug   Apply a small amount of antiseize thread compound to the spark plug threads (Figure 11–26). This application is recommended for cast iron cylinder heads and is mandatory for aluminum cylinder heads. The antiseize compound will prevent galling and seizing of the threads and make the spark plug easier to remove the next time this maintenance is needed. Since spark plugs last a long time before they need to be replaced, it is more likely that the threads in the cylinder heads will suffer this damage if no antiseize is used.

When the new plug is gapped properly, screw it into the threaded hole by hand, as far as it will go. Do not use a wrench to start the plug into the hole. An old spark plug wire boot makes an inexpensive plug starter tool.

Figure 11–26: A tune-up kit containing silicone grease and antiseize. Antiseize is being applied to a spark plug prior to installation.

After the plug is threaded into the hole, use a torque wrench to tighten each spark plug to the manufacturer's specifications. If you don't have a torque wrench, there is a general rule you can follow: Tighten new spark plugs that have ringlike gaskets on their base one-quarter turn past the point at which they begin to feel snug. Tighten new taper-seat plugs without gaskets only one-sixteenth turn past snug. This guideline will ensure that the plug is neither too loose nor too tight. Most spark plugs are of the taper-seat design.

Push the cable boot firmly onto the spark plug or reinstall the ignition coil. Repeat the procedures for each spark plug. Keep the plugs in order to aid in diagnosing problems.

Spark Plug Wire Replacement Spark plug wires should be replaced one at a time between the distributor cap or ignition coils and the spark plugs to avoid misconnecting the wires. If you misconnect the wires, the engine will not run. Refer to a service manual for the correct spark plug wire arrangement.

You should be aware that spark plug wires could be defective without showing any physical signs or damage. The conductor in a spark plug cable can increase in resistance or become "open," or completely nonconductive. If you find any cracking of insulation or evidence of high-voltage deterioration on the wires, replace them. Also examine the spark plug wires for evidence of swelling. An oil leak, even a small one, can cause the insulation of the spark plug wire to swell and damage the cable in a short period of time. The source of the oil leak must be fixed, or the new wires will be ruined as well.

It's a good idea to replace spark plug cables (including the coil wire) every 50,000 miles or 4 years–even if nothing seems wrong with them. If problems with engine operation occur, such as an engine miss, cylinder misfire, or noticeable bucking or jerking while accelerating, replace the spark plug wires and the spark plugs.

When replacing spark plug wires it is a good idea to use a small amount of silicone grease on each boot end. This grease prevents the rubber boot from sticking to the spark plug, makes removal much easier in the future, and acts as a moisture barrier. (Don't confuse silicone grease with silicone sealer, as they are two different products.)

Inspecting the Distributor

The distributor cap, rotor, and spark plug wires handle and distribute high voltage (20,000 to 50,000 volts) from the ignition coil to the individual spark plugs. A defective cap or rotor can cause the engine to misfire or may prevent it from running. The inside of the distributor cap and the rotor should be inspected for cracking, carbon tracking, or evidence of high-voltage deterioration. The distributor cap and rotor should be replaced along with the spark plug wires and spark plugs.

Some caps are attached with flat metal clips that hold the cap through spring tension. Screws that thread into holes in the distributor base hold down some distributor caps. To remove a cap held by screws, just loosen the screws until the cap can be lifted off. Another type of distributor cap has hooks that swing under the metal base of the distributors to hold it in place.

After you remove a distributor cap, look for a specially designed alignment notch or tab on the cap or base. These tabs and notches must be re-aligned to replace the cap correctly on the distributor body. Examine the inside of the distributor cap carefully. You will notice metal terminals around the inside edge at the top of the cap. A fragile carbon terminal is located in the center of the cap.

The spark plug cables must be replaced in a new cap in the same relative locations as they occupied on the old cap. Use the alignment tabs and notches on the old and new caps to judge where the cables should go. Transfer the wires from the old cap to the new cap one at a time. If you misconnect the wires, the engine will not run.

Examine the plastic of the distributor cap carefully. Even the smallest hairline cracks in a distributor cap will allow moisture to collect. The moisture can cause the high-voltage electricity to travel to ground instead of to a spark plug gap. A cracked distributor cap should be replaced as soon as possible.

Another problem that you may see inside the cap is carbon tracking. Carbon tracking produces a fine, sooty line of carbon along the inside surface of the distributor cap plastic. Any evidence of carbon tracking means that you should replace the distributor cap.

Also look for evidence of high-voltage deterioration of the plastic cap. High voltage can cause a discoloration in the plastic. Sometimes this discoloration looks whitish, and the plastic becomes roughened. At other times, it takes on a rainbow-colored shine. The discoloration or roughened area is evidence that high-voltage electricity has found a path to ground through the plastic. If you find any evidence of unusual coloration on the plastic, it's a good idea to replace the distributor cap.

Finally, inspect the metal terminals inside the cap. Examine them for deep grooving. Deep grooving indicates that the gap, or space, between the tip of the rotor and the metal terminals has widened because the metal has been removed through high-voltage deterioration. The cap should be replaced.

With the cap off, you will see a distributor rotor in the lower part of the distributor. The distributor rotor is a plastic part with a metal tip. High-voltage electricity passes through the tip from the carbon center terminal to the metal terminals of the distributor cap.

Some rotors are held on with screws. Loosen the screws or remove the bolt to lift the rotor from the distributor. On other distributors, the rotor simply fits snugly over the shaft on which it turns. Lift the rotor firmly but gently off the shaft.

When you pull the rotor off, turn it over. You will notice that there are tabs or grooves in the central mounting hole. These tabs or grooves fit corresponding notches or flat spots on the distributor shaft. Notice how the rotor must be aligned to fit on the shaft.

Examine the rotor carefully for the same type of problems that you may have found inside the distributor cap. If you find any cracks or evidence of high-voltage deterioration (discoloration), replace the rotor. Be sure to check the underside of the rotor as well.

Metal rotor tips can become burnt by high voltage. If the tip is burnt, clean it. You can scrape it lightly or use sandpaper. However, some rotors have a special brownish anti-arcing coating on the tip. Do not mistake this for burning, and do not scrape or sand it off.

Your inspection of the distributor and spark plug cables is now complete. Align the tabs on the rotor and distributor shaft. Push the rotor back on the shaft. Replace any fasteners that held the rotor securely. If you have trouble fitting the rotor in place, don't force it. Remove the rotor and look at the alignment tabs. The rotor should fit easily on the shaft.

Next, align the placement tabs on the distributor cap and base and put the cap back on. If you have trouble positioning the cap just right, try rotating it slightly until it drops into place. Fasten the cap in place with the clips, screws, or hooks.

Automatic Transmission Service

For maximum reliability and a long service life, the fluid and filter of an automatic transmission should be changed on a regular basis. The interval for automatic transmission service is normally every two years or approximately 25,000 miles. However, if the vehicle is used for trailer towing, continuous stop-and-start driving, or off-road use, this interval should be cut in half.

Regular monitoring of the automatic transmission fluid (ATF), described earlier in this chapter, may alert you to potential problems. Regular fluid and filter changes provide further insurance that the transmission will continue to perform well.

Most automatic transmissions typically have no drain plug like that of the engine oil pan. Thus, to change automatic transmission fluid on these vehicles, the transmission oil pan must be removed. There are a few that do provide a drain plug and use an external filter similar to the filter used on the engine for the lubrication system. On cars equipped with an external filter and drain plug, the procedure to drain and refill is the same as it is for changing engine oil. In most other cases, the filter is located inside the transmission, just above the oil pan area. To do this job, you will need to raise the vehicle. Use a jack to raise the front of the vehicle, and place jack stands under the recommended frame or chassis support points, or drive the car onto a set of ramps to raise the front end. If the pan is removed suddenly, all of the ATF will come flooding out. However, there is a technique that will allow you to drain the oil slowly, in a controlled manner. The following procedure will provide the general steps for removing the pan and replacing the internal filter.

CAUTION

Never work under a vehicle supported only by a jack. Set the parking brake, place the transmission selector lever in PARK, and block the rear wheels securely before you begin.

Drain ATF

To drain the oil slowly, start by removing the row of bolts on the end of the transmission that is tilting toward the ground. Leave the bolts on the other three edges in place; position a drain pan under this end. Then, slowly begin loosening the bolts on either side, closest to the removed bolts, one at a time. With this technique, the pan can be tilted downward slowly.

As the pan tilts and separates from the bottom of the transmission, the oil will begin to come out. It may be necessary to use a putty knife or other tool to break the pan free from the gasket and allow it to begin tilting downward.

As the oil begins to flow into the drain pan, gradually loosen more bolts along the two edges to allow the pan to tilt farther. Finally, remove the last row of bolts and remove the transmission pan.

Inspect Interior of Pan and Transmission   Inspect the inside of the transmission pan for any thick, blackish or brownish deposits or metallic particles. These deposits are the residue from damaged clutches and bands within the transmission. A dark film or small amount of deposits can be considered normal wear. However, if there are large amounts of fiber deposits or metal particles, the transmission probably needs to be overhauled.

Clean Pan and Gasket Surfaces   Clean the inside of the pan thoroughly. Scrape off any gasket residue from the pan and the transmission housing above it with the putty knife. Be careful not to gouge the soft aluminum of the transmission housing when you scrape the gasket surface.

Replace Filter or Clean Screen   As you look up into the bottom of the transmission, you will notice a screen or filter, which is held in place with screws, bolts, or metal clips (Figure 11–27). Some screens can be cleaned with solvent; all filters must be replaced. Clean the screen or replace the filter.

Inspect Pan Flange   When you install the new pan gasket, pay particular attention to the flange on the pan—the part that comes in direct contact with the gasket. Any time a gasket is tightened down, the metal is distorted. This metal will protrude through the gasket and hit the surface on the other side. Thus, the sealing qualities are lost.

Each time you take a pan down, the bolt holes should be checked for indentations. Before any bolts are put in place, place the pan on a flat surface and carefully flatten the areas around the bolt holes with a hammer.

Install New Gasket   Some vehicles have a reuseable pan gasket. Simply inspect the gasket and reuse it if it is in good condition. If the gasket is not in good condition or the gasket is not a reuseable type, replace the gasket. Do not put any sealing material on the new pan gasket. Place the gasket carefully on top of the pan. Insert two bolts through holes near opposite corners of the pan. Lift the pan into place and turn the bolts with your fingers until they begin to thread into the transmission housing. Do not tighten the bolts.

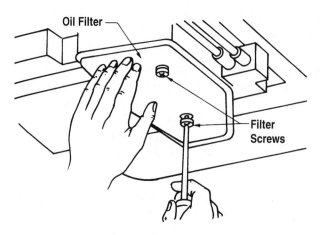

Figure 11–27: **Replace the filter and oil at the recommended intervals. Always replace the filter and gasket in the correct and proper positions.**

Insert the rest of the bolts into place carefully, lining up the holes in the gasket so that the bolts pass through the holes. Tighten the bolts evenly in a crisscross pattern, a little at a time, to the manufacturer's recommended torque specification.

In general, 4 to 6 quarts of ATF will be needed to refill the transmission. Some fluid remains in the converter, unless it can be drained separately. If you want to replace the fluid, refer to a service manual for your particular vehicle. Few modern automobiles have drain plugs on the converter.

Check the owner's manual for the fluid capacity of the transmission. Add a portion of that amount to the transmission. Start the engine and allow time for the transmission to warm up. Then check the fluid level and add additional fluid until it reaches the correct level on the dipstick.

Change Hoses

Hoses may last more than 60,000 miles. However, it takes about five years for the average car owner to drive 60,000 miles. Old, dried, and cracked hoses may rupture at any time. Thus, it is a good policy to change all important hoses at this interval, including the upper and lower radiator hoses, heater hoses, and vacuum hoses.

If you have followed the normal maintenance procedure outlined earlier, the chances are that many of your hoses have been replaced before this time. If any hoses have not been replaced, it is now time to do so.

If you have spring-type clamps, it is recommended that you replace them with the worm-drive screw clamps as shown in Figure 11–28. For all practical purposes, these stainless steel clamps can be tightened and loosened indefinitely without losing their clamping capacity.

To replace cooling system hoses, the radiator should be drained. Thus, it may be convenient to flush the cooling system, drain it, and replace the radiator and heater hoses.

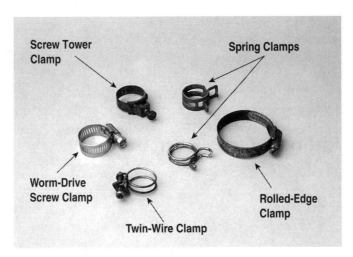

Figure 11–28: Whenever you replace hoses, replace other types of clamps with the worm-drive screw clamps to prevent leaks

Vacuum hoses should be inspected and replaced if they show signs of cracking or deterioration. Make sure you change vacuum hoses one at a time to avoid misconnections and related problems.

Change Belts

The normal procedure is to replace belts when they show signs of cracking or deterioration. However, if your original belts still are intact after 60,000 miles or five years of driving, they should be replaced.

To remove the belts, the units that they drive and an idler pulley or a belt tensioner must be loosened and moved so that the belts are no longer stretched tight. It can take a novice many hours to change all the belts on an engine. Thus, you may want to have this job done by a professional. Most cars have a single multigroove **serpentine belt** that may drive most or all accessories on the engine.

To replace a belt that is close to the engine, all belts in front of it must first be removed. Thus, it is a wise idea to install all new belts at one time. Belts are replaced in the reverse order of removal. In other words, the last belt to come off goes back on first.

Belts often are difficult to position around pulleys. Here is a tip to help you replace belts more easily: Always position the last portion of the belt over the largest pulley.

Multigroove belts almost always use a self-tensioner to maintain proper tension on the belt. The tensioner will need to be rotated to allow the belt to be removed. Be sure to make a drawing of how the belt is routed. Sometimes a belt routing diagram is found under the hood on a sticker. Verify that the routing is correct and that you are familiar with the proper routing.

Inspect the pulleys for any nicks or other damage and correct if needed. Check the pulley grooves for any debris and clean if necessary. After the new belt has been properly routed, release the tensioner or allow it to return to its position against the belt. Make sure that the belt is properly positioned on the pulleys.

If the belt does not use a tensioner, use a pry bar to push on the unit for adjustment. Make sure the belt is tensioned correctly—there should be no more than a half-inch of movement. After about 100 miles of driving, a new belt stretches and must be retightened. After this first stretching, the length of the belt remains relatively stable over the rest of its life.

Change Camshaft Belt

As described in Chapter 3, many engines have belt-driven overhead camshaft valvetrains. The timing belt is enclosed within a protective cover at the front end of the engine. Thus, there is no easy way to inspect or change the belt. It must be replaced at the recommended interval.

The timing belts have teeth to engage the crankshaft and camshaft sprocket wheels, and are reinforced with steel or fiberglass strands. If the belt breaks, the engine cannot run. Worse, if the belt breaks when a valve is in the downward position and a piston is moving to the top of an upward stroke, severe engine damage can result.

Although timing belts may last longer than 50,000 miles or five years, it is in your best interest to have them changed at approximately this interval. Because valve timing must be correct for the engine to run, the positioning of the camshaft and crankshaft sprocket wheels must be exact when a new belt is installed. It is best to leave this job to a professional technician.

Check Coolant Pump

Coolant pumps often fail unexpectedly. However, there are signs that can indicate when a pump is nearing the end of its life. Loosen the belt that turns the coolant pump and attempt to rock the pump pulley. Any noticeable movement indicates that the bearings in the water pump are beginning to fail.

Another way to check a coolant pump is to hold a small mirror near the underside of the pump. All pumps have what is called a "weep hole" on the underside. If you find seeping coolant or rust deposits, the coolant pump seals are beginning to fail. Think about having the pump changed immediately.

On some vehicles, the coolant pump is driven by the camshaft drive belt described in the preceding section. A cover encloses the belt and the coolant pump, so that there is no easy way to check for wear or leakage. However, as mentioned, the camshaft timing belt should be changed by a certified technician at approximately 60,000 miles. When the timing belt is changed, the pump can be checked or replaced.

INTERIOR CARE

It's important to keep your car's interior clean and well maintained. Regular vacuuming removes much of the dirt that can contribute to wear and tear of the carpet and fabric and can reduce the likelihood of stains. Always vacuum the car's upholstery and floor mats each time you wash your car. Cars with a leather interior should have the leather periodically cleaned. A leather treatment should also be applied to keep the leather conditioned for long life.

Spills and other mishaps should be cleaned up as soon as possible to minimize the possibility of a stain becoming permanent. Use a carpet or upholstery spot cleaner according to the manufacturer's recommendations. Test all cleaners for compatibility in an inconspicuous spot first before using. After cleaning, protect your car's fabric upholstery and carpeting with an application of a fabric guard. This type of product repels both water and stains.

Here are tips for interior care:

- Start in the front of the car and work toward the rear.
- Always remember to clean out your glove compartment and consoles of excess documents and materials. Keep your owner's manual and insurance cards.
- Clean the dash and the vents.
- When cleaning the interior, always pull out your floor mats and vacuum them, and also vacuum the inside of your car regularly.
- If you notice a spot on your carpet or fabric, spot clean it with a gentle spot cleaner.
- Clean the doors, moldings, weather stripping, and door area.
- Clean and protect any leather.
- Apply protectants, making sure that you use and follow manufacturer's instructions.
- Clean the glass last. Use a gentle glass cleaner, preferably alcohol based.

EXTERIOR CARE

The most obvious sign of a well-maintained car is a clean and shiny exterior. It is a common misconception that clear coats serve as a coat of armor that needs little or no care at all. What most people fail to realize is that clear coat on your car is basically unpigmented paint. You should wash your car regularly. A common mistake people make when washing their cars is not using a product that's specifically formulated for automotive paint. Many times people will use dishwashing detergent reasoning that it does an excellent job in the kitchen sink. Dishwashing detergent is too harsh for automotive finishes, especially clear coats. It works so well that it can strip the wax from the paint and even attack the paint itself. You will need to choose a car-wash product that is made specifically for use on your car's paint.

Start by rinsing any loose dirt or debris off the body of the car. Wash your car from the top and work down. The wheels and tires should be washed last. Spray water up inside the wheel wells and under the rocker panels. Dirt, salt, and debris can get trapped and cause rusting. Here are a few washing tips.

- Always park the vehicle in the shade. Allow the body to cool to air temperature.
- Hose off loose debris with a spray of water.
- Mix the car-wash detergent according to the product's directions.

- Apply the washing solution with a wash mitt or an open-cell foam sponge.
- Wash from the top down, and in sections. Wash the tires and rims last.
- Rinse often before the detergent dries, working section to section and keeping the car wet.
- Final rinse with a sheeting action to minimize water spots.
- Apply a quality wax following the product's directions.

Cleaning and protecting your car on a regular basis will make it easier to keep clean and provides long life. A clean car is more comfortable to drive in, and it also increases the value of your car.

The tasks described in this chapter are the elements of a regular preventive maintenance program. Chapter 12 presents an overview of some of the most common and popular do-it-yourself repairs. These repairs typically require a greater degree of commitment to working on your own car.

1. What are some of the factors that affect the maintenance needed for a car?

2. What are the additional concerns for a cold climate?

3. What are the additional concerns for a hot climate?

4. List the different driving types.

5. What are the additional concerns for each of the driving types?

6. What checks should be done on a daily basis?

7. What checks should be done on a weekly basis?

8. What maintenance should be done at 3,000- to 7,500-mile intervals?

9. Explain the general steps in changing oil.

10. Why should tires be rotated?

11. Why might some tires not be able to be rotated to the opposite side of the vehicle?

12. What maintenance should be done yearly or at 12,000- to 15,000-mile intervals?

13. What are the different types of antifreeze? What are their maintenance intervals?

14. What is the major maintenance for the cooling system?

15. Why is it important to flush the cooling system?

16. What is the major maintenance for the ignition system?

17. What are some of the differences in spark plugs? What are their maintenance intervals?

18. What should be applied to the spark plug before replacement? Why?

19. Describe the necessary maintenance for the automatic transmission.

20. What is the life expectancy of hoses and belts?

21. Why should a camshaft belt be replaced?

22. What are some signs the coolant pump is nearing the end of its service life?

23. List some interior cleaning steps.

24. List some exterior cleaning steps.

25. Why is a clean car important?

A LOOK AT REPAIRS

After reading this chapter, the reader should be able to:

- List some common repairs a car owner might make.
- List the steps for changing a tire.
- List the steps for changing an alternator.
- List the steps for changing valve cover gaskets.
- List the steps for changing a thermostat.
- Explain some of the factors to look for when deciding on a repair facility.
- List some tips when deciding on a repair facility.

RECOGNIZING REPAIR NEEDS

Earlier sections of this book have been aimed at increasing your awareness and understanding of the major parts of your vehicle. In the course of your checks and maintenance tasks, you may have noticed problems that require repair. Vehicles today are complex and sophisticated; however, there are some common repairs that you may wish to do yourself. Many other repairs will require you to take the vehicle to a repair center for the services of a technician. This chapter discusses repairs that you may be able to perform. Also, it discusses what to do if you decide to seek a technician to perform your repairs.

The instructions in this chapter are intended as general guides for the work involved. If you feel you need more detailed information, you should obtain a service manual for your car. Service manuals are available through auto dealerships, auto parts stores, and bookstores. These manuals present specific repair information that may be needed to complete the repair you wish to do.

SHOULD YOU DO IT YOURSELF?

When it comes to repairs, the questions often arise: Should you do it yourself? Is it safe? Can you handle it? Before you undertake any repair job, the answer to all three questions should be yes. However, it is not recommended that you attempt repairs involving safety components such as brake systems unless you are thoroughly familiar with the tools, procedures, and safety precautions to be used.

CHANGING TIRES

If you can change your own tire, you don't have to depend on others or wait for help. You can be on your way and also have the satisfaction of having done the job yourself.

Be Prepared

Individuals with limited strength might find changing tires a somewhat more difficult job than the more robust person. Thus, it might be wise to carry a few items to aid in the process of changing tires. For example, a 3-foot length of $1\frac{1}{2}$-inch diameter water pipe can be used as an extension for the lug wrench. This simple extension, available at any hardware or plumbing supply shop, will increase your leverage greatly when you loosen and tighten lug nuts with the lug wrench. If it is combined with a block of wood, the pipe can also act as a lever to help lift the tire.

As discussed previously, the wheels that will remain on the ground always should be blocked before you use a jack. Thus, it is also a good idea to carry two or more good-size wooden blocks or wheel chocks in your trunk. If you don't have any blocks in your trunk, try to find a couple of substantial rocks or other heavy objects to place on both sides of the tire.

At a minimum, block the front and back of the tire diagonally opposite the one you are going to change. If you have more blocks, block the other wheels as well. This precaution is especially important when the car is not on a perfectly level surface. Of course, you always should try to maneuver your vehicle to as level an area as you can find. Also, your working area should be as far from the flow of traffic as possible.

Follow These Steps

Figure 12–1 through Figure 12–5 illustrate a step-by-step procedure for changing a tire. Refer to the photographs as you read the following descriptions. Note that specific instructions on use of the jack and changing tires can be found in your owner's manual or on a sticker on the inside of the trunk lid.

Step 1. *Remove wheel covers.* Use the end of the lug wrench or a large screwdriver to pry the wheel cover from the wheel (Figure 12–1). Insert the tool and pry or twist at several locations around the cover to remove it.

Step 2. *Loosen wheel lugs.* Place the lug wrench on a wheel lug and turn it counterclockwise (Figure 12–2). If the lugs are extremely tight, try using your body weight to help turn the lug wrench: Brace yourself against the car so that you won't fall and, carefully, stand and bounce on the end of the lug wrench. If you have a length of pipe, place the pipe over the end of the lug wrench to add leverage. *Don't remove the lugs.* Simply loosen them one or two turns, or until they can be turned by hand.

Figure 12–1: Tire changing step 1: Block the diagonally opposite wheel, then remove the wheel cover of the tire to be changed.

Figure 12–2: Tire changing step 2: Loosen each wheel lug approximately one turn.

Figure 12–3: Tire changing step 3: Use the jack to raise the tire about an inch off the ground.

Step 3. *Raise car.* Refer to your owner's manual or the placard near the jack storage area or spare tire compartment for proper jack positioning and operation. Most cars have jacks that fit into cutouts under the rocker panels near the wheels. Typical locations for placing the jack are in front of the rear wheel and behind the front wheel. Operate the jack to raise the tire about an inch off the ground (Figure 12–3).

Step 4. *Remove wheel lugs and tire.* Brace your foot against the tire so that it cannot fall on you. Then loosen and remove the lugs (Figure 12–4). Remove the tire and place it away from the work area.

Step 5. *Mount spare tire.* Lift the wheel and align it on the studs, or threaded shafts, on the wheel hub. If you car has wheel bolts instead of

Figure 12–4: Tire changing step 4: Remove the wheel lugs and the tire.

nuts, align the tire with one of the holes on the wheel hub. Screw one wheel lug in by hand to hold the tire in place. Then replace the rest of the wheel nuts or bolts (Figure 12–5). Use the lug wrench to tighten the lugs snugly.

Step 6. *Lower the car and tighten wheel lugs.* Lower the jack until the tire rests on the ground. Use the techniques previously mentioned (using your body weight or a pipe) to tighten the lugs securely with the lug wrench (Figure 12–6).

Figure 12–5: Tire changing step 5: Position the spare wheel over the wheel studs. Screw the lugs on by hand and then tighten them with the lug wrench.

Figure 12–6: Tire changing step 6: Lower the tire onto the ground and remove the jack. Then tighten the wheel lugs securely with the wrench.

Figure 12–7: Be aware that the spare tires for most cars are smaller than normal tires and should not be driven more than 50 miles per hour, and only for an hour at a time.

Observe Precautions When Using a Temporary Spare

Temporary spare tires are not to be used as regular tires (Figure 12–7). They may only be driven at speeds up to 50 mph for a maximum of one hour. If you have not reached a location where you can have your regular tire repaired by this time, stop and allow at least 30 minutes for the spare to cool before you continue. Have the regular tire repaired and replaced on the car as soon as possible.

CHANGING THE ALTERNATOR

If a charging system check has revealed that your alternator needs replacement, you may want to try to remove and replace the unit yourself. Gaining access to the alternator may be your most difficult chore if many other parts are in the way.

Following is a general description for changing an alternator. Refer to a service manual to determine the specific procedures and precautions that are necessary to remove and replace the alternator in your car.

⚠ CAUTION

Before you start any electrical work, disconnect the battery negative (ground) cable to prevent sparks, shocks, and possible fires from accidental grounds or short circuits.

Step 1. *Remove the old alternator.* Remember to disconnect the battery, and then start by removing the belt on the alternator. Bolts attach the alternator to the engine (Figure 12–8). Loosen them with a wrench or socket. There is a set of wires that allow the alternator to provide power for

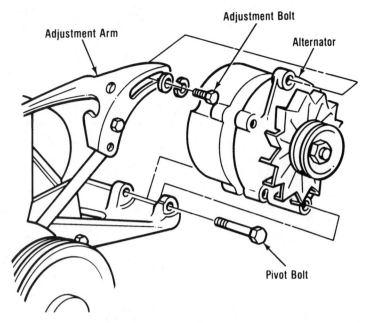

Adjustment Arm

Adjustment Bolt

Alternator

Pivot Bolt

Figure 12–8: Alternators are mounted with several bolts. Make sure the battery is disconnected and that all alternator bolts have been removed before removing the alternator.

the car and that control the alternator. Note the position of all wires and connectors on the alternator so that they can be replaced properly. You can use pieces of masking tape to mark each wire and its corresponding connection on the alternator. Remove bolts, nuts, screws, and electrical connectors and remove all connecting wires. The alternator may need to be partially removed and turned to gain access to the connectors.

If you try to remove the alternator and it won't move, double check to make sure you have removed all of the attaching bolts. Extra bolts that are difficult to see on first inspection may be used to hold some alternators in position. Look at the replacement alternator to see if there are some other possible places for bolts or brackets.

Step 2. *Install the new alternator.* Installing the new alternator is basically the reverse of removal. Align the new unit with the bolt holes on the engine. Thread the bolt in with your fingers to hold the unit roughly in position. Reconnect the wires in their proper locations. Tighten the bolts and replace the belt. Check the belt tension if necessary. Recheck the electrical connections to make sure they are correct.

CHANGING VALVE COVER GASKETS

If you notice oil leakage from the valve cover gasket, first try tightening the cover bolts. Clean the oil leaks from the engine. If the gasket continues to leak, it must be replaced. Figure 12–9 shows how one type of valve cover gasket fits between valve cover and cylinder head. Two valve covers are used on V-type engines.

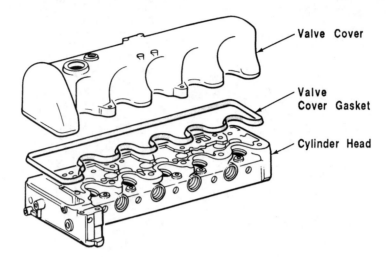

Valve Cover

Valve Cover Gasket

Cylinder Head

Figure 12–9: The valve cover gasket is positioned between the cylinder head and the valve cover.

Step 1. *Clear your access to the valve cover.* Move hoses, wires, spark plug cables, and other items away from the valve cover and attaching bolts. In some cases, you may have to disconnect and remove some or all of these parts that prevent access to the cover. Identify the items that you remove and their connection points with pieces of masking tape so they can be properly replaced.

Step 2. *Remove the valve cover bolts.* Start near the center and work outward in a crisscross pattern to prevent warping the cover. As you remove the bolts, notice if any of them are of different lengths. Make a note of where longer or shorter bolts are to be replaced.

Step 3. *Remove the valve cover and the old gasket.* After you have removed the valve cover, the gasket material must be removed from the cover and the sealing area on the cylinder head. Many cars have one-piece rubber gaskets that simply pull off. Other cars use cork materials or beads of silicone sealant as gaskets. Typically, cork gaskets and silicone sealant must be scraped from the cover and cylinder head with a gasket scraper or putty knife. It is important to prevent any of the gasket or sealant material from getting into the engine and oil supply. Place a paper towel or rag around the valvetrain parts to catch dirt and broken bits of gasket.

Step 4. *Examine the cover.* Some cars have cast aluminum valve covers while others may use plastic or composite materials. Examine these covers for warping by placing them on a flat surface. Use a droplight to check if the valve cover lies flat against the surface. If a valve cover is warped badly, it must be replaced.

Examine the bolt holes in a pressed sheet-metal valve cover. If the areas around the holes are dimpled inward, the new gasket will not seal properly. To remove dimpling from a sheet-metal cover, take a small block of metal and place it on the outer surface of the cover, over the bolt hole. With a ball peen hammer, carefully flatten the dimpled areas against the block of metal. This procedure will assure a good seal.

Step 5. *Use a sealant as recommended.* Some manufacturer's service manuals recommend using a sealant between a pressed sheet-metal valve cover and its gasket. If you use a sealer, use it on the valve cover side, just to hold the gasket in place. This sealant is only for convenience, so that the gasket doesn't slip. If you use a sealer on the engine side of the gasket, the gasket will adhere to the cylinder head and be extremely difficult to remove the next time.

There were concerns with older room-temperature vulcanizing (RTV) synthetic sealants that they could damage the oxygen sensor in the computer and fuel injection system. RTV sealants, sometimes called silicone sealer, are almost always safe for oxygen sensors. Look for labeling on the package that indicates "sensor safe" or "oxygen sensor safe." There are many sealants that are available. Use the sealants recommended by the vehicle manufacturer to seal valve covers and other engine gaskets.

For valve covers that are sealed with beads of sealant material, apply a small bead (approximately ⅛-inch wide) of sealant in the middle of the valve cover channel and around the inner edge of the bolt holes.

Step 6. *Install the new gasket and replace the valve cover.* Tighten the bolts carefully, starting from the center and working outward in a crisscross pattern. Use just a slight amount of pressure. Tighten the bolts evenly until you feel a little resistance; then stop. If the gasket material starts to bulge outward, you are tightening the bolts too much—back off a little.

CHANGING THE THERMOSTAT

Thermostats are usually changed for one of two reasons: (1) Either the engine is overheating or (2) it is running too cold and the heater is not putting out enough heat. However, on today's cars with intricate emissions-control systems that depend on engine temperature sensors, a malfunctioning thermostat can lead to many unusual running problems. Some of these symptoms include poor fuel mileage, slow engine warmup, and faster than normal idle. Thus, it is a good idea to change the thermostat approximately every three to five years, or whenever a heating, cooling, or engine-running problem is noticed.

Step 1. *Lower the coolant level.* When the engine is cold, drain approximately 1–2 gallons of coolant from the radiator drain. Draining the radiator will lower the coolant level in the system below the thermostat housing

CAUTION

Avoid removing or changing a thermostat when the engine is hot. Hot coolant can scald skin and cause painful physical injuries. For maximum safety, perform work on the cooling system only when the engine is cold.

area. If the coolant is less than a year old, you can collect it in a clean bucket or deep pan and replace it when you are finished.

To drain the coolant, look for a small drain faucet near the bottom of the radiator outlet tank. Open the drain faucet by turning it counterclockwise. Allow the coolant to drain into your pan or bucket.

Some radiators don't have drain faucets; thus, it is necessary to loosen and remove the lower hose from the radiator to drain the coolant. Unscrew the hose clamp screw several turns until the clamp is loose around the hose. Then grasp the hose and rotate it on the radiator outlet connection to loosen it. Pull the hose off the connection to drain the coolant. Special care should be taken to avoid damage when removing hoses from plastic or composition radiator tanks. Remove the radiator pressure cap to vent the top of the radiator and allow the coolant to flow out rapidly.

No special tool is needed to open the radiator cap; simply place the palm of one hand over the cap to grasp it. Press down on the back of this hand with your other hand. Then lean your body weight onto the cap and rotate it counterclockwise. The cap will rotate, then stop at a safety notch. Press a little harder and continue turning the cap counterclockwise. At the next stop, the cap can be lifted off.

Step 2. *Remove the thermostat.* Loosen both clamps on the upper radiator hose. Remove the upper hose leading into the thermostat housing. In most cases, there will be two bolts holding the thermostat housing to the engine block. Remove the bolts by loosening them equally, first on one side, then the other.

After the bolts have been removed, it may be necessary to use a putty knife to remove the thermostat housing from the cylinder head. Pry gently between the thermostat housing and the head to loosen the housing. As you remove the housing, note the position of the thermostat. The temperature-sensing bulb is at one end, and the opening mechanism is at the other end. The temperature-sensing bulb must always face the engine. Typically, the words TOWARD RADIATOR and an arrow are stamped into the opening mechanism end. Here again, if you identify which end is which as you remove the thermostat, you will have no problem installing the replacement correctly. A thermostat design is illustrated in Figure 12–10.

Step 3. *Clean the housing and the cylinder head area, where it mounts, of old gasket material and sealant.* Use a scraper, putty knife, and wire brush to remove these materials. Take care not to gouge or deeply scratch the housing cylinder head sealing surfaces.

Step 4. *Install the new thermostat.* When the sealing surfaces are clean, apply a small amount of nonhardening gasket sealer to both sides of the new thermostat housing gasket. Install the new thermostat with the arrow pointing toward the housing outlet. Place the gasket over the thermostat to help hold it in place as you replace the housing.

Insert the bolts through the housing and then align the bolts with the threaded holes in the cylinder head. Screw the bolts into place with your

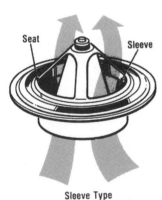

Seat   Sleeve

Sleeve Type

Figure 12–10: Thermostats use a wax pellet that melts at a specific temperature. When the wax melts, it expands and opens the sleeve in the thermostat allowing coolant flow.

fingers. Make sure that the thermostat remains in its recess as you tighten the bolts. Use a wrench or socket to tighten the bolts alternately until they are snug.

Step 5. *Replenish the coolant.* Reinstall the upper hose on the thermostat housing and radiator. Tighten the clamps securely. Close the radiator drain or replace the lower radiator hose and clamp securely. Refill the cooling system, using either the coolant you removed earlier or a fresh mixture of the proper proportions for your climate. Check for leaks at the upper and lower radiator hoses and around the thermostat housing. Tighten clamps or bolts as required to stop leaks.

Step 6. *Check for proper operation.* Reinstall the pressure cap on the radiator and start the engine. When it has reached normal operating temperature, again check for and correct leaks if necessary. After you have replaced a thermostat, there may be air bubbles in the cooling system. Let the engine run for a few minutes; then check to see if more coolant is needed. Also remember that it may be necessary to raise the front end of FWD cars to purge air from the cooling system after it has been drained (refer to Chapter 10 for details).

SEARCHING FOR REPAIR SERVICES

Searching for a repair facility can sometimes be confusing and time consuming, but it doesn't need to be. Know what to look for and your chances of finding it will be greater than if you jump in unaware. The best time to check for services is before they are needed. It might sound odd to look for something you don't need, but we do it all the time. We buy insurance beforehand to protect ourselves and we hope not to need it. We might examine schools, medical services, and other items before a move to another community. So, too, should the availability of repair services be checked out before they are needed.

New cars typically have a warranty. Generally, you need only find a dealership that has a service department that handles the same make of vehicle as you have. If your car is not covered under warranty, you have a number of options. Your choices will include dealerships, repair chains, approved repair facilities, and independent repair facilities.

Dealerships have the latest technical information, use factory original parts, and may offer a nationwide warranty on repairs. Repair chains typically have stores in various locations either in a region of the country or nationwide. They offer consistent repair facilities and up-to-date equipment and training of their technicians. Approved repair facilities are screened by a sanctioning organization to assure that strict repair standards and criteria are maintained. For example, AAA (the American Automobile Association) has perhaps the best-known network of approved automotive repair facilities. There are also independent repair facilities available. Some of these may specialize in a particular make of vehicle while others may specialize in a particular system, such as fuel injection or computer diagnostics.

There are a number of things to look for when choosing a repair facility (Figure 12–11). The facility should be clean and well lit. It need not be in a brand-new building, but it should present a professional image and show attention to detail. Amenities should be available for customer convenience, and waiting rooms should be clean and comfortable. Some facilities may offer pick-up and drop-off service for convenience. The equipment in the shop should be up-to-date, well organized, and appear well maintained. The technicians should have access to repair information either with service manuals or, more commonly, on computer.

Ask if the technicians are familiar with your type of vehicle and about their level of experience. Determine if the facility specializes in a particular area of repair, for example, transmissions, exhaust, or brakes. Check to see if certifications or qualifications are displayed. Technicians will usually be certified by a professional organization such as the National Institute for Automotive Service Excellence (ASE). ASE certifies technicians in the areas of engines, drivetrain, suspension and steering, brakes, electrical, heating and air conditioning, performance, and advanced engine specialist. It also certifies other professional positions including service consultant, collision repair/refinishing, parts specialist, medium and heavy truck, al-

Figure 12–11: Is the service consultant listening to your service needs? Is the facility well lit and clean? These are some of the things to look for when deciding on a repair facility.

ternate fuel, and many others. Dealership technicians may also be certified by the manufacturers on the specific systems used on their vehicles.

Ask around to see if other people have had service work done at the facility you are considering. The facility should be courteous and provide good communication to the customer regarding repairs. The vehicle should have been delivered when promised and at the quoted price for repairs performed. You may also want to check with the State Attorney General and the local Better Business Bureau for any complaints and reports of unresolved disputes with customers. Also find out how long the business has been in operation.

Another item to research is the type of warranty the facility offers. Warranties will vary greatly, so it's best to check out what is available. Some offer a warranty that expires as soon as you drive out of the parking lot, while others may offer a "parts-only" warranty. Look for a 12,000-mile or a 12-month warranty. Check to see if it is honored in other areas of the country if you travel. Many dealerships, repair chains, and approved repair centers can provide guarantees on repairs.

When choosing a repair facility, the deciding factor is often your comfort level with the employees. They should be courteous and able to answer your questions. Most people have good instincts. You will decide within a few minutes if you feel comfortable dealing with a repair facility. If they perform adequate repairs, you still need to deal with your service consultant so you must be able to trust his or her advice.

Once you have decided on a repair facility, the next step will be to communicate effectively your repair needs to the service consultant. The service consultant might be called the service writer or the service advisor. It is this person's job to listen to your service and repair needs, prepare estimates and service orders, schedule work, and work out the details for the return of your car. Basically, the service consultant acts as the intermediary between you and the technician. He or she can help communicate your vehicle concerns to the technician.

When communicating with the service consultant, you need to provide the clearest and most comprehensive information possible. You have the best information about your car and its maintenance and repair history. You will be the first to notice any changes in how the car drives. Make this information available to the technician. If your car has been worked on recently or has an unusual repair history, be sure to share this with the technician. Some people distort or lie about the facts concerning a vehicle problem, but doing so only serves to delay correct diagnosis and repair of the problem.

The more precise you are in describing the problem, the more you help the technician focus and zero in on the repair. The best way to help communicate with the repair facility would be to put your observations in writing. This method of communicating gives you a chance to clarify what you are thinking. It also provides you with a list of items so that none are forgotten. Include all of your observations, no matter how trivial you may think they are. It is up to the technician to decide what may or may not be relevant. Not providing information might cost more to diagnose, in the best case, or may lead to repairs that are not needed. Holding back information could endanger yourself and your passengers should the technician not check a system or area of the car that you noticed had a problem. The technician is not a mind reader and generally only completes repairs that the customer requests. Note if the problem occurs when the engine is warm or cold, at a particular speed, a certain gear, accelerating or decelerating, going up a hill or on level ground, going over bumps, during turning or braking. Include observations concerning any unusual warning lights, vibrations, leaks, noises, smells, or smoke. Be as complete in your observations as you can. Share your information.

There are a few things that you don't want to tell the repair facility. Avoid making a diagnosis. Trying to seem knowledgeable may get you a repair you didn't need or want. Don't go in and say it might be an engine problem. Chances are that is where the technician is going to start looking for a problem. Don't tell the service writer what you are willing to spend. You might just as well announce how much you want your bill to be. You will get a cost estimate, so don't get too anxious.

When you get the estimate for repairs, keep in mind that there may be some charges for diagnostic time. You should understand everything that the estimate contains. It should include parts, labor, and any other charges. All items on the estimate should be clearly spelled out, including the exact nature of the work to be performed, what parts are to be replaced and their brand, and the time expected for the repair. Ask if the parts are factory original, parts that meet the original manufacturers specifications, or parts that are remanufactured. If any of the items on the estimate are not clear, ask for clarification or that the estimate be rewritten. Be sure to get the estimate in writing.

You may want your old parts back. Tell the repair facility that you want your old parts. Keep in mind that if remanufactured parts are used, frequently there is a core charge for the old part. The charge for a remanufactured part is separate from the charge to obtain a used part for remanufacturing. The remanufacturer will take your old part back for disassembly and remanufacture for resale. If you keep your old part, the remanufacturer will need to purchase a replacement core, usually from a salvage yard. In some cases if the core is badly damaged, you may need to pay the core charge. The remanufacturer is expecting a part that can be rebuilt in return.

Labor rates can vary greatly across the country from $45 to $100 per hour. The technician does not receive the hourly rate listed on the estimate. The hourly rate does include the pay the technician receives, plus other business expenses. These expenses include the cost of overhead, utilities, lighting, taxes, equipment costs, training, and so on. A repair facility is not an inexpensive business to operate. A wheel alignment machine can cost $40,000 or more. And that's just one piece of equipment.

Diagnostic time can vary greatly as finding a problem can be elusive. Most labor is charged at a flat rate, meaning that the same job should take the same amount of time no matter who is doing the repair. A water pump for a specific vehicle, for example, may take 1.2 hours to replace. The estimate will take the 1.2 hours and multiply it by the shop labor rate to determine the labor charge. The technician will be paid for 1.2 hours of labor, but the technician will receive only 20% to 40% of the labor charge.

As your car gets older, parts wear out. Expensive repairs may be necessary to keep the car in a reliable and safe operating condition. You must balance the cost of repairs against the value and use you expect to get out of the car. In Chapter 13, you will learn what factors should be considered in making a decision about fixing your old car or getting a replacement vehicle.

1. What are some questions you should ask yourself before undertaking a repair job?
2. Describe how to change a tire.
3. What is a major precaution before beginning work on the electrical system?
4. Outline the steps for an alternator replacement.
5. Describe how to change a valve cover gaskets.
6. Why would a thermostat need to be changed?
7. What are some things to look for when searching for repair services?
8. Who can certify technicians?
9. How should you communicate with the service advisor and technician?
10. What are some things to avoid telling the service advisor or technician?
11. What is the labor rate in your area?
12. What is flat rate?
13. Describe how flat rate benefits the customer and the technician.

THE FIX-IT/DUMP-IT DECISION

After reading this chapter, the reader should be able to:

- List and explain the motivations for owning a vehicle.
- Explain the importance of evaluating repair and maintenance costs.
- Compare real and false economy.
- List sources of information in estimating the value of your car.
- Explain the importance of comparing the costs of repair to the cost of a replacement car when deciding to keep or purchase a new vehicle.
- Explain the importance of setting a budget when deciding on a new vehicle.

UNDERSTAND YOUR MOTIVATIONS

Sooner or later, you must make a major decision about fixing your old car or getting a new or used model to replace it. For most car owners, this is a financial decision. You weigh the costs of fixing an older car against the benefits that you expect to receive. The benefits that might be important to you include performance, appearance, economy, reliability, safety, and comfort. Try to decide among these qualities which are most important, and how much they are worth to you.

Performance

Performance is the ability of a car to do work on demand. The work that the vehicle is expected to do is dependent on the driver. Some drivers want a car with dramatic acceleration. Others feel agile handling is a prime concern. Still others look for better performance for towing purposes or during adverse road conditions.

To some, performance may mean the ability to cruise all day long on a single tank of fuel. If you live in a large city, driving range may not be a concern. However, if you live in a sparsely populated area, it may be one of your most important considerations.

Performance may mean the ability to carry several passengers, as well as large amounts of luggage or cargo. Or, you may have an interest in off-road and camping activities for which a 4WD vehicle might be best. You must define your own concept of performance, and then decide how well your present car, or a different car, might fulfill these desires and needs.

Appearance

Appearance may be the most subjective of all evaluations. Style and appearance have always been major factors in consumer decisions about car buying and is largely based on a personal opinion of the vehicle. Even in the early days of the automobile, a certain body style could attract thousands of buyers, as it certainly does so today. A variety of exterior and interior appearance options confront the modern car buyer.

Auto shows provide the best examples of appearance-conscious vehicle design. International auto shows can host hundreds of cars, SUVs, trucks, and vans representing the work of over 40 manufacturers for the future styling of automobiles. These car shows can draw tens of thousands of spectators. They are great places to see what the manufacturers may be offering in terms of styling and appearance, as well as other features.

Presumably, style and appearance played some part when you selected your current vehicle. Does that attraction still apply? Or is there a different make or model on the market that attracts you more?

Economy

The area of economy includes the consideration of fuel mileage, but fuel consumption is only one aspect of overall economy. Economy also includes the costs of operating and maintaining a car. Are you on a restricted budget? How much maintenance are you willing or able to do for yourself? If repairs are needed, are they more costly for some makes and models than for others? For example, some expensive European vehicles do not require many repairs. However, the price for those few repairs may be much greater than the cost of more-frequent repairs on less-expensive cars. Consumer magazines often publish yearly guides that provide a comparison of repair needs and repair costs. You should be able to determine the average expected costs to own and operate your vehicle.

Perhaps the most basic economic decision involves the down payment or monthly payments for a replacement vehicle. You may consider leasing a car to avoid a large down payment. In that case, the monthly payments will probably be higher. In most cases—unless you use your car for business purposes—it costs more in the long run to lease a car than to buy it. Check Chapters 14 and 15 for information about buying and leasing.

If you are thinking about buying a used car, necessary repair or restoration costs might be involved. Even a brand-new car may develop mechanical problems that might occur soon after the warranty expires. In that case, you could be saddled with big monthly payments and a hefty repair bill.

Other costs are involved with new cars. Insurance costs are higher because a new car costs more to replace if it should be damaged beyond repair in an accident. In most states, registration and license costs are higher for new cars than for older models.

Think about the economy of your present car. It may require extensive work to get it to, or keep it at, a safe and reliable running condition. Try to estimate the costs of this work.

Make yourself a list of estimated costs of operating and maintaining your present vehicle over the next two or three years (see the next section, "Evaluate Maintenance and Repair Costs"). Compare these costs with those of a new or newer car. Decide what you can afford. Decide how much of your budget you wish to devote to car ownership. Now you can relate all your other decisions to this critical factor.

Reliability

As with the other qualities, your individual driving needs will determine how important reliability is to you. If you must use your car every day to commute long distances to work, an unreliable car could be a disaster. If your car is used primarily for pleasure driving, reliability may not rank as high on your list. Consult the same sources you did for determining the economy of your vehicle to help determine its reliability.

Do you live in a very cold or very hot climate? A breakdown during a snowstorm or in the searing heat of a desert might represent a physical and personal danger.

In addition, you need to consider the availability of parts and service for your present car or for a replacement vehicle. Is it the kind of car for which parts and service are widely available at reasonable costs?

Comfort

Comfort is another subjective area and again, dependent on the personal preference and opinion of the driver. To some drivers, comfort means reclining power bucket seats. Others prefer bench seats. Some people may want a variety of power-assist options. Others prefer to do the work of rolling windows down and adjusting seats manually.

Your circumstances might change. For example, you might move to a warmer or colder climate, in which case air conditioning, heating/defrosting capacity, or other options such as heated mirrors and seats might seem like necessities, not options.

Some accessories that enhance your driving comfort can be added to an older car for a price. Adding air conditioning, for example, usually requires installing a heavier-duty radiator and making sure that the entire cooling system is in top condition. There are three basic questions related to adding aftermarket accessories:

1. How much will it cost to bring an old car up to your comfort standards?
2. Will the addition of new equipment detract from the car's performance or economy?
3. Will you be able to recover some of the cost of added accessories when you finally do sell the car?

Comfort, like appearance, is highly personal. In fact, all of these areas involve making personal decisions. You may seek advice, but you must reach your own conclusions.

EVALUATE MAINTENANCE AND REPAIR COSTS

In previous chapters, you have learned to detect a number of symptoms that can help you to analyze performance and safety. Now, in evaluating the continuing costs of vehicle upkeep, you should analyze projected repair costs over the next two years (Figure 13–1).

If you aren't sure about your car's overall condition, diagnostic services are available in many locations. For example, the best-known national automobile club will conduct diagnostic examinations of its members' cars. Dealerships and independent service shops will perform inspections and evaluations of major car systems. These diagnostic services usually cost about $50 to $75. However, you should beware of shops that may use diagnosis as a pretext to sell unneeded repairs.

Other types of repairs, such as body work and paint, require estimates from specialists. First, decide how important your car's appearance is to you. For some people, a shiny new paint job or new interior upholstery can be well worth the money in terms of owner satisfaction and pride. But if your car is used primarily for transportation, its appearance might be of minor importance to you.

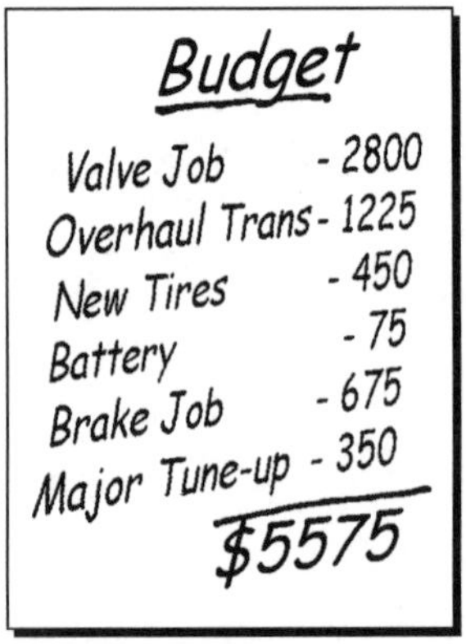

False Economy versus Real Economy

Among the continuing maintenance costs are such things as fuel and oil. If your older car is a large model, are your fuel costs excessive? Perhaps the engine has been burning oil. Try to be realistic about such costs. If the engine burns a quart of oil every 600 miles, and you drive 12,000 miles a year, that's 20 quarts of oil, or about $40 a year. Even if the engine burns a quart of oil every 300 miles, the annual cost amounts to only about $80. These costs are minor in comparison with the monthly payments, registration, and insurance costs for a new car. Also keep in mind that many states require emission testing of vehicles. If the vehicle does not pass the test, repairs will need to be made to bring the vehicle into compliance with the

Budget

Valve Job	- 2800
Overhaul Trans	- 1225
New Tires	- 450
Battery	- 75
Brake Job	- 675
Major Tune-up	- 350
	$5575

Figure 13–1: You should prepare a projected repair cost for your car for the next 2 years.

emission regulations. If an engine is burning oil to the point where it fails the emissions test, then a decision will need to be made about repairing the engine at great expense or getting another vehicle.

Major Repair Costs versus New Car Prices

Another way to compare maintenance costs with the costs of new car ownership is to consider the prices of new cars. New car prices have gone up dramatically in recent years. Repair costs have risen, too, but they haven't kept pace with the prices of new cars.

Remember, as a car gets older, more and more parts will have to be replaced. Chapter 11 covers what you can expect to go wrong in 50,000-mile cycles. By the time a car reaches 200,000 miles or more, there is a good chance that you may face some sizable repair bills.

To illustrate, consider the cost of rebuilding an engine. You may be looking at repair bills of $3,000 to $5,000, or even more. Some years ago, when new cars cost an average of $15,000, this level of expense used to warrant a decision to buy a new car. Although some smaller models sell for less than $14,000, the average price of a new car today is approximately $22,000.

You may be facing total expenses of $2,000 to $3,000 to keep your car operating over the next several years. The engine, transmission, brakes, suspension, and steering may need major work. You also may need new tires and other expensive replacement parts. You may want to have a new paint job and new interior upholstery. Try to analyze maintenance and repair costs in light of new car prices before you decide.

ESTIMATE MARKET VALUE OF YOUR PRESENT VEHICLE

There are various ways to estimate the market value of your car or of a new or used replacement car. Price guides are an important source of information. The guides will give you complete car information on manufacturer's suggested retail price (MSRP). The pricing guides are usually more accurate and concise than the other publications for a variety of reasons. They specialize only in the pricing of cars and take into account other factors that affect the pricing of vehicles. The two most common price guides are the National Automobile Dealers Association's *N.A.D.A. Official Used Car Guide* and the Kelley *Blue Book*. Price guides can be found on the Internet. For example, the Kelly Blue Book price guide is available at http://www.kbb.com. This handy site provides used car pricing as well as new car pricing. Also check your local newspaper used car section to see the advertised price for used cars. Check the used car dealerships. This resource is often overlooked as you are trying to sell your car, not buy another one like it, but you can find out what the selling price is of cars similar to yours.

More than price is involved in a decision to sell or trade your car. You also must consider convenience. On the one hand, the simplest and easiest method, of course, is just to trade in your old car on a new one at a dealership. Unfortunately, that is usually the least economical method for the owner. On the other hand, are you willing to go to the trouble of advertising the car and selling it yourself? How much time and effort are you willing to devote? It may not be easy to get a fair price for the car. A good rule of thumb is not to be optimistic. It is safer to be pessimistic in your figuring.

SET YOUR BUDGET

Decide what you can afford or what you are willing to spend for transportation over the next two to three years. Then, figure realistically what the cost of owning a car will be during that period. How will these costs fit in with your budget?

Be realistic in evaluating some of the main elements involved. How many miles will you be driving each year? Mileage will determine such factors as fuel costs, regular maintenance costs, and replacement of tires, filters, and so on. Relate these expenses to those of your present car. Decide how much transportation is worth to you. Come up with realistic figures and commit yourself to them. Otherwise, you may risk financial difficulties.

Decide realistically and firmly what your total expenditure will be. Include the costs of registration, insurance, gasoline, and regular maintenance services. These costs can vary dramatically among models and types of cars.

Some people feel they need a new or sporty car to enhance their egos or to reflect their status or a status they would like to have in life. Other people may look at a car as merely a way to get from home to a job and to pick up the kids from school. You need to consider the purchase of maintenance of a car in light of your own needs and what you can truly afford to spend.

CONSIDER TOTAL OWNERSHIP COSTS

Much more than the purchase price of a car must be considered before you decide whether to buy. Other significant costs include finance charges, insurance, license fees, and, in most states, sales taxes.

Under some economic situations, the greatest cost can be the expense of borrowing money. If interest rates are high, the eventual cost of a car can be more than twice the sticker price if you pay it off over five or six years.

Under other economic conditions, some manufacturers and dealers have offered cars at less than 2% or even at 0% interest, if the loan is paid off in full in one to two years.

Insurance costs for a new car will be considerably higher than those for an older car. License fees in most states will be considerably higher, and in most states you will have sales taxes to pay on a new car. Some states also charge a use tax, usually at the same rate as sales tax, on transactions between private parties. States that don't levy sales taxes generally have very high new car license fees.

Figure all these costs into the equation. Then, determine what a new car will cost to own over the projected life span of your present vehicle. At this point, you have a realistic basis for comparison. You aren't comparing apples and oranges any more. If you are considering a new car for reasons of vanity, you know what your vanity will cost.

UNDERSTAND THE IMPACT OF YOUR DECISION

It is important to understand the full impact of your two- to three-year total cost projection. You may be saving money on repairs with a new car, but your payments may be $350 a month or much more.

You don't want to purchase the answer to your automotive dreams, only to hate it in six months because you can't afford to maintain it. Worse yet, you don't want to get in over your head financially. If you cannot make your monthly payments, the car may be repossessed, and you would still be financially responsible for making the payments. Your credit rating can be harmed. In addition, if you wind up without a car, you may be reduced to looking for a low-priced wreck just to get to work.

Buying a new car involves some financial risk. Evaluate your situation. Be sure to understand why you want to buy, what you are getting, and how you can handle the costs. Buying a new car is the second largest investment most people ever make (the first being their home), and they do it repeatedly. The move should be pleasing and satisfying to you. Satisfaction is what car buying should be about. If there is some question about your satisfaction, you should have some doubt about your decision to buy.

The important rule is to approach the fix-it/dump-it decision on the basis of intelligent information gathering and selection of alternatives. Chapter 14 deals with the steps you need to take after you have made your decision.

1. List the motivations for owning an automobile.
2. Describe each of the motivations briefly.
3. Which of the motivations are objective?
4. Which of the motivations are subjective?
5. What is false economy?
6. What is real economy?
7. How might you estimate the market value of your car?
8. Why is it important to set your budget?
9. List some items that should be included in the total coat of ownership.
10. What may happen if you don't make all of the payments on a loan?
11. Can you be held responsible for loan payments after a repossession? Why?

BUYING TIPS FOR PARTS AND INSURANCE

After reading this chapter, the reader should be able to:

- List and describe the places to buy parts for your car.
- List and explain the different types of insurance coverages.
- Analyze the value of an extended warranty.
- Explain what to do in case of an automobile accident.
- List some things that can to help you avoid an accident.

BUYING PARTS

If you've decided to keep your present car or to buy a used car, you may realize that some repairs will have to be done. To save the most money possible, you may want to do some of your own repairs. If you know how and where to shop, there are ways to save money on necessary parts.

Automotive parts are sold in a competitive world generally known in the trade as the automotive *aftermarket*. Many companies—other than the original manufacturer—specialize in making replacement parts. Be aware that replacement parts vary widely in both price and quality. All auto manufacturers purchase parts from subcontractors. Some aftermarket parts are original-equipment type parts made by the subcontractor and sold either under the subcontractor's brand or a special brand name.

Tires are a good example of aftermarket quality. For many years, the original equipment tires on new cars were considered minimal quality in the industry. This situation is no longer true with many car models, but it is still possible to buy tires of higher quality in the aftermarket.

There are three basic places you can go to buy parts for your car:

1. Authorized dealers
2. Auto supply and chain stores
3. Used-parts and salvage yards

Some auto parts are popular and common to more than one vehicle. These items move, or are sold, quickly. Shock absorbers, brake linings, spark plugs, belts, and hoses are examples of fast-moving items. These are easy to find in most auto parts stores. Other parts, such as specific mechanical parts or trim items, may stay on the shelves for years before they are purchased. It is necessary to go to a dealer to obtain such parts.

At one time, the only source for body parts—fenders, doors, and so on—was a dealership parts department. In recent years, aftermarket manufacturers have begun to produce body parts as well as other items. In some cases, these body parts may have an inferior fit and less corrosion resistance as compared with factory units. If you are having body parts replaced at a body and paint shop, you should be aware of your right to choose between original manufacturer body parts and aftermarket parts.

It is important to note that parts are not usually purchased at swap meets or garage sales because these are not reliable sources for parts. In addition, the parts can vary greatly in quality and are not readily available.

Authorized Dealers

Authorized dealers are licensed to sell specific brands of new cars. Under their contracts with the manufacturer, they must invest a certain amount of money to buy and stock **OEM parts** (that is, parts made by the original equipment manufacturer). Dealers must carry or be willing to order any part that can be found on specific models within a manufacturer's line (Figure 14–1).

Another part of the manufacturer/dealer commitment is that the dealer must submit complete inventory reports periodically and replenish the parts stock to the level specified by the manufacturer. Dealer parts departments often are the only source for unusual or slow-moving (low sales volume) parts. Dealer parts departments also have the ability to send emergency orders to the factory or to a regional parts warehouse.

Some parts, as mentioned, can be obtained only through dealership parts departments. As a private purchaser, you do not receive a discount at a dealer's parts department. Repair facilities and technicians get discounts

Figure 14–1: Dealerships are authorized by the manufacturer to sell OEM parts.

of about 15–20%. That is a point to remember about buying parts from a dealer. On the one hand, you know you are getting the factory-authorized part. On the other hand, you are paying the full retail price. There is no shopping around to be done because all dealers charge the full list price. Some dealers, in fact, offer no discounts at all—even to other shops or trade sources.

Auto Supply and Chain Stores

Some of the parts sold in auto supply and chain stores may exceed the quality of OEM parts. In some cases, OEM parts are of medium quality; in others, the OEM replacements are of the highest quality. For example, OEM MacPherson strut cartridges and shock absorbers are generally medium-quality items. Auto parts stores typically stock several grades of strut cartridges and shock absorbers. Thus, you have a choice of good-, better-, or best-quality units. In addition, the replacement-parts business is highly competitive. If you spend a little time to shop around at different auto parts stores, you may be able to save significant amounts of money. For example, you might find the best-quality shock absorber at a price that another store charges for a medium-quality unit. It always pays to shop around.

Major retail chain stores often have complete automotive parts and service facilities. In addition, franchised chain auto parts stores often provide low prices and excellent values. For the cost-conscious do-it-yourselfer, the aftermarket can have both price and quality advantages (Figure 14–2).

Used-Parts and Salvage Yards

Used-parts and salvage yards typically offer a selection of body parts and most mechanical parts. Salvage companies buy wrecked vehicles, mostly from insurance companies that have declared the vehicles to be "totaled"

Figure 14–2: An aftermarket parts store is usually well stocked with parts.

and paid the owners the fair market value (Figure 14–3). Salvage operators also acquire old cars that are no longer running.

A late-model car is much more valuable for its salvageable parts than as scrap metal. Successful salvage operators know which parts of a car are usable. For example, a car that has been struck from behind may have a perfectly good engine, radiator, and transmission. In some cases, the salvage operator will disassemble such a car and separate the good parts for individual sale. In other cases, the entire front end of the car will be sold as a unit to an auto body shop.

Be aware that this sector of the marketplace cannot offer the types of guarantees or warranties you might expect of a factory or a major retailer.

Figure 14–3: A modern savage yard buys wrecked or old vehicles, disassembles them, and sells the parts to various customers.

Salvage yard guarantees are usually very limited, if offered at all. A reputable salvage yard might offer a free replacement guarantee for a limited time. This offer sounds good, but remember that you are investing your time and labor (or paying someone else) to install the part. It is not the same as taking your car to a dealer or garage for repair work. Salvage yards cannot afford to offer money-back guarantees.

There are two relatively recent developments in auto salvage yards. The first is a type of operation in which the customer brings his or her own tools, then disassembles and removes the desired part from a vehicle. These operations offer lower prices than salvage yards that do the removal labor for the customer. However, some expertise and judgment are required on the part of the customer. It is possible to spend time, money, and energy on a part that turns out to be more badly worn than the unit on the car to be repaired. It is also possible to save considerable money on expensive parts, if you know what you are doing.

A second type of operation completely disassembles cars, then stores the parts on organized shelves and tables and catalogs them. Finding a specific part for a specific car becomes a matter of simply consulting a computerized inventory. As you might expect, the prices for parts in these operations are typically higher than in other salvage-yard type operations. The convenience of being able to locate a needed part quickly makes up for the additional price. Generally, these yards also offer an exchange guarantee if the part is returned within a few days.

If you deal in the used parts market, compare the prices of new, used, and reputably rebuilt parts. Some large chains, for example, have major engine rebuilding operations with reputations to protect, and so their guarantees are usually more meaningful. But check carefully, because in areas where there is no pricing competition, there is no standard of comparison. The price of a used engine or other part may be "whatever the market will bear." At times, this price will not be a bargain. So, be very careful in buying engines or drivetrain components in the used market. It is possible to save money, but it also is possible to end up paying nearly as much for a used component as for a new unit.

The reason many people go to the used parts market is that vehicle manufacturers are not required to continue making parts for discontinued models forever. After approximately 10 years, many parts may become increasingly hard to find. Thus, salvage yards often are the best sources for parts for older cars.

Rebuilt or Remanufactured Parts

You also should understand that, with certain parts, rebuilt items might be your best option. Factory-new alternators and starters, for example, may be prohibitively expensive. A major source of parts for authorized dealers and parts stores is the remanufacturer or rebuilder. These companies purchase used parts and refurbish them with new components. Rebuilding such equipment is straightforward and reliable enough that you should be able to buy a rebuilt alternator or starter with a reasonable assurance of its quality. After the unit is disassembled, only a few components need to be replaced to return the unit to good working order. In many cases, the rebuilt item is virtually as good as new. Just try to make sure you are dealing with a reputable firm.

Generally, the parts store will provide and service the warranty if needed. If a dispute over a part cannot be resolved, which is unusual, governmental consumer or business standards departments exist in most areas. In addition, the Better Business Bureau (BBB) in your vicinity can provide evidence of bad performance; a blank BBB reference usually means no complaints have been filed. Be aware that some rebuilt parts are of inferior quality. You may have to remove and replace a unit several times until you find one that works properly. Deal with a reputable parts supplier.

BUYING INSURANCE

It should be clear that the bank, finance, or lease company really owns your new car until you have made the last payment. You may be willing to take risks on your liability, but the financial institutions are not. Therefore, they will require that you carry insurance to protect their equity in your car.

If you finance the car, your insurance protects the financial institution as well as yourself. Should an insurance claim be settled for more than the lender's equity in the vehicle, you would get the difference. The basic protection against a car's being totaled generally benefits the lender—unless your loan balance is quite low.

Most states have financial responsibility laws. These laws state that you, as the registered owner/driver of a vehicle, are responsible up to certain limits for any accidents causing property damage or personal injury. It is important that you learn what liabilities your state places on the owner/driver of a motor vehicle. Then, you have to decide how vulnerable you or your family might be should you become involved in an accident that is your fault. In many states, a serious auto accident can destroy a person financially if that individual does not have adequate insurance coverage.

No one pulls out of the driveway with the intention of getting into an accident. But there are more than 6 million traffic accidents in the United States every year resulting in 3 million injuries, and more than 40,000 persons killed on the nation's streets and highways. Property damage and liability due to these mishaps run into the billions of dollars. You don't have to be completely in the wrong to suffer liability in an auto accident.

Liability coverage is the foundation of any auto insurance policy and is required in most states. If you are at fault in an accident, your liability insurance will pay for the bodily injury and property damage expenses caused to others in the accident, including your legal bills. Bodily injury coverage pays for medical bills and lost wages. Property damage coverage pays for the repair or replacement of things you wrecked other than your own car. The other party may also decide to sue you to collect so-called pain and suffering damages. The following are the common types of coverages that may be found in an automobile insurance policy:

- Liability coverage
- Bodily injury liability
- Property damage liability
- Medical payment coverage
- Personal injury protection
- Uninsured and underinsured motorists
- Collision coverage
- Car rental expense
- Extended warranty protection

Because required and minimum coverages will vary from state to state, it is prudent to consult an insurance agent for coverages in your state.

Liability Coverages

The foundation of an auto insurance policy is liability insurance. Most states require auto liability, so your insurance minimum will depend on where you live. Liability coverage is the most important to your financial security. In the face of ever-increasing medical costs, you want to make sure you're choosing the right amount of liability coverage to make your policy worthwhile. If you don't have enough coverage, a serious accident could wipe out the insurance money very quickly and you would be financially responsible for the remaining bills. The minimum coverage limits required by law may not be enough in the event of such an occurrence and serious property damage. Choosing a low liability limit is unwise and a poor way to save money on insurance.

The insurance numbers are usually expressed as a set of numbers; for example, (25, 50, 20). This notation means $25,000 worth of bodily injury coverage per person; $50,000 worth of bodily injury coverage per accident; and $20,000 worth of property damage coverage.

Bodily Injury Liability

If you injure other people—passengers in your car, people in other vehicles, or pedestrians— through the use of your insured vehicle, bodily injury coverage would apply. Bodily injury coverage pays for medical bills and lost wages. Bodily injury coverage can provide for your legal defense if a lawsuit is brought against you. This coverage pays the damages for bodily injury assessed against you, up to the dollar limits in your policy, if you are found legally liable by a judge.

Property Damage Liability

Property damage coverage pays for the damage done to the property of others. It does not pay for the damage to *your* property. If claims or lawsuits are brought against you, property damage liability coverage provides for your legal defense. It also pays the damages for property damage assessed against you, up to the dollar limits in your policy, if you are found legally liable by a judge.

Medical Payment Coverage

Medical payment coverage pays for the medical expenses of you and your passengers, no matter who's at fault. If you and relatives living in your household are pedestrians or passengers in someone else's car, and are injured by another car, the coverage applies. Any reasonable and necessary medical expenses are covered up to the amounts and time limits specified in your policy. If you do not have health insurance, payments will be made while liability is being determined. If you already have health insurance, you only need to purchase coverage for expenses beyond that provided by your health insurance. Also, any medical expenses not covered by your primary insurer, including deductibles and copayments, are also provided.

Personal Injury Protection

Personal injury protection is an optional coverage for you. It will cover you, relatives living in your household, and your passengers who have been injured by a vehicle. It allows medical, loss of income, funeral expenses, and other qualifying expenses up to the limits specified in the policy. These expenses are usually paid without regard to fault. The most common limits are $10,000 to $35,000.

Uninsured and Underinsured Motorist

Uninsured and underinsured motorist coverage provides extra protection for bodily injury or death that results from an accident when the other driver is uninsured or underinsured. If the bodily injury liability limits of the person at fault are not sufficient to cover your damages, this coverage will pay the difference up to your policy limits. In some states this coverage will also pay for property damages. Also, if you and relatives living in your household are pedestrians or passengers in someone else's car, and are injured by another car, the coverage applies. This coverage can also include damages for pain and suffering, as well as medical expenses and loss of earnings.

Collision Coverage

Collision coverage will pay to fix or replace your car if it's damaged in a collision with another car. The insurance company pays for the cost of your repairs or for the value of your car, whichever is less, minus your deductible. Collision coverage is normally the most expensive part of your auto insurance coverage. If you choose a higher deductible, say $500 or $1,000, you can lower your insurance costs. Collision coverage is available only with a deductible. Keep in mind that you must pay the amount of your deductible before the insurance company pays any money after an accident. The finance company may require collision coverage.

Each person's insurance company pays for certain financial losses, such as medical or lost wages, regardless of fault. In exchange for these benefits, the right to sue may be restricted.

Car Rental Expense

There are other items you may want to look at in your insurance coverage. A provision for rental car expenses is one such option. Car rental expense coverage will reimburse you for the cost of renting a car while your car is repaired after an accident.

Look at the entire scope of coverage. Do you want to have enough coverage so that you can get a replacement if your car is totaled in an accident? Do you want to provide for your own medical care and that of your passengers in case of a collision? Do you want to protect your car against theft, vandalism, and other noncollision damages? Make sure you understand what types of insurance coverages are available and analyze what you need.

Buying insurance is like buying a car or financing. First, decide what you need; second, shop around for the best value. Insurance premiums can vary widely for the same coverage. There are several avenues available to most people for obtaining insurance. You may have an agent who handles your life insurance or the fire insurance on your home. Because you are already a customer, that agent might work harder to obtain the best coverage for you at the least cost (Figure 14–4). Credit unions usually have good insurance sources. The company that underwrites group coverage where you work may be another source. The idea is to shop around before you buy.

Some mutual insurance companies and auto club insurance exchanges offer rebates based on claims experience during the year. If the company has a good year—if claims are relatively low—this arrangement can save you money on your next year's premium. If claims payouts are high, of course, no rebate is made.

There are direct-selling and agent-selling companies. Direct-selling companies claim they can offer lower rates because they don't have to pay

Figure 14–4: The insurance agent works for you. The agent knows the insurance requirements for your area and can recommend what coverage you will need.

commissions to agents. You have to decide for yourself, and the best way to do this is to shop around and get several quotes.

BUYING EXTENDED WARRANTY PROTECTION

Most new car dealers now offer so-called extended warranty contracts. These are special insurance policies that cover the costs of repairs or replacement of parts that break down or wear out. These policies typically provide coverage from three to five years or for a specified number of miles, whichever comes first. The policies are not factory warranties, although the underwriting company may be a wholly owned subsidiary of the auto manufacturer.

In some cases, extended warranties are available either from the car manufacturer or from private insurance firms. One advantage of a manufacturer's extended warranty is that the manufacturer is much more likely to stay in business than is a private insurance firm. Another advantage is that paperwork and claims procedures are simplified.

If you are considering buying extended maintenance protection, be sure your analysis of the cost is based on actual coverage. Subtract the original factory warranty period (months or mileage). The remainder is the true period of coverage. Divide that number of months or miles into the total insurance policy premium to calculate the cost of useful coverage. In some cases, you can buy an extended warranty at a later time; for example. just before your factory warranty expires. Buying an extended warranty at the same time as a new car may mean that the first few years represent wasted money, because you're already covered by the manufacturer's warranty.

As in the other buying situations covered in this chapter, here are some final words of advice about buying any kind of insurance or extended warranty: Read everything carefully. Be sure you understand what coverage is provided by an insurance policy. Make sure it is what you want and need *before* you sign the papers.

WHAT TO DO IN AN ACCIDENT

According to the National Safety Council, one in every eight drivers will be involved in a motor vehicle accident this year. You may or not be at fault. If you've ever been involved in an accident, you know it can be very stressful. Many people are overwhelmed with emotions and worries. You will be concerned with everyone's safety and anxious about your vehicle. You might be angry with other drivers. You may be worried how the accident will affect your driving record and insurance. It may be hard to think clearly and respond properly. And if there are injuries, the stress can be amplified and panic may occur. But that's when a clear head and quick action are really crucial. It is vitally important to remain as calm as possible (Figure 14–5). Here are some tips to follow should you be involved in an accident:

- Stay calm, stop the vehicle, and turn off the ignition. Turn on your hazard flashers. Survey the accident scene and check for injured people.
- Contact the police regardless of the circumstances. If necessary, contact an ambulance or medical services.
- Make the accident scene as safe as possible. Move the vehicle out of the roadway, if it is clear, safe, and legal. In some states it is against the law to move the vehicle from the accident scene. Check the ordinance in your area or ask the police when you call.
- Mark the scene of the accident if you have retro reflective triangles in your car. Keep people away from any dangerous items resulting from the accident, such as battery acid or fuel leaks.
- Record the name, address, and phone number of the other driver. Write down the make and license number of all vehicles involved. Gather the names of all the people who witnessed the accident. You'll want to get as much information as you can about the other driver's insurance agent, policy, and insurance company. Since many cases end up with the parties blaming each other, witnesses can be important. Don't hesitate to approach anyone who may have seen the accident.

Figure 14–5: Accidents can happen at any time. Stay calm and call the police.

- Do not discuss fault or place blame; this is not the time or place. Be careful about what you say. Even an offhand or casual remark could be used against you in court.
- Make a diagram of the accident scene and where drivers and passengers were seated. Indicate the direction of travel, lane used, date, time, and weather conditions.
- Get a copy of the police report from the police department (Figure 14–6).

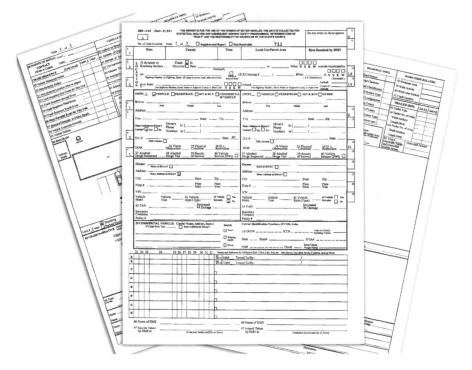

Figure 14–6: Should you be involved in an accident, be sure to get a copy of the police report for your insurance agent.

- Contact your insurance agent as soon as possible The faster your agent gets the information, the faster he or she can act to help you. Discuss the accident only with your insurance agent and the police.
- Examine the damage to your vehicle carefully. Carry a disposable camera in the glove box. It will be handy for taking photos for any situation that comes up at a moment's notice, even if the accident is not your own. Take photos of the vehicles and of the accident scene. These photographs could easily become an important witness to what actually occurred.
- Without being overly suspicious, observe the other driver's actions. Should the other driver or passengers later claim to have a serious injury, what you notice could be important.
- Do not drive an unworthy vehicle on the road.

The whole insurance process will be less stressful following your accident if you know the details of your insurance coverage. Don't wait until after an accident to find out that your insurance doesn't cover costs for towing or a rental car. You can add coverage for rental car reimbursement, which provides a rental car for little or no money while your car is in the repair shop or if it is stolen. This coverage usually only costs a few dollars a month to add. Check with your agent for specifics on your policy.

Always report an accident to the police. The other driver may agree to pay for the damage to your car on the day of the accident. Later, when the repair bills arrive and the other driver decides they're too high, it is too late. At this point, time has passed. Your insurance company will have a difficult time assembling the evidence should you then file a claim.

Remember, you have no way of knowing whether the other driver will change his or her mind and report the accident. The other driver may even claim damages or injuries that were not visible at the accident, and your insurance company might have to pay a hefty settlement. A lawsuit could be the result. Make sure that your agent and insurance company has your version of what happened.

Automobile accidents take a toll on everyone involved, both financially and emotionally. The chances are high enough that at some point you will be involved in a minor accident. Just keep your head, remain calm, and make safety your primary concern, both for yourself and others. You can deal with the consequences later.

Here are a few tips to help you avoid an accident:

- Carefully look both ways before crossing an intersection.
- Don't speed into an intersection the instant the light turns green.
- Be aware of the green light. Green lights turn to yellow eventually and this should not be unexpected.
- Look twice at intersections, especially when turning.
- Anticipate unexpected changes in traffic.
- Always check your blind spot before changing lanes. Don't just rely on your rearview mirrors.
- Practice extra caution when passing large vehicles, such as tractor trailers, which have large blind spots.
- Pay special attention to posted speed limits, especially around schools and in residential areas.
- Leave enough space between your car and the one in front of you.
- Regularly perform safety checks, maintenance, and repairs on your car, especially for your brakes and tires.

Parts and insurance are necessary for the operation of your vehicle. An older care may require more parts, but has a lower cost to insure. A newer car usually requires fewer parts but has a higher insurance cost. While you decide which is right for you, we will take a look at buying a new vehicle in the next chapter.

1. List places from which to buy parts.

2. Why do you think parts aren't usually purchased at swap meets or garage sales?

3. What are the advantages and disadvantages of purchasing parts at an authorized dealer?

4. What are the advantages and disadvantages of purchasing parts at a parts store?

5. What are the advantages and disadvantages of purchasing used parts at a salvage yard?

6. What are rebuilt or remanufactured parts?

7. Why do you need insurance?

8. What is liability coverage?

9. What is bodily injury coverage?

10. What is property damage coverage?

11. What is personal injury protection?

12. What is uninsured/underinsured motorist coverage?

13. What is collision coverage?

14. What is car rental expense coverage?

15. Describe extended warranty protection.

16. What should you do in case of an accident?

17. How many accidents occur in the United States per year?

18. How many injuries from accidents occur in the United States per year?

19. How many fatalities from accidents occur in the United States per year?

20. What are some tips for avoiding an accident?

BUYING A NEW CAR

After reading this chapter, the reader should be able to:

- List the steps for preparing to buy a car.
- List sources of information for researching the purchase of a car.
- Explain the importance of the test drive.
- Describe how to determine the value of a car.
- Describe the various conditions of a vehicle.
- Explain the different methods to sell or dispose of your old car.
- Compare leasing and buying costs.
- Explain how the cost of a new vehicle is determined.
- Describe dealer selling tactics.

DECIDING TO BUY A CAR

Getting a new car involves a number of steps:

1. Decide what kind of vehicle you want.
2. Determine the value of the vehicle you presently own.
3. Decide if you want to buy or lease.
4. Become familiar with financing options.
5. Determine the true cost of the vehicle that you're interested in.
6. Plan on how to close the deal.

This process may seem long and, at times, never ending, but each step is manageable and relatively easy to understand. The key is to make a plan and follow the steps in order. If you get out of order on the steps, then you are vulnerable to confusion and manipulation.

BE A PREPARED SHOPPER

Before you can begin to shop for a vehicle you must choose what kind of vehicle you want (Figure 15–1). Make a list of what you want in a car. Your list should include the following items: brand, features, vehicle type (sports car, compact car, midsize car, minivan, luxury car, sport utility vehicle, pickup truck), options, price range, maintenance costs, and other characteristics. When you get this list compiled, don't forget to check on the insurance costs of the vehicle. Many buyers are sometimes surprised to learn the cost of insurance for their car. Some vehicles have higher insurance costs than others.

Figure 15–1: A dealership sells new cars as well as used cars. Be prepared before you go to the dealership to buy.

Narrow down your choices to a few or perhaps even a single vehicle. If you have already chosen the vehicle you want, you have passed the first obstacle in getting a new car.

INFORMATION SOURCES

There are a number of information sources that you can use when deciding on a car. You may want to research consumer resources such as enthusiast resources. Many of these can be found in your library in the periodicals or magazines that are available. Another resource is the Internet, which has today a wide variety of offerings, ranging from photos and specifications, to a virtual 3D tour of the car.

Good sources for an objective view of a car include buying guides, cost of ownership guides, price guides, and periodicals or magazines.

Buying Guides

Buying guides can be found at your local AAA (American Automobile Association; Figure 15–2) office and other sources. The AAA *New Cars and Trucks* provides information on all aspects of ownership of a vehicle including evaluation, pricing, financing, leasing, buying, insurance services, safety evaluations, and other information. The publication also lists an often-overlooked option for car buyers, the automotive "twin." Automotive twins are vehicles from the same manufacturer that have different nameplates but the same engine, drivetrain, and chassis. Many are manufactured on the same assembly line. These are worth checking out, as a twin may have different equipment and options and offer a better deal in

Figure 15–2: **AAA** publishes buying guides that are excellent sources of information.

the market than its counterpart. Additional information may be found at http://www.aaa.com.

Cost of Ownership Guides

Cost of ownership guides are another resource available. A good example is *IntelliChoice*, which includes the costs of operating a vehicle such as fuel, maintenance, insurance, and depreciation. Also found in these guides are estimated financing costs, state fees, and repair costs. Having these details makes it easy to spot the best value for each category and allows you to compare several different models or brands quickly and easily. It also allows you to focus on the category or categories that interest you the most and find the best value. *IntelliChoice* usually highlights the best value in each category for you. It may be found at your local library, in larger book stores, or may be ordered on the Internet at http://www.intellichoice.com.

Price Guides

Price guides are an important source of information and should always be included in your research. They can give you complete car information on the MSRP, which is the manufacturer's suggested retail price, better known as the *sticker price* of the car. The dealer invoice price is given, which is the price the dealer will pay the manufacturer for the car. Prices are also given for the base model of the car, as well as the prices for the different options available. The pricing guides are usually more accurate and concise than the other publications for a variety of reasons. They specialize only in the pricing of cars and take into account manufacturer, regional, and seasonal factors that affect the pricing of vehicles. These price guides may include the NADA books and the Kelley *Blue Book*. Another place to find a price guide is on the Internet. One of these is http://www.kbb.com, which is the Kelly *Blue Book* price guide site. This is a handy site, as it provides used car pricing as well as new car pricing.

Periodicals

Most popular among the periodicals are *Consumer Reports* and *Consumer Guide*. These magazines offer an objective analysis and review of a vehicle from the consumers' point of view. They analyze a car with regard to safety, reliability, performance, and value. Specific reviews of cars are often available in an automotive buyer's guide from these publishers. You may not find a review for the car that you want in the current issue, so take a look and find the most recent review that most closely matches the car in which you are interested. It also may be helpful to look up reviews of the previous year's model to see how the initial review compares with a more recent review. Some guides will direct you to a review from a past issue.

Other periodicals that may be of interest are the enthusiast magazines such as *Car and Driver* or *Motor Trend*. These periodicals tend to take a slightly different viewpoint than those in the consumer category. Their main viewpoint is that of the driver, with a focus on performance, styling, and technology available in the car. Usually a similar group of cars will be evaluated at the same time, and they will all be ranked against one another. These magazines provide a good way to find out which car has the best performance or the best styling, but, to be sure, you will need to look at the numbers closely. If the car you are interested in isn't ranked at the top, then your choice might be a very close finisher to the top ranked car. Read carefully and with an objective eye.

Other Sources of Information

Another way to help narrow your car choices is to talk to friends, relatives, and coworkers. They may own a vehicle that you are interested in and can provide you with valuable firsthand information about their experiences. If you value their opinions, then ask them. Their insights as to the car's operation, performance, and maintenance requirements, both good and bad, can be of great benefit. Try to get a range of opinions from several of these sources, if possible. It will help to average out the true experiences of the car owners and their experiences with their cars.

THE TEST DRIVE

One of the final steps before purchasing a car should be a test drive. Actually driving the car gives you an opportunity to see and feel how the car performs. Locating the car should not be too difficult, but you will need to find a dealer that has the car you are interested in available for a test drive. Most dealerships have a number of cars designated for such test drives by potential customers. Try to select the car that is the closest to the one you want to buy with respect to your wish list. It may not be a perfect match, but it will be close enough for a test drive. These test drive cars are sometimes referred to as *demonstrator* or *demo* cars (not to be confused with demolition derby cars).

If you are unsure about how they will affect the handling and ride characteristics of the vehicle, be sure to ask about special characteristics in the car such as traction control, active suspension, or four-wheel drive. Try to drive the vehicle as closely to the way you plan to routinely drive the vehicle. A little time to plan your test drive route may be beneficial. You will likely face some pressure to buy the car while you are on your test drive. Don't let a pushy or high-pressure salesperson talk you into a car. You certainly have the option to test drive a car before you buy. Or if you want to avoid the dealership for a little while longer, a visit to a friend, relative, or coworker might result in a private test drive. If you know someone who has the type of car you are interested in, ask him or her for a test drive. It could be less hassle and result in a longer test drive. For an extended test drive, an option is to rent a similar car. This alternative will give you a lot longer time to drive the car, but it will cost you some money. You may want to rent the car for a short trip to get a more complete feel for driving. In any event, spend some time with the car.

Remember your personal preferences will ultimately help decide what you like or dislike in a car. When shopping for a vehicle, try to meet as many of the needs on your list as possible. Doing so will help avoid finding out later that the car is not meeting your expectations and minimize the chance that you end up buying a car that you become bored with quickly.

FEATURES AND OPTIONS

Features and options are the next items to consider for your car. Sometimes features and options may get confused or mixed together. An easy way to remember the difference is features are more common in nature, like air bags or antilock brakes that can be found on various brands of cars, while options tend to be less common, like heated seats or a moon roof. Whether a feature or an option, just list what you want in your vehicle.

Economy considerations may include the car's mileage, its insurance costs, a lower vehicle cost, maintenance costs, depreciation, and expected vehicle lifetime. Depreciation is the decline in the car's value over a period of time. Features like these may sound difficult to accurately list, but it's not as bad as it sounds. Safety features can include front air bags, side air bags, antilock brakes, traction control, all-wheel drive, and active restraints or active seatbelts.

Options may include more specific items such as power steering, power windows, rear window defrost, keyless entry, and so on. The list of options may seem endless and sometimes confusing. Many vehicles are equipped with optional equipment as a matter of course. It is expected to see options such as a tilt steering column, an AM/FM stereo radio, cruise control, and air conditioning on vehicles today. These options are sometimes called *standard options*. Take a close look at these as not all vehicles have the same standard options. A good example is antilock brakes. Most but not all vehicles have them. Be sure to note what options your vehicle may have that you did not have on your list; so as you shop around for a car, you can compare the cars against one another more accurately. This attention to detail will help you further narrow down your choices and may point out some option you don't want or need. Other things being equal, a car with more options costs more money. And the so-called free options are options whose cost is hidden in the overall cost of the vehicle.

DETERMINING THE VALUE OF YOUR CAR

After a decision has been made as to the car you want, the next item to be dealt with is how to get rid of your old car. If you are lucky enough to not have to deal with this step, or you are keeping your current car, then "no problem." Just skip this step and concentrate on purchasing your new car. But if you aren't keeping your old car, then how you get rid of your old car can impact how you buy your new one. The four options for disposing of your old car are sell it yourself, trade it in, sell it to a salvage yard, or donate it to charity. But before you decide on how to dispose of your car, you need to determine its value.

As mentioned previously, there are a number of factors that affect the value of your used car. In addition, mileage, geographical location, time of the year, and vehicle condition can affect its value. The more mileage a vehicle has, the less it is worth. Geographical locations and the time of the year are considerations in that, for example, convertibles *may* be more valued in warmer climates and summertime while four-wheel drive vehicles *may* be more valued in northern climates and in the wintertime. Luckily, the price guides will take these factors into account for you, so there is little need to worry about how to calculate them. The one factor that you will need to determine personally is the condition of your car. Condition is usually categorized as excellent, good, fair, and poor. Most of us would like to think our cars are in better shape than they are, but a quick look at the descriptions for the four types will help clear things up.

Excellent Condition

Excellent condition means that the vehicle looks great, is in excellent mechanical condition, and needs no reconditioning. It should pass a smog inspection. The engine compartment should be clean, and there should be no fluid leaks. The paint is glossy and free from scratches, dents, or dings. The body and interior are free of any visible wear or defects. No rust should be visible, and there should be no obvious repairs made. The tires are the proper size, and they are all the same or match one another and look new or nearly new. The vehicle is in excellent shape and it has a clean title. This is the type of vehicle most would want to buy if purchasing a used vehicle.

Good Condition

Good condition means that the vehicle is free of any major defects. The paint, body and interior have only minor defects, and there should be no major mechanical problems. Any rust present on the vehicle should be minimal, and a deduction should be made to correct for it. The tires should match and have 50% or more tread wear left. A vehicle in good condition will need some reconditioning to be sold at retail. Any major reconditioning needed should be deducted from the value of the car. The vehicle should have a clean title. Most cars owned by consumers fall into this category and are probably what most of us have if we have taken good care of our cars.

Fair Condition

Fair condition means that the vehicle probably has some mechanical, body, or interior defects, but is still in safe running condition. The paint, body, or interior need some reconditioning or work before it is sold. The tires may need to be replaced. There may be some repairable rust damage. The vehicle should have a clean title. The value of fair condition will vary quite a bit. Even if the vehicle undergoes significant reconditioning, this vehicle may not qualify for the full used car retail price.

Poor Condition

Poor condition means that the vehicle has severe mechanical, body, or interior defects and may be in safe running condition. The vehicle may have problems that cannot be readily fixed, such as a damaged frame or a rusted-through body. It may have already undergone substantial reconditioning. An example might be a vehicle that was made from two totaled or severely damaged vehicles. A vehicle with a salvage or flood title, or with excessive or unsubstantiated mileage, should be considered in poor condition because of potential problems. Any vehicles in this category should be looked at very closely and appraised independently to assure its condition, value, and safety.

Interior

Some specific areas to look at while determining the value of the car are its interior, exterior and tires. The headliner is the lining above your head on the underside of the roof. It is usually not a problem, but if it is faded or has stains or tears, it should be replaced (which requires time and expense). The upholstery should be clean, not faded, and free from rips and tears. Leather, if it is in good shape, usually has some added value; however, it is much more expensive than cloth or vinyl to repair. Some carpet wear will be evident in older cars and will need to be replaced if the vehicle is to bring top dollar. Stains, tears, and heavily worn areas will lower the value. The dash does not present many problems. But if it is scratched, faded, cracked or shows other damage, then its condition, too, will lower the vehicle's value. Sun damage and fading are the most common of dashboard problems.

Exterior

Exterior items are subjective items, which means the seller is a little more forgiving of imperfections or problems, while the buyer tends to want the vehicle to be in better shape and is, therefore, more critical of any blemishes. Any car will be subject to more scratches and dents the older it gets and the more miles put on it. Many times the driver may not notice these problems. Any damage should be closely looked at, as other subtle damage may have also occurred. Any repairs or reconditioning for damage to the body or paint that may have been made may call for a deduction in value, although this deduction should be much less than if the damage had not been repaired. If two cars look the same, the one with some damage repair will have a little less value than the one that did not have any initial damage.

Be sure to look in and around moldings, wheel wells, headlamp and tail-lamp moldings, and all the hinge areas of the hood, doors, and trunk. You are looking for any signs of rust and corrosion. Rust under the paint will raise the surface of the paint and it may look like the paint is bubbling up. The windows should be free from cracks and scratches, and they should also operate properly. The windshield and rear window are usually the most expensive to replace. If the front windshield is cracked or shows sand damage or scratches from windshield wipers, it should be replaced. Some states allow minor cracks in the windshield if they are not in the sightline of the driver, while other states do not allow any cracks. In most cases a damaged windshield will need to be replaced. The tires must also match and be properly sized for the car. They should be new or nearly new to avoid a deduction. If the tires are worn, then one of the first things that needs to be done is the purchase of a new set. Bear this in mind if you are buying or selling a used car. New tire tread depth for passenger cars is from $^{10}/_{32}$ to $^{12}/_{32}$ (tires are considered unsafe at $^{2}/_{32}$. A tire with $^{6}/_{32}$ or $^{7}/_{32}$ tread depth is considered to have 50% tread remaining. Anything less than 50% tread wear can result in a deduction.

Title

A check on the title to see if it is clean can affect the value of the vehicle. The best way to check a vehicle's title to see about its background is to use a service such as Carfax (http://www.carfax.com). You can find out a lot about the title history of a used car and see how many owners it had and where it came from. The title type, such as a salvage title, can even be indicated. However, these services cannot guarantee that a vehicle has not been flooded or wrecked. It is possible for a vehicle to be sold from state to state in order to clean up its title, or for an owner to buy back a wrecked or flooded vehicle from the insurance company before it has a salvage title assigned to it. But a service can tell you if it has been assigned a salvage title.

If a vehicle is found to have a salvage title, an industry acceptable deduction is 50% after the condition has been determined.

GETTING THE VALUE OUT OF YOUR CAR

When you decide to get rid of your vehicle, there are several ways in which you can do this. Your options include selling it yourself, trading it in, selling it to a salvage company, or donating it to charity.

Selling It Yourself

Of the options mentioned, selling the vehicle yourself will usually result in the most money, yielding somewhere between trade-in value and used car retail value. This value is referred to as the *private party value* because it represents the value you might expect to receive when selling your own used car to another private party. It also represents what you might expect to pay for a used car when purchasing from an individual or private party. The *N.A.D.A. Blue Book* or the Kelly *Blue Book* and its website http://www.kbb.com are excellent resources in determining the price of your car. Other sources are the newspaper classifieds, automotive magazine classifieds, and other sites on the Internet. Another place to check is to see what a similar car on a dealer's used car lot is selling for, but remember that the dealer is asking used car retail, and your price will usually be a little lower than that. The dealer usually has cleaned and prepared the car for resale and has some other overhead costs associated with the car, such as financing, advertising, and general expenses. You will need to take some time to get your car ready for resale and perhaps some time and money to advertise your car for sale. If you take the time to prepare your car for sale, you should be able to get more money for it in the end.

Be honest about your car and upfront with any problems it may have. Include any records you may have for its maintenance and repairs. Set your price a few hundred dollars above what you want, to give some room for negotiating the price. Be sure to sell it as is and make a note on the bill of sale that you are selling it "AS-IS," and that you are not offering a warranty of any kind. Ask the buyer for a cashier's check. A personal check can bounce and you should not be walking around with a large sum of cash. A next best option is to go to the bank of the buyer and get a cashier's check and sign the title over all at once.

Trading In

An approach that will get you less money for your car, but also is less hassle, is to trade in your car during the purchase of your new car. This trade-in price is often negotiable and the trade-in vehicle condition open to interpretation. The dealer is, in effect, buying your car, so is interested in its condition. Your trade-in will shortly end up for sale on the dealer's lot. Again, the cleaner your car and the better the condition, the more value it will have. Use your resources to help you determine the trade-in value of your car. Using your car as a trade-in is a better option if it is not in good or excellent condition. It rarely pays to make major repairs to a vehicle prior to a sale or trading it in.

Vehicle Salvage Companies

Another option for cars in poor or fair condition is to contact a vehicle salvage company. It will buy vehicles for salvage, disassemble them, and offer the components for sale. It's strange to think that a car is worth more as parts than as a whole. Bear in mind that a salvage yard won't give you a lot of money for your vehicle, and many people are shocked or feel insulted to only get 5% or 10% of what they thought it was worth. A salvage yard will almost always use a salvage price for a vehicle, which will fall in between scrap metal price and poor condition price. The salvage yard's primary source for vehicles comes from insurance companies that "total" a vehicle and contract with the salvage company for sale of these wrecked vehicles. If you have a vehicle in poor condition, a salvage company may be an option, but make it your last option.

Donating to Charity

If the condition of your car is poor or fair, donating it to charity may be a viable option. Organizations such as the Salvation Army may be able to put the car to use. You may also find that the tax benefit received might be comparable to the trade-in value. This is not always the case, but the donation is for a good cause. Other places to donate your car include churches and schools.

If you have a used car to dispose of, research the value of the car and determine which option for getting that value works best for you. After you have made this decision, then comes your next big decision: to buy or to lease the new car.

TO BUY OR TO LEASE

There are a number of factors to consider in deciding whether to buy or lease. When you decide to buy a car, your payments tend to be higher because in the end you will eventually own the car. But don't you own it right away? You only own the car if you pay cash or have made all the payments. Most people will take a loan or finance the car through a bank or other financial institution. So is leasing better? Most people who lease cars tend to have lower payments than if they purchased the car. A lease can be thought of as renting the car from the lessor. The lessor is the company, usually a bank, that buys the vehicle and leases it to you. You pay the lessor for the right to drive the vehicle during the term of the lease. At the end of the lease, you return the car to the dealership.

When investigating the alternatives, be sure to use the resources mentioned before, as several of them have current tips on buying and leasing.

The Advantages of Leasing

Let's look at the positive aspects of leasing. On a new car the down payment is usually lower, or sometimes there is no down payment involved. Parting with a substantial amount of cash as part of a transaction is something a lot of people avoid. Many times we don't have a lump sum of cash to use as a down payment. The monthly payments are usually lower compared to purchasing, and sales tax is usually paid monthly, based on your lease payment. Leases usually involve new cars, so if you are leasing, it will likely be a new car. At the end of the lease period you return the car to the dealership, which means the disposal of the car is not a concern. You won't have to hassle with the decision to sell, trade, or donate the car. You then decide if you want to lease another new car or buy one when the lease has ended. There are also some tax benefits if you use the leased vehicle for business purposes. Many businesses lease vehicles because they get to deduct the cost of the lease as an expense.

The Disadvantages of Leasing

There are some limitations to leasing. It is important to know them and to see if you fit into the best profile for leasing a car. There is almost always a mileage limitation on a lease car. It's not an unreasonable limit, but is around 15,000 miles per year for the lease. If the lease period is two years, then your mileage limit is 30,000 miles in this case. Any miles over the limit will incur a penalty in a cost per mile over the limit. The planned mileage for the car should be around the mileage limit. If the planned miles are more, you might want to purchase additional miles when you negotiate the lease, but don't go too far over. Leasing is not a good idea for individuals who plan on exceeding the mileage limit, or for those who will not come near it. The lease does not allow for a mileage credit. If you drive the car 20,000 miles and the lease limit is 30,000, you are not going to get a credit for 10,000 unused miles when the lease is up.

You may also be penalized for excess wear and tear on a vehicle. Any damage or wear in excess of normal for the given mileage and age will need to be paid. When the vehicle is returned, it needs to be in a near sale-ready condition. Modifications to the vehicle are usually not permitted. Any that are permitted and made will need to remain on the vehicle. An example is the addition of bug deflectors or running boards. These are usually made at the dealership at the time of lease, and these items will remain with the vehicle. Any modifications made or accessories added to the vehicle and then removed are subject to a penalty.

Leasing also tends to be more complex and confusing than buying, mainly because there are a lot of confusing terms. Not understanding these can lead to opportunities for you to become the victim of fraud or misrepresentation. The same thing could happen if you were buying a car, but it seems more prevalent in leasing, again due to the confusing terms. While lease payments are lower than buying payments, remember, you are not building equity. Equity is the value or ownership that you build up in a car over a period time as you make payments. While you are leasing, you do not own any equity in the car; the lease company has the equity. Therefore, they are assuming all of the risk of investing in the vehicle. For example, suppose the car you lease is really popular right now, and it looks like the value of the car two years from now will be really good. But, in the meantime, gas prices go up or down substantially. In the case of higher gas prices, a low mileage vehicle will drop drastically in value, while a vehicle with good mileage may not depreciate as much. The lease company assumes the risk of depreciation during the lease period.

Don't forget that at the end of the lease, you return the car with nothing to show for your payments. Leasing a car is similar to leasing or renting an apartment or house. When you are done, you can leave with no obligations as long as there is no excess wear or damage. Chapter 16 covers leasing in more detail.

The Advantages of Buying

Buying also has its pros and cons. Let's look at the advantages first. Buying a car is a rather straightforward approach with two major options for purchase: You can pay cash or you can finance until you have made all the payments and own the car. Buying works well for those who plan on keeping the car for three to four years or more. The vehicle generally will cost less, and you are building equity in the car, which can be used to help buy another car in the future, either from a private sale, or as value during a trade-in. It is a fairly simple process to buy a car. You don't have to worry about any limitations such as mileage, wear and tear, or modifications as you would if you had leased the vehicle.

The Disadvantages of Buying

There are some disadvantages to buying a car. If you need to sell the car in the first year or two, you will probably not be able to sell it for what you owe on the car. Two factors combine to create this situation. The first is depreciation. Even though it is a new car, most of the decline in the value of the car occurs in the first few years. So the value of the car starts to decline right from day one. The second factor is the loan itself. In the first years of a loan, you are paying mostly interest and not much on the principal. The principal is what you actually owe on the car. You could find yourself still owing on the car even after it has been sold. This situation in which you owe more on the car than what it is worth is often called being "upside down." A way to avoid this situation is to use a down payment or to increase your down payment, which can sometimes get you a lower interest rate for a loan. A bank does not want to see a buyer in this "upside down" situation. Let's face it, if a buyer has to sell the car in the first couple of years, the bank is at risk for the outstanding balance and trying to collect. A bank wants the lowest-hassle method and will sometimes reward a buyer who has a larger down payment with a lower interest rate because it is less risky for the bank. So, without a hefty down payment, you may find that your monthly payments are too high.

PAYING FOR THE CAR

Generally, it is important to determine how to pay for a vehicle before going in to make the purchase. Research your options and make this decision beforehand. Knowing how you will pay for the vehicle will reduce the chance to fall victim to manipulation at the dealership. Planning your financing options is a very important step and must be mentioned again. The three routes to financing are paying cash, using dealer financing, and borrowing from a bank or credit union.

Paying Cash

Paying cash is definitely low hassle and gives you the flexibility to avoid financing all together. Paying cash does not literally mean having the actual cash in your hand to pay the dealer. Having money in the bank from which to write a check, or other financial transfer of funds will be sufficient. Most of us don't have readily available amounts of cash to pay for a new car, but if you do, then take a look at financing the car and using the cash to invest and see if this strategy will save you money in the long run. Financing a car, either in whole or in part, and making regular payments will also help establish a good credit history. It might also be good to consider keeping some cash for emergencies that may pop up. Even if you have the cash, you should still look at all of your options.

Using Dealership Financing

Dealership financing is easy to obtain, especially when special interest rates or promotional rates are available. However, you will need to watch the terms of these agreements and read the fine print, as you should on any contract or document you are going to sign. Some of these special rates are for a limited time, meaning that the low interest rate might be good only for a grace period, or the first year or two or three. After this grace period, then the rate jumps to a much higher rate. You might find a "balloon" clause in which if the car is not paid for in a set amount of time, the whole loan is recalculated starting from the beginning at a higher rate. In general, regular-rate dealership financing tends to be higher than that obtained from a bank or credit union. Dealership financing may be the only option for those people with no credit history or even a bad credit history.

Borrowing from Banks and Credit Unions

Banks and credit unions are good places to shop for money. It may seem strange, but when you take out a loan, you are actually buying money. But for the purpose of buying a car, getting money from a bank or credit union is a loan. Shop around at various local banks to get the best rate that you can. If you belong to a credit union, don't forget to check with it as well, as it may have better rates than banks. An important step if using a bank or credit union is to get your loan preapproved. This status has several advantages. You can find out if the car you want is affordable, the monthly payment can be estimated for your car, and, once the car purchase is finalized, you just need to go into the bank and get the check to conclude the deal.

DETERMINING THE VEHICLE COST

So just how much do you need to give the dealer for the car of your dreams? A little cost research for the vehicle you want will tell you that. To many people the research consists of little more than looking at the vehicle sticker MSRP (manufacturer's suggested retail price) and guessing what they should pay (Figure 15–3). The MSRP is the suggested price that the dealer should charge the public. In most cases the actual amount paid for the car will be below the MSRP. In fact, it is unusual to pay the MSRP for a car, and rare that one would need to pay over the MSRP. In the case of low-production and high-in-demand vehicles, a customer might pay over the MSRP, but for most cars, paying below the MSRP is the norm.

Some dealers mark up the MSRP and display a higher price. It is important to remember that these prices are meant to be negotiated and haggled over. Most people who pay the MSRP have not done their research or are uninformed. You can use the fact that the MSRP is open to negotiation to your advantage. The dealer knows most people will try to negotiate a price lower than the MSRP. So how much is open to negotiation? More research!

You will need to find out what the dealer paid the manufacturer for the car and a few other prices. Use the same resources mentioned before, and

Figure 15–3: The MSRP is displayed on the vehicle along with the options and their retail cost.

Dealer Invoice
+ Advertising Cost
– Hold Back
– Factory to Dealer Incentives

= Vehicle Cost to Dealer

Figure 15–4: The formula for calculating the dealer's cost of a vehicle.

you can find out what the dealer needs you to pay based on a simple formula: dealer invoice + advertising cost – holdback – factory to dealer incentives = vehicle cost to the dealer (Figure 15–4).

So now let us take a quick look at how to calculate the dealer invoice. The *dealer invoice* is the vehicle's base model cost to the dealer, plus the cost of the options, and the cost of shipping and destination, a fee the manufacturer charges for the delivery of the vehicle to the dealership from a given area of the country. Some manufacturers charge the *advertising* fees to a dealer based on the dealer's sales region. Other manufacturers may include them in the base cost of the vehicle. This fee is reasonable and should not present a concern for the buyer other than to include it in the cost calculation. Figure 15–5 shows how a typical dealer invoice must be calculated. In this example, the dealer invoice is $21,500.00. The *holdback* is money returned to the dealer at set intervals either monthly or quarterly. The money returned is based on a percentage of the MSRP, or on the dealer invoice once the dealer sells the vehicle. Holdback reduces the cost

$19,000	Invoice price
750	Air conditioning
850	Automatic transmission
200	Premium sound system
500	Shipping and freight
200	Advertising
$21,500	Dealer invoice

Figure 15–5: This example shows how to calculate the dealer invoice cost

to the dealer and allows it to sell a vehicle at or even below invoice and still make money. So when a dealer offers to sell you a car at $100.00 over invoice, there is still a good profit being made. Other money to the dealer comes under the category of *factory to dealer incentives*. These are usually unadvertised and are paid directly to the dealer when the vehicle is sold. The dealer may or may not tell the buyer about these incentives, but either way they lower the cost to the dealer and need to be included in the formula to calculate the true cost of the vehicle. Look for this information in your resources or check the Internet for the latest prices.

Let's work through an example. After some research, you find that this car has a 2% dealer holdback based on the MSRP. The MSRP of the vehicle is $24,500. Take 2% of the $24,500, and you find the holdback is $490. You also find the factory to dealer incentive is $750 for the vehicle. So now do the math: $21,500 (dealer invoice) − $490 (cash back) − $750 (factory to dealer incentives) = $20,260, the true cost to the dealer. Keep in mind that the factory to dealer incentives can change, depending on the time of year, model popularity, and sales rates. There may be no incentives. There may not be an advertising fee, or the holdback may be based on invoice cost instead of MSRP. Also be on the lookout for options that have been bundled for a discount. Often the manufacturer will bundle popular options together and offer them at a discounted price compared to purchasing them separately. Calculate the dealer cost, and then allow for a 2–5% profit over the dealer cost. The dealer needs to make a profit in order to stay in business.

ADDITIONAL COSTS (ADD-ONS)

Be aware that there are a number of so-called add-ons, or additional fees, at the dealership when closing the deal. There may be considerable pressure to purchase or pay for these add-ons. And be aware that there are a number of tactics and strategies that may come into play when getting close to the end of the deal. The dealer may "forget" to factor the cost of options until the closing of the deal.

Don't get caught off balance. Just make sure you have done your research on the cost of the options. Then be sure to look at your wants and needs and see if they match up to what you are willing to pay.

Taxes

In addition to the vehicle and your chosen options, there are some other items that will need to be paid for. State, local, and business sales taxes must be paid on the purchase price at the time of purchase. When leasing, these taxes are based on the monthly lease payment and are paid each month. State and local motor vehicle taxes for tags and registration must be paid whether you are purchasing or leasing. These fees are required and outside of the control of the dealer.

Fees

Other fees may include the processing fee and the prep fee. The dealer may charge a processing fee for doing the paperwork associated with the purchase or lease. The dealer may also charge a fee for preparing the car for sale. These fees do little more than enhance the profit for the dealer; they aren't really necessary. So if the dealer presents these fees, just ask the dealer to include them in the overall cost of the vehicle. If the dealer does not include them in the price, then keep them in mind when negotiating on the final price.

You may encounter a market adjustment fee, which is a dealer markup fee usually added to the MSRP. The dealer may actually get this fee for a vehicle that is in high demand. You may try to get the dealer to forgo this fee as part of negotiations. If you are after a high-demand vehicle, then you may have to pay it to get the car you want. In the majority of cases, it is simply a dealer tactic to get a larger profit on the sale.

LOW-HASSLE OPTIONS

There are a number of low-hassle approaches to getting a car. You can use a buying service or a broker, or you can go to a one-price dealership. Just be sure to first check out the dealer cost on a car to make sure you are getting a good deal.

Buying Services

Buying services use their personnel to purchase a car for you. They vary in their exact approach but most take your vehicle request and contract a price for you. Some of these services charge a fee, while others get a commission from the dealership once the sale is concluded. Some buying services have a set price that is available when you call. They have already negotiated the prices. Other services will call you back with a quote. Usually, they have sent your request to a number of dealerships and are collecting the best prices available. Most of these buying services are after serious buyers interested in purchasing a car immediately and this time constraint may not give you much time to do your research.

Most local AAA offices offer car buying help or can refer you to a car buying service. Credit unions are another good source for buying service information. Even if you don't use the buying service when making your final purchase, it can give you an idea of the most you should pay for the car.

Brokers

Another low-hassle method is the use of a broker. A broker can negotiate a deal for an automobile at a fixed rate. You tell the broker what you want and how much you want to pay, and the broker will negotiate a deal that the broker feels is fair. While low hassle, this is a service for which you need to pay. If the price that comes back from the broker is not a good deal, then you will be out the fee for the broker. Check the broker's reputation before choosing this method.

One-Price Dealerships

One-price dealerships have been steadily increasing in numbers due to their popularity with a growing number of people. These are also known as "no-haggle" or "no-dicker-sticker" dealerships. Their main appeal is the low-pressure, no-hype sales approach. The dealer posts the discounted price, and there is no negotiation. But you still need to know if the discounted price is fair. Find out the dealer's cost for accurate comparison. If the advertised price is too high, then it may not be worth the extra cost just to avoid haggling.

NEGOTIATING YOURSELF

The method of choice still is negotiating the price yourself. If you research the costs and are properly prepared, many times the price you can negotiate will be as good or better than what you could get through the low-hassle alternatives.

Most dealerships aren't difficult to deal with when it comes to negotiating. Again, use your resources if you feel the need to check out a particular dealer's reputation. Working with a friendly dealership makes the negotiating process a little easier. Even if you run across a not so friendly dealership, you have done your research and know what the costs are. Know the power of "NO." Some sales personnel will put a lot of pressure on you to buy after you've said "no." If at any time you don't feel comfortable or get upset, say no and leave. The fact that you can leave is one of your best negotiating tools.

Remember to negotiate your deal in a specific order. First, the price of the new car, second, the price of the trade-in if you have one, and last, the financing. Conclude each step before going on to the next step. When negotiating the price of the new car, make sure you allow for the fees that you must pay, like taxes, tags, and title. Allow a fair profit for the dealer. The dealer, like any other business, needs to make a profit. That's what they are in business to do—make money just like the rest of us. Compare the price the dealer is offering to the prices that you have researched. Don't be afraid to show the prices that you have. If a dealer cannot meet the price you want to pay, you can leave and keep shopping somewhere else.

GOING TO THE DEALERSHIP

There are no really best times for going to a dealership, but some are better than others. The last few days of the month are better because sales personnel may be trying to meet sales quotas. The last hour of the day is better because the salesperson may want to go home. During a sale is another good time. Sales may occur during September because the dealer will need to clear out inventory to make room before the new models are delivered in October. The dealer and the manufacturer may also offer extra incentives for the sale of these cars. Winter months usually mean higher utility and holiday bills. Coupled with a sales slowdown, sales personnel may be willing to compromise more on the price of the car during winter.

When you do decide on a time to go, find a dealer that has the car you want. Call the dealer and ask about a test drive. Test drive the vehicle to see if it has any problems. Inspect the vehicle for any damage. Look at the mileage. If the mileage is more than 200 miles on a new car, then it may have had several test drives. You may wonder why. If it is the vehicle you want, ask about a discount for the extra miles. Keep in mind the warranty may be affected by the mileage. The initial mileage should not affect a lease for mileage limitations. Be sure it is the car you want before going any further in the negotiations.

Ordering a vehicle is one way to get what you want in a car. Ordering may get you a little better deal, as the dealer has a sure sale and doesn't have to pay interest on a car while it is sitting around. Drawbacks include waiting for delivery, lack of a test drive, and a selection limited to vehicles built domestically. Most people don't order cars. You can usually find a vehicle on a dealer's lot, unless you want some special options, such as a vehicle with limited slip differential, a particular engine, or a towing or off-road package, or are particular in your choice of options. Likewise, vehicles for special uses may need to be ordered.

When you go to a dealer, take along a friend or relative to help you in the buying process. He or she can give you moral support and can bring with them an objective eye, alerting you to some of the tactics that may be played against you.

TACTICS AND TRICKS OF THE TRADE

There are a number of tactics and tricks of the trade that dealers can use when negotiating with a buyer. After all, they play the game every day and, quite honestly, are quite good at it. You can level the playing field somewhat if you know what to expect. So let's look at some of the more common practices that might be used.

Mixing Transactions

A dealer or salesperson may try to mix transactions. The purpose of this technique is to rush you and confuse you into agreeing to a higher price than you originally intended. A popular tactic is to ask what you want for your trade-in vehicle before you agree on the purchase price. By satisfying the buyer with a higher trade-in price, the dealer can then ask for a higher price on the vehicle. Sometimes using the retail-to-retail prices of both cars as a guideline, the buyer feels that the price is "fair." Or another transaction mixer is when the dealer asks what you want your monthly payments to be. What your monthly payment will be is based on the price of the vehicle. In this case the dealer is mixing financing with negotiating the purchase price. Any mixing of transactions will make the price more difficult to determine, which takes away your buying leverage.

Good Cop, Bad Cop

While the expression "good cop, bad cop" is a cliché, the relationship it describes does exist. Its success as a negotiating tactic depends on your willingness to be a buddy to the salesperson. The salesperson will act like your partner—you and him against the sales manager. He'll tell you that he will try to get the sales manager to come down on the price if only you'll come up on what you're willing to pay. Many times you will have come up more than they have gone down. What happened to your research? When exactly did the salesperson become your buddy? Don't fall for this tactic. Stand firm and show your salesperson that the prices you have researched are fair.

Dealer-Installed Options

A sure way to enhance the dealer's profit is the use of dealer-installed options. These are options you usually don't want that the dealer will try to add in after the final price is set. Should the vehicle have these dealer-installed options, ask the dealer to remove them. If the dealer cannot or will not remove the items, then say that you are not paying for these extra items or ask for a vehicle that does not have them. Otherwise, the best thing to do is to leave and find another dealer.

Good Faith

Sometimes the buyer may be asked to show a sign of "good faith." When a buyer is interested in a car, the salesperson asks for an amount of money to hold the vehicle. This amounts to little more than an early down payment and makes the negotiating more difficult. Once they have your money, there is little incentive for them to give it back should you decide to leave and go elsewhere. You should not give the dealer anything until the decision is made to purchase—not a check, or credit card, or a driver's license. Your presence at the dealership should be good faith enough.

Trade-In

If you have a vehicle to trade in, it is not unreasonable for the dealer to examine the vehicle to appraise its value. But make sure you give the dealer a spare set of keys or have an extra set made for this purpose. Sometimes a dealer will hold your trade-in to prevent you from leaving, telling you that your old vehicle is on its way over from wherever it is. Meanwhile, you will be under pressure to buy the new car. Should this delay tactic be used, demand your car. Another excuse is that your keys got misplaced and they are looking for them. Giving them a spare set of keys makes that ploy useless to them.

Emotional Plays

Emotional plays are another favorite selling strategy. Intimidation, high-pressure, and "poor me" are examples of emotional tactics that dealers might use. Dealers may say that they are not making a profit. They may understate or flat out deny dealer holdback or factory to dealer incentives. Or they may even question the prices you have researched. Stand firm. You know their costs. Remember, if you are not comfortable or are upset, then leave. Don't lose sight of the fact that you are buying a car at a fair price. You have the ultimate last card, the power of "NO."

Let's review some things to keep in mind when buying a vehicle:

- Know what you want.
- Do your research.
- Take someone with you.
- Allow plenty of time to negotiate.
- Thoroughly inspect the vehicle.
- Know the power of "NO."
- Negotiate in a specific order.
- Leave if you become uncomfortable or upset.

And here are some final words of advice: Read everything carefully. Be sure you understand the terms of any contract or agreement—purchase or lease (see Chapter 16)—that you sign. Make sure it is what you want and what you are willing to pay for before signing the papers.

1. What are some decisions that should be made before you buy a car?
2. List some sources of information to help you when buying a car.
3. What are buying guides? What are pricing guides?
4. Why is the test drive important?
5. How do you determine the value of a trade-in?
6. What factors affect the value of a trade-in?
7. Describe the different categories of the condition of a used car.
8. What are some areas of the car to inspect?
9. Why might it be a good idea to do a title check on a car?
10. Describe some ways to dispose of your used car.
11. What are some advantages and disadvantages of buying?
12. What are some advantages and disadvantages of leasing?
13. Describe some financing options for purchasing a car.
14. What is MSRP?
15. What is the formula for calculating dealer cost?
16. Explain each factor used in the formula.
17. What are some required fees that must be paid when buying a car?
18. What are some fees that might appear when buying a car?
19. What are some of the low-hassle methods of buying a car?
20. What is the correct order in which to negotiate a deal?
21. What are some of the better times to buy a car?
22. List and explain some dealer tricks and tactics.

REVIEW QUESTIONS

LEASING

After reading this chapter, the reader should be able to:

- List some advantages and disadvantages of leasing.
- Compare leasing and taking out a loan.
- Describe the contents of the Federal Lease Disclosure.
- Explain the lease end procedure.

LEASING CONSIDERATIONS

Financing an automobile can be done through a loan or a lease. A loan finances the *purchase* of a vehicle and a lease finances the *use* of a vehicle (Figure 16–1). Each method has its own benefits and drawbacks. It's not possible to simply say that one is always better than the other because, well, it depends on your own particular situation and preferences. (See "To Buy or Lease" in Chapter 15.) You will need to calculate both methods and see which is financially better for you. Some factors that will affect buying versus leasing include interest rates, lease incentives, and the term lengths.

Advantages

There are a number of advantages to leasing. Some or all of these may apply to you should you decide to lease a vehicle. Some people may only be interested in specific benefits of leasing, especially if a vehicle is to be used in a business.

- Most leases require little or no down payment, making it easier to lease a new car and frees up cash for other things. However, a down payment can be made on a lease to lower your monthly payment amount.
- In most states you pay sales tax on the entire value of a vehicle when you purchase it. However, in a lease, you are only taxed on the portion of the value that you use during your lease. The tax is spread out and paid along with your monthly lease payment.
- You are only paying for the portion of the car or truck that you actually use, your monthly payments are 30–60% lower than for a purchase loan of the same term.
- Because your payments are lower, you may be able to get more car for your money and drive a brand-new vehicle every two to four

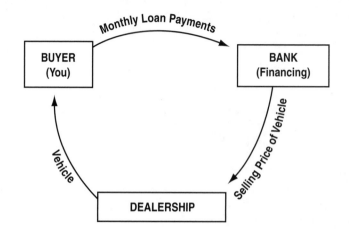

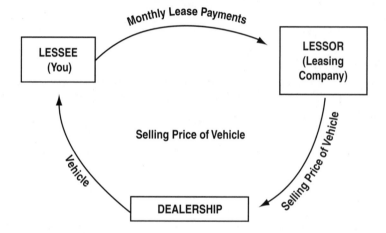

Figure 16–1: Buying and leasing are almost the same. In either one, the dealership sells the vehicle to a bank or lease company and you drive the vehicle in exchange for monthly payments.

years, depending on the length of your lease. (Most people like to lease for a length that matches the length of the manufacturer's warranty. Then if something goes wrong with the car, it is covered.)

- The headaches of selling a used car are eliminated. When your lease ends, you simply turn it back to the dealership or lessor and walk away, unless you decide to buy it.

Disadvantage

Though there are many advantages to leasing, it is wise to consider if you meet the best fit for a lease. If not, then a lease would be a poor choice and could wind up costing you more than purchasing.

- Lease contracts are purposely written to discourage and, in many cases, even prevent early termination. If you terminate a lease early, you will pay termination charges and all remaining payments. You will generally wind up making all the lease payments one way or the other. Ending a lease early is the most common leasing problem people have.

- While leasing allows you drive a new vehicle always under warranty every two to four years, you will always be making payments.
- Because leases typically require a smaller down payment and lower monthly payments, there is a higher risk to the lessor. So you generally must have a better credit rating than would be required for a loan to purchase. (Obtaining a lease is easier if you have a history of making credit payments on time and have minimal debt.) You may have to pay a higher finance rate to lease or be denied a lease.

LEASE VERSUS LOAN

Loan Payments

Loan payments have two parts: a *principal charge* and a *finance charge*. The principal pays off the vehicle purchase price. The finance charge is loan interest. When buying, the entire cost of a vehicle must be paid for, regardless of how many miles it is driven. Typically a down payment, sales taxes, and interest are rolled into the cost of the car and thus into the loan amount. You make your first payment a month after you sign your contract.

Lease Payments

Lease payments are made up of two parts: a *depreciation charge* and a *finance charge*. The depreciation part of each monthly payment compensates the lessor for the portion of the vehicle's value that is lost during your lease. The finance part is interest on the money the lessor has in the car during the lease period.

When you lease, you pay the portion of the purchase price that represents the vehicle's depreciation or decline in value. In addition to the depreciation, you must also pay for any other items you rolled into your lease, such as extended warranties or other dealer options. You have the option of not making a down payment. In most states the sales tax is paid, based on the amount of the monthly payment, and you pay a *money factor* that is similar to the interest rate on a loan. With leases, you may also pay extra fees and possibly a security deposit that you don't pay when you buy. You make your first payment at the time you sign your contract.

Negotiate Price

Many people think that the price of leasing a vehicle cannot be negotiated. This mistaken belief allows dealers to make a larger profit. Often they will imply, or state outright, that price isn't negotiable in a lease. They may try to show that leases are different because you aren't buying the car, which is simply not true. Always negotiate the lowest price possible. Remember that the dealer is still selling the car—it will be sold to the lease company rather than a bank. The price you finally negotiate for a vehicle is called the *capitalized cost*. Just as in buying a car, this cost would be lower than the MSRP. The lower the capitalized cost, the lower the monthly lease payments will be. The capitalized cost is sometimes called the *lease price*. The capitalized cost, or lease price, can be reduced by rebates, factory to dealer incentives, trade-in credit, or a down payment just as if you were calculating the final price of the vehicle for a purchase. Reducing the capitalized cost, such as by making a down payment, can create smaller monthly lease payments.

The difference between monthly lease and loan payments is simply the amount you are paying. In a vehicle loan, you repay the total cost of the car, which is a combination of the down payment and monthly payments. Leasing can free up cash for other things. Every dollar you don't spend buying a vehicle instead can be used for something else like paying off debt, investing, or putting into savings.

If the vehicle was purchased and then sold, and the money put in the bank, then the overall cost of leasing compared to buying, over the short term, is about the same. Comparisons sometimes show buying to cost a little less than leasing due to fewer fees and the assumption that a purchased vehicle will return full market value if it is sold or traded at the end of the loan (often a bad assumption). However, if you factor in the benefits of wisely investing your monthly lease savings, the net cost of leasing can, in the short term, easily be less than buying.

But if a buyer keeps the car after the loan has been paid off and drives it for several more years, the cost is spread out over a longer term. It doesn't take a math major to figure out that the cost of buying *one* car and driving it for ten years is less expensive than leasing *four* or *five* different vehicles for the same time period.

SOME THINGS YOU NEED TO KNOW ABOUT A LEASE

While buying a car is a relatively straightforward financial transaction, a lease is more complex because there are more factors that can affect your lease payment.

Leasing enables you to drive a more expensive vehicle than you could afford to purchase. But you need to know what leasing is all about before you choose that option. In the excitement of leasing a vehicle, it can be easy to overlook the total costs. The following items should be looked at closely on any lease contract.

Residual Value

The **residual value,** also called *lease-end value, buyback,* or *purchase option,* should be a reasonable estimation of what the vehicle will be worth at the end of the lease, with mileage and normal wear and tear factored in to the cost. If the lease is from a new car dealership, traditionally a residual value is based on a percentage of the manufacturer's retail price. Residual values decline with time as the car gets older.

Manufacturers and dealerships will sometimes inflate the residual value as a way to keep the lessee's payments lower. The higher the residual value, the lower your monthly payments. This factor is a bonus when planning to lease, but may not work in your favor if you plan on purchasing the vehicle at the end of the lease. The residual value is often ignored. It becomes important if a customer wants out of the lease early or if the allowable mileage is exceeded. Dealerships may mark up the residual value to allow a greater profit should you buy the vehicle back at the end or break the lease early.

Vehicles that change styles frequently or are fad vehicles usually lose resale value quickly, which means their lease residual value is lower and your monthly payments are higher. These kinds of vehicles could easily have higher payments than a more expensive vehicle with a better resale value.

If the residual value is lowered, your monthly payments will increase. The dealer stands to make a little more profit at the beginning of the lease. However, a lower residual value will make the vehicle cheaper to buy at the end of the lease. A low residual value is a disadvantage to the person who turns the vehicle in and leases another new one.

To see if the residual value figure is reasonable, check out the price of a two- or three-year-old vehicle similar to the one you are planning to lease. Check with the dealer to see if the residual value is negotiable. Sometimes it may be lowered without changing the monthly payments if the dealership is willing to give up some of its markup. Check to see if it is necessary to purchase the vehicle at the end of the lease, or if you are able to re-lease the vehicle or extend the lease at the end of its term. Also check to see if financing is available should you decide to purchase the vehicle.

At the time you lease, the lease company estimates the vehicle's residual value and, if the vehicle is actually worth less than the residual when you turn it in, the leasing company takes the hit, not you. Alternately, if the vehicle is worth more than the residual and you have the option to purchase, you may want to buy it, then sell it and make a profit. Although selling means some work on your part, you might know ahead of time that someone wants to buy the vehicle you are leasing.

Gap Coverage

Gap coverage, or *gap insurance,* pays the difference between what you owe on your loan or lease and what your vehicle is actually worth should the vehicle be stolen or destroyed. It is common to owe more than your car is worth for most of the life of the lease. If the car is totaled, you could still owe hundreds or thousands of dollars to the finance company even after your insurance has paid off a car you no longer have. This fact surprises most people caught in this unfortunate circumstance. Most leases have gap insurance, while most loans do not. You are better protected with a lease, unless you purchase the gap insurance separately at extra cost for a loan.

Mileage Limits

A lease limits the number of miles you can drive without being penalized. The most common mileage limit is 15,000 miles per year, although other limits may be used. If you exceed the mileage limit, you have to pay excess mileage charges at the end of the lease. When you lease, you may be able to select the limit that best fits your needs. Additional miles may be purchased when the lease is negotiated and are cheaper at the front end of the lease than paying excess charges at the closing end. Only purchase extra miles if you are going to use them. If you don't use the extra miles, you lose them at the end of the lease. The penalty for excess mileage can add up quickly, ranging from $0.10 to $0.25 per mile.

Vehicle Maintenance, Insurance, and Residency

You must return the car at lease-end with no more than normal wear and tear. Anything more than the normal wear and tear will be subject to you paying for excess "damages." Generally, the vehicle is expected to be in ready to resell condition, with little if any preparation. You are responsible for insurance, upkeep, and maintenance just as with a purchased vehicle. Some people mistakenly believe the leasing company is responsible. A leased car belongs to the leasing company. Therefore, you cannot make modifications and install custom equipment that alters the car. If you do, you will be charged for the cost of repairs to undo the modifications. This requirement is usually specified in the lease contract and is explained fairly well in a disclosure statement. Generally, you have to take good care of the car and keep it properly maintained.

Lease agreements require you to maintain full coverage insurance coverage. You will likely want full coverage for your new vehicle regardless of whether you're leasing or buying.

Be aware that many lease contracts may restrict you from moving in some way. Most prohibit moving their vehicles out of the country. Some leases may prohibit moving out of state. Check the lease contract if you may need to move during the lease period. Moving your vehicle out of state is usually not a problem. Your payments may change slightly due to different sales tax rates between states. When moving, always notify your lease company so they can make the necessary changes.

Closed-End or Open-End

Automobile leases are either closed-end or open-end. There's a big difference between the two types, and you should understand that difference. The type of lease must be clearly indicated on the lease contract as required by federal regulations.

The closed-end lease is the most common for consumers. It allows you to return your vehicle at the end of the lease and have only to be responsible for excess mileage and excess wear and tear, should they exist.

Open-end leases are used mostly by businesses. With an open-end lease, the lease company has no risk, but, rather, you assume the risk in the lease. When the lease ends, you are responsible for paying any difference between the residual value and the actual market value. This could be a significant amount of money should the market value of the vehicle drop. Check your lease to make sure it is a closed-end lease.

Money Factor

Now, since the lease company has used its money to buy the vehicle while you drive it, it is bound to have an interest in some sort of compensation for its money. When buying a car, the bank charges interest. A lease involves almost the same thing as interest, but it is called the *money factor*. The money factor, sometimes called *lease factor*, is specified as a small number such as .00285. Sometimes the dealer may quote this as a larger number like 2.84 that might make it sound more like an interest rate. The money factor can be converted to an interest rate by multiplying by 2,400. To convert the money factor to an interest rate, 2,400 is always used regardless of the length of the lease. In this example a money factor of .00285 multiplied by 2,400 = 6.84%. After you get an interest rate, you should be able to compare the monthly payment the dealer gives you to the monthly payment from your bank or credit union. You should check the dealers numbers against your own. The two sets of numbers should match very closely. If the two sets are not close, then ask the dealer for the numbers he is using for capitalized cost, residual value, money factor, and other terms.

Security Deposit

Some leases require a security deposit. The amount of the security deposit will usually be equal to or a little more than a monthly payment. It may be refunded at the end of the lease or held for any excess mileage or excess wear and tear that may be present. Or, in some leases, it will be used to make the last lease payment. A lease may refer to this fee as the "last payment."

It is not unusual to make the first and the last payment when starting a lease. Don't confuse this fee with a down payment.

Other Fees

The leasing company may charge an administration fee. This fee is often not explicitly specified in your contract, but is included in your capitalized cost when calculating monthly payments. Ask about this fee if you don't see it mentioned. The range for this fee is typically $200 to $850. While this is a legitimate fee on the low end, it amounts to a profit enhancer on the high end.

Sometimes the leasing company will charge a fee due at the end of the lease to compensate for the expenses of selling the vehicle. Sometimes this fee is required even if you decide to purchase your vehicle at the end of the lease. Try to negotiate this fee to a minimum or eliminate it if possible. See if it can be rolled into the administration fee to keep it to a minimum. If the fee is charged, a typical range would be from $200 to $500.

You will pay sales tax on the purchase amount of the vehicle if you buy it, usually having it added to the loan when financing. However, in a lease you only pay tax on the part of the car you lease. Usually, you pay tax on the monthly lease payment at the state and local sales tax rate. In most states if you make a down payment on your lease, you will be charged state and local sales tax on the amount of the down payment. It is paid when you sign your lease.

The registration, license, tag, and title fees are the same fees you would normally expect to pay in your state whether you lease or buy your new car.

Federal Lease Disclosure

On January 1, 1998, a new regulation went into effect that made a number of significant changes to vehicle leasing disclosures and advertising. This section in your lease contract should be titled, "Federal Consumer Leasing Act Disclosures" (Figure 16–2).

The main sections of the this regulation are as follows.

- A four-cell lease box with the four main cost disclosures that look like disclosures on a loan installment contract.
- A two column disclosure of the amount due at lease signing and the source of those funds.
- A ten-line worksheet disclosure of the calculation of the monthly payment including disclosure of the full capitalized cost, disclosure of the value of the vehicle, and a description of each.
- Additional disclosures regarding early termination, excessive wear and use, and the lessee's purchase option.

Amount Due at Lease Signing or Delivery (Itemized below)* $_____	Monthly Payments Your first monthly payment of $____ is due on the_____, followed by ___ payments of $ _____ due on the __ of each month. The total of your monthly payments is $_____	Other Charges (Not part of your monthly payment) Disposition fee (if you do not purchase the vehicle) $_____ _____ _____ Total $_____	Total of Payments (The amount you will have paid by the end of the lease) $_____

*Itemization of Amount Due at Lease Signing or Delivery

Amount Due at Lease Signing or Delivery:	How the Amount Due at Lease Signing or Delivery will be Paid:
Capitalized cost reduction $_____	Net trade-in allowance $_____
First monthly payment _____	Rebates and noncash credits _____
Refundable security _____	Amounts to be paid in cash _____
deposit	
Title fees _____	
Registration fees _____	
_____ _____	
Total $_____	Total $_____

Your Monthly Payment Is Determined as Shown Below:

Gross capitalized cost. The agreed upon value of the vehicle ($_____) and any items you pay over the lease term (such as service contracts, insurance, and any outstanding prior to credit or lease balance). .... $_____

If you want an itemization of this amount, please check this box ☐

Capitalized cost reduction. The amount of any net trade-in allowance, rebate, noncash credit or cash you pay that reduces the gross capitalized cost... −_____

Adjusted capitalized cost. The amount used in calculating your base monthly payment............................... =_____

Residual value. The value of the vehicle at the end of the lease used in calculating your base monthly payment.. −_____

Depreciation and any amortized amounts. The amount charged for the vehicle's decline in value through normal use and for other items paid over the lease term.. =_____

Rent charge. The amount charged in addition to the depreciation and any amortized amounts.................... +_____

Total of base monthly payments. The depreciation and any amortized amounts plus the rent charge........... =_____

Lease term. The number of months in your lease .... ÷_____

Base monthly payment.. =_____

Monthly sales/use tax... +_____

_____.. +_____

Total monthly payment ....=$_____

Early Termination. You may have to pay a substantial charge if you end this lease early. <u>The charge may be up to several thousand dollars.</u> The actual charge will depend on when the lease is terminated. The earlier you end the lease, the greater the charge is likely to be.

Excessive Wear And Use. You may be charged for excessive wear based on our standards for normal use (and for mileage in excess of _____miles per year at the rate of _____per mile).

Purchase Option At End Of Lease Term. You have the option to purchase the vehicle at the end of the lease term for $_____(and a purchase option fee of $_____). (You do not have an option to purchase the vehicle at the end of the lease term.)

Other Important Terms. See your lease documents for additional information on early termination, purchase options and maintenance responsibilities, warranties, late and default charges, insurance, and any security interest, if applicable.

Figure 16–2: An example of the Federal Consumer Leasing Act Disclosure. It contains key payment items, amounts due at signing, determination of monthly payments, early termination, and other disclosures.

The Federal Lease Disclosure is an excellent presentation of the information you need to know without so much detail that you get lost in the particulars. The key items that will interest most consumers are as follows:

- Amount due at signing
- Monthly payment
- Other charges
- Total of payments
- How monthly payment is determined
- Early termination statement
- Wear and tear explanation.

WHAT TO EXPECT AT THE END OF THE LEASE

About one to two months before the end of the lease, the leasing company will contact you. It will instruct you where to return the vehicle—usually at the dealer where the car was leased—and provide details regarding the inspection of the vehicle. You may be reminded of your option to purchase your vehicle, and you will be provided a purchase price different than the residual value, which may be a better price. It may also offer to extend your lease. Usually the dealer that leased the vehicle will simply inspect the vehicle and take it back. You then need to decide which option you wish to pursue. If you decide to buy the vehicle, no inspection is needed, and you simply finance the purchase exactly like buying a car.

Many people fear that the leasing company will pick apart the car and examine every missing fleck of paint and minor scratch at the end of the lease. In reality, the lease company is looking at the vehicle from the perspective that it will need to be ready for resale. Damaged windows or windshields, excessively worn tires, deep scratches, dents, damaged interior are items that will need to be repaired or paid for after the vehicle is returned.

You should handle all matters regarding the end of your lease with your lease company, not the dealer, unless specifically directed to the dealer by the lease company. Being directed to the dealer is common if the lease company is a captive lease company. A captive lease company is a finance company related to a particular automobile manufacturer or distributor. Bear in mind a dealer's lease-end advice may lead you to make poor decisions. The dealers' goal is to sell or lease you another new vehicle.

At the normal end of a lease, you may have some or all of the following options:

- Return your vehicle
- Extend your lease
- Re-lease your vehicle long-term as a used vehicle

- Purchase your vehicle
- Arrange for a third-party to purchase your vehicle

Make your decision ahead of time so that you'll be less likely to become a victim of the dealer's pressure.

SOME FINAL TIPS

Whether you choose to lease or buy, the following tips that apply to both choices and may help you get the best deal:

- Don't let the dealer tell you your source of invoice prices is wrong.
- Don't give the dealer a deposit on a car during negotiations.
- Don't reveal your attraction to a vehicle.
- Don't accept an offer to take a car home overnight.
- Don't let the dealer tell you that lease prices are not negotiable.
- Don't agree to extended warranties, credit insurance, or add-on services.
- Don't tell the dealer what monthly payments you can afford.
- Always negotiate price, never monthly payments.
- Always negotiate the price from dealer's cost, not from MSRP.
- Always be prepared with the dealer invoice price.
- Always know what your trade-in is worth.
- Don't sign any kind of agreement or contract until the deal is settled.
- If you're not comfortable with the salesperson, ask for another, or leave.
- Always give yourself the option of leaving if you become tired, confused, intimidated, or pressured.
- If you feel the salesperson is playing games with you, ask him or her to stop, or you always have the option to leave.
- Always check the dealer's monthly payment figures against your own figures.

Remember, if you decide you're not happy with the vehicle after you have signed the lease contract, it is too late to change your mind. You have signed a legally binding contract and will need to adhere to its terms. Whether you buy or lease, there is no three-day grace period in which you can cancel the deal. Once you sign the contract, it is legal and binding.

Give your options some thought, do your research, and don't make a hasty decision.

1. What is the difference between a loan and a lease?
2. Explain the components of a loan.
3. Explain the components of a lease.
4. What are some advantages of leasing?
5. What is residual value?
6. What is gap coverage?
7. What are the two types of leases? Which is better for consumers?
8. Explain the money factor.
9. How are lease payments different than loan payments?
10. What is contained in the Federal Consumer Leasing Act Disclosures?
11. What are some lease end options?
12. What are some tips to help you in your lease negotiation?

accelerator   The gas pedal.

accelerator pump   A carburetor pump that provides a squirt of extra fuel for added pickup when you depress the accelerator quickly.

accidental ground   A condition that exists when a wire connected to the positive battery terminal contacts a grounded metal part of the car. *See also* ground; short circuit.

adhesive   A substance that causes two materials to stick together.

aeration   The process of mixing air with a liquid.

air bag   A safety system that includes impact sensors, an inflation module, and nylon bags that will deploy during a collision.

air cleaner assembly   A metal or plastic housing that directs outside air into the fuel delivery system; contains the air filter element and may act as flame arrestor in case of backfire.

air ducts   Tubes or channels used to carry air to specific locations in the automobile. *See also* plenum.

air filter   The paper element in the air cleaner that filters dust and dirt from the air before it enters the fuel delivery system.

air pump   A belt-driven or electric device that pumps air into the exhaust manifold or oxidizing catalytic converter to lower the amount of unburned fuel in the exhaust gases.

air spring   A rubber cylinder filled with air that functions as a spring in a suspension system.

alignment   *See* wheel alignment.

all-wheel drive (AWD)   A system of powering all four wheels on a vehicle. Also known as four-wheel drive.

alternator   *See* generator.

American Automobile Association (AAA)   An organization that offers a wide a variety of services for motorists from roadside assistance to financial and insurance services.

antifreeze   An ethylene glycol-based or propylene glycol-based liquid that, when mixed with water in a car's cooling system, raises the boiling point and lowers the freezing point. It also helps to inhibit rust and corrosion in the cooling system. *See also* coolant.

atomization   The dispersion of fuel into tiny droplets to form a fuel/air mixture that can be burned in an engine's combustion chambers. Atomization takes place at the nozzle of a fuel injector.

automatic transmission   A transmission that selects and changes gears automatically through three mechanisms: a torque converter, a planetary gearbox, and a hydraulic control system. *See also* transmission.

automatic transmission fluid (ATF)   A special light oil used in automatic transmissions. The most common is Dexron III.

axle   A metal shaft on which a wheel rotates; also, a shaft that transfers torque directly from a differential unit to a wheel.

ball joint   A movable suspension or steering joint that allows three-dimensional motion. Ball-type joints are used to connect control arms to steering knuckles and also to connect tie rods to steering linkage.

band   A holding device in an automatic transmission that is used to engage and disengage planetary gear mechanisms for shifting.

battery   A device that produces electricity on demand through chemical reaction. The battery is charged by the electricity-generating action of the alternator.

bearing   A device that supports a rotating part and reduces friction.

blowby   A condition in which combustion gases are forced past worn piston rings.

Blue Book   The Kelley *Blue Book*, a bimonthly publication that lists used car wholesale and retail prices. It may be accessed on the Internet at http://www.kbb.com.

body   The parts of a vehicle that form the enclosure for passengers, cargo, and other parts.

body-over-frame   A type of vehicle construction in which the frame is the foundation, with the body and all major parts of the vehicle attached to the frame. *See also* unibody.

boot   A protective rubber or plastic cover.

bore   The diameter of a cylinder; also, to enlarge a cylinder.

bottom dead center (BDC)   The lowest position that a piston can reach in its stroke within a cylinder.

brake disc A flat, round metal part in a disc brake unit against which friction pads can be pressed to create braking action; also known as a brake rotor.

brake drum A hollow, cylindrical metal part in a drum brake unit against the inner surface of which semicircular brake shoes can be pressed to create braking action.

brake fluid The liquid used in a hydraulic brake system to activate pistons in wheel brake caliper or cylinder units; also used in hydraulic clutch linkage systems.

brake lines The reinforced rubber hoses or metal tubes through which brake fluid flows between the master cylinder and wheel brake caliper and cylinder units.

brake lining A heat-resistant friction material attached to a disc brake pad or drum brake shoe.

brake pad *See* pad.

brake rotor A disc shaped component that rotates with the wheel on which friction pads can be clamped against the spinning disc to slow and stop the attached wheel.

brake shoe A semicircular piece of metal to which a drum brake lining is attached.

brake system The components that enable a car to slow or stop. Consists of a brake pedal, linkage, master cylinder, brake lines, wheel calipers, cylinders, and either disc or drum brakes at the wheels.

break-in The initial period of operation of a mechanism, during which moving parts wear slightly against each other so that the surfaces become smoothly matched.

bushing A protective plastic, rubber, or metal liner that acts as a bearing to reduce friction and allow free movement.

CAFE standards *See* Corporate Average Fuel Economy.

caliper The portion of a disc brake that holds the pads and forces them against the disc to slow or stop the car. The caliper may be movable or fixed in place.

cam A projection on a metal shaft that provides a pushing force against other parts as the shaft rotates.

camber The adjustable angle at which the front wheels tilt away from the vertical, as viewed from the front of the car. *See also* wheel alignment.

cam lobe   The raised projection of a cam that extends outward from a camshaft.

camshaft   A metal shaft onto which cams have been cast or machined.

carbon dioxide (CO_2)   A relatively harmless gas exhaled in human breathing and absorbed by plants; also created when a hydrocarbon fuel is burned completely.

carbon monoxide (CO)   A poisonous gas created when the burning of a hydrocarbon fuel is incomplete.

carburetor   A mechanical device that atomizes fuel into a stream of air that passes through the barrel, or opening, formed within it.

caster   The forward or backward tilt of the steering knuckle from true vertical. *See also* wheel alignment.

catalytic converter   A mufflerlike emissions control device that chemically converts harmful exhaust gases into relatively harmless products.

cavitation   A condition in which a fluid pump attempts to pump air.

charging system   The electrical parts and connections that generate electricity and store it for use on demand. Consists of the alternator, voltage regulator, battery, and wiring connections. *See also* electrical system.

chassis   The underlying structure of a car, on which other parts are mounted.

circuit   A complete path through which electrical current flows from a source, through a load, and back to the source.

circuit breaker   An electrical device used to protect a circuit against an overload or malfunction. These devices are reusable and many will automatically reset themselves after a short period of time.

closed loop   An operating condition in a computer-controlled system in which signals from sensors are fed back to the computer, which monitors and adjusts a process to maintain a stable condition. *See also* computer; electronic control unit; open loop.

clutch   An assembly that can engage or disengage the flow of torque between rotating parts. *See also* clutch disc; clutch linkage; flywheel; pressure plate; throwout bearing.

clutch disc   The part of the clutch that takes power from the engine flywheel and transfers it to the input shaft of a manual transmission.

clutch linkage   The system parts that provide force to press a clutch into contact with a rotating part or to withdraw the clutch from contact. Includes the clutch pedal, mechanical or hydraulic linkage, clutch fork, throwout bearing, and pressure plate.

coil   A series of connected spirals or concentric rings formed by winding. For example, an electrical device formed from a length of wire wrapped into a circular pattern creates a magnetic force when electrical current flows through it. *See also* ignition coil.

coil-on-plug (COP) system   A system in which each cylinder has its own coil delivering spark to the spark plug. It is controlled by the computer. Sometimes called a distributorless ignition system. *See also* direct ignition.

coil spring   A thick steel bar that is coiled to expand and compress under force. *See also* suspension system.

cold cranking amps (CCA)   A rating of a battery's ability to deliver electrical current at 0 degrees Fahrenheit.

combustion   A process in which a fuel combines rapidly with oxygen; burning.

combustion chamber   The space at the top of a cylinder where combustion occurs; formed by the top of a piston, the cylinder walls, and the lower surface of the cylinder head.

compression ratio   A numerical comparison of the cubic volume in a cylinder when a piston is at bottom dead center (BDC) and top dead center (TDC). The higher the compression ratio, the higher the octane rating of fuel needed to prevent detonation problems during engine operation.

compression stroke   The upward movement of a piston that compresses an air/fuel mixture prior to ignition. *See also* four-stroke cycle.

compression test   A check of the amount of pressure created during a piston's compression stroke. A low compression reading can indicate valve or piston ring sealing problems.

computer   An electronic device that accepts input data in the form of electrical signals and produces output signals in response to a program, or set of instructions. *See also* electronic control unit; powertrain control module.

conductor   A device that allows current flow.

connecting rod   The metal rod that connects a piston to the crankshaft. Helps convert up-and-down motion of the piston into rotating motion of the crankshaft.

constant-velocity (CV) joint   A flexible joint used at the inner and outer ends of driving axles on front-wheel drive vehicles.

continuously variable transmission (CVT)   A transmission that has an infinite number of "gear ratios" to vary the torque needed at the wheels.

control arm   A part in a suspension system that locates and controls the movement of a wheel. *See also* suspension system.

coolant   A mixture of antifreeze and water that is circulated through the cooling system to transfer heat generated by the engine to the outside air. A proper coolant mixture protects against rust, corrosion, freezing, and boiling. *See also* antifreeze.

coolant pump   A belt-driven pump that draws coolant from the outlet tank of the radiator and forces it to circulate through the water jacket passages in the engine block and head and back to the radiator inlet. Formerly known as a water pump.

coolant recovery system   An addition to a cooling system that recycles heated coolant that has been expelled from the cooling system. As the liquid in the radiator cools and contracts after the engine is shut off, suction draws the coolant from a recovery bottle back into the radiator.

cooling system   A system that stores and circulates liquid to transfer heat from the engine into the outside air. A liquid coolant is heated as it circulates through the water jacket spaces in the engine head and block. As the coolant flows through the system, heat is transferred into the metal tubes and fins of the radiator. The heat is transferred into the air that passes through the radiator core. *See also* coolant pump; fan; fan belt; pressure cap; thermostat.

core   A radiator or heater part consisting of tubes through which coolant flows to transfer heat to air flowing around the tubes. *See also* heater core; radiator core.

Corporate Average Fuel Economy (CAFE) standards   Federal legislation requiring auto manufacturers to produce vehicles with engines that meet efficiency standards.

crankcase   The lower portion of the engine block that forms the housing for the crankshaft. The oil pan, or oil sump, is bolted to the bottom of the crankcase.

crankshaft   A metal shaft with offset journals to which the connecting rods are attached at the bottom of the engine. When the engine runs, the crankshaft rotates and converts the up-and-down motion of the pistons into rotary motion.

crankshaft pulley   The grooved wheel attached to the front end of the crankshaft, to which belts are connected to operate the fan, alternator, and other belt-driven accessories. Also called the power pulley.

crude oil   The natural state of unprocessed oil as it is pumped from the ground.

cylinder   A tube-shaped, hollow area in which a piston moves.

cylinder block   The lower part of an automotive engine that contains cylinder holes and which houses the pistons, connecting rods, and crankshaft. Also called the engine block.

cylinder head   The upper part of the engine that contains intake and exhaust ports, valves, and threaded holes for inserting spark plugs.

D

desiccant   A special substance that absorbs moisture in the air conditioning system.

detonation   The uncontrolled combustion of fuel that occurs too early in the four-stroke cycle; can damage pistons and other internal engine parts.

diesel engine   An internal-combustion engine that uses the heat of compression, rather than a spark, to ignite injected diesel fuel.

dieseling   A condition in which a gasoline engine continues to run after the ignition is switched off. Caused by fuel being ignited by heat within the engine.

differential   The system of gears that transfers different amounts of torque to driving wheels in turns.

dipstick   A flat metal rod that is used to measure the level of a liquid in a reservoir, using reference marks on the stick.

direct ignition   A distributorless ignition system in which the spark distribution is controlled by the computer. *See also* coil-on-plug system; distributorless ignition system.

disc brake   A brake assembly in which a caliper with pads exerts pinching pressure on a rotating disc to slow or stop a vehicle. *See also* brake system; power brake.

displacement   The volume displaced by a piston as it moves up the cylinder from bottom dead center (BDC) to top dead center (TDC). An engine's total displacement, or "size," is calculated by multiplying an individual cylinder's displacement by the number of cylinders. Displacement can be expressed in cubic inches, cubic centimeters (cc), or liters.

distributor   The device within the ignition system that distributes high-voltage current to each spark plug at the right moment for efficient combustion. *See also* electronic ignition; rotor.

distributor cap   The cap that covers the distributor mechanism and protects it from dirt and moisture. Contains towers to which each spark plug wire is connected, and a center tower through which current from the coil is conducted to the rotor.

distributorless ignition system (DIS)   A system in which each pair of cylinders has a coil delivering spark to the spark plugs for those two cylinders. It is controlled by the computer. *See also* direct ignition.

double A-arm suspension   A form of front suspension in which two A-shaped control arms locate and position a wheel and allow it to move up and down, as well as to pivot for steering. *See also* MacPherson strut.

drive belt   A flexible belt that turns driven units such as alternators, mechanical fans, air conditioning compressors, or air pumps. Also known as a V-belt, or multigroove or serpentine belt.

driveline   The assembly of a drive shaft and universal joints used on a rear-wheel drive or four-wheel drive vehicle to transfer torque from a transmission to a differential.

drive shaft   A metal shaft that is turned to transfer torque in a drivetrain.

drivetrain   A series of components that transfers engine power to the driving wheels. It includes a clutch or torque converter, manual or automatic transmission, drive shafts, and a differential.

driving wheels   The wheels on a vehicle, front, rear, or all four, that receive torque from the drivetrain.

drum brake   A brake that exerts pressure to force brake shoes against the inside of a rotating brake drum to slow and stop a vehicle.

dual overhead camshaft (DOHC)   An engine design in which two camshafts are located above the cylinder head to operate separate sets of intake and exhaust valves.

electrical system   Several separate systems that combine to store, generate, and distribute electrical current needed to start and run a car's engine and electrically operated accessories. *See also* charging system; ignition system; starting system.

electrolyte   The liquid in the battery that conducts electrical current between the terminals. Consists of a mixture of sulfuric acid and water.

electronic control unit (ECU)   A computer used to monitor and adjust an operating process. *See also* computer; powertrain control module.

electronic ignition   An ignition system that uses signal-generating components to operate a control module; the control module interrupts the current flow to the ignition coil or coils to create a high-voltage spark. Also called high-energy ignition.

emergency brake   *See* parking brake.

energy   The capacity to do work. *See also* force; power; work.

engine   An internal combustion device that intakes air and fuel and converts them to rotary motion.

engine block   *See* cylinder block.

ethanol   An alcohol-based alternative fuel produced by processing starch crops that have been converted into simple sugars. It is a renewable resource.

evaporative emission controls   Devices used to help control gasoline vapors from escaping into the atmosphere.

evaporator   The part of an air-conditioning system in which refrigerant is expanded to produce cold; air is circulated through the evaporator and into the passenger compartment.

exhaust gases   Residue from the burned air/fuel mixture that has passed through the exhaust system.

exhaust gas recirculation (EGR)   An emissions-control system that introduces exhaust gases into the intake air/fuel mixture to lower the burning temperature and reduce oxides of nitrogen (NO_X) in the exhaust gases.

exhaust manifold   The hollow structure that carries burned exhaust gases from the cylinders into the emissions control and exhaust systems.

exhaust stroke   The final stroke of the four-stroke cycle in which the piston moves upward and pushes the hot burned gases out through the open exhaust valve.

exhaust valve   The valve that opens to permit exhaust gases to escape from the cylinder into the exhaust manifold.

expansion   An increase in size or volume.

F

fan   A rotating device in the cooling system that helps to push or draw air through the radiator to increase cooling capacity. On most modern cars, the fan operates mainly when the car is idling or operating at low speeds.

fan belt   The belt that drives a mechanical fan on a rear-wheel drive vehicle. *See also* drive belt.

fast-idle speed   The adjustable, faster speed (rpm) at which the engine runs during warmup.

fault code   In a computer-controlled system, a self-diagnostic feature that can produce a sequence of flashes that corresponds with a number. The number is a code for a specific problem condition.

feeler gauge   A device that measures the gap between two surfaces. A wire (round) gauge is used to check spark plug gap; a flat gauge is used to check valves and breaker point gaps.

firewall   The insulated panel that separates the engine and passenger compartments. It protects passengers from engine noise, heat, and fumes.

firing order   A sequence in which ignition occurs in the cylinders of an engine.

float bowl   A chamber in the carburetor that holds a supply of fuel to be atomized and mixed with air in the barrel. A small float and needle valve regulate the level of fuel in the bowl.

flywheel   A heavy, circular metal part connected to the end of the crankshaft to smooth out an engine's power strokes. On manual-transmission cars, the clutch can be pressed against the turning flywheel to transfer power from the engine to the input shaft of the transmission. *See also* clutch.

force   The ability to cause movement. *See also* energy; power; work.

four-stroke cycle   A power cycle in which the piston moves within a cylinder four times—down, up, down, up—to deliver driving torque. The first down stroke (intake) draws the air/fuel mixture into the cylinder. The first up stroke (compression) compresses the mixture. The next down stroke (power) forces the crankshaft to turn. The final up stroke (exhaust) purges the cylinder of burned gases. *See also* compression stroke; exhaust stroke; intake stroke; power stroke.

four-wheel drive (4WD)   A powertrain layout in which engine torque can be transferred to all four wheels to move the vehicle.

free travel   The distance a clutch pedal moves before it begins to take up the slack in the clutch mechanism.

freeze plug   *See* soft plug.

friction   The resistance caused by the rubbing of moving parts against one another or against the nonmoving housings in which they are contained. Friction creates heat and causes parts to wear. Bearings and lubricants (oil and grease) help to reduce friction.

front-wheel drive (FWD)   A drivetrain layout in which engine torque is transferred to the front wheels through a transaxle and drive shafts. Typically, the engine in a FWD vehicle is mounted transversely, at right angles to the axis of the chassis.

fuel filter   A device in the fuel system that removes solid particles and small amounts of moisture from fuel before it reaches the carburetor or fuel injection system.

fuel injection system   A fuel delivery system that sprays fuel through injectors into the air passing through the intake manifold. Port-type fuel injection includes separate injectors for each cylinder, mounted in the intake manifold. Throttle-body injection (TBI) has one or more injectors located at a central point above a throttle plate that leads to the intake manifold. Fuel injection systems are controlled electronically by computers.

fuel injector   A device that sprays fuel in response to signals from a fuel injection computer.

fuel line   A synthetic rubber hose or metal tube through which fuel passes from the tank to the engine's fuel delivery system.

fuel pump   A pump that draws fuel from the tank and sends it through fuel lines to the fuel delivery system. May be operated mechanically or electrically.

fuel system   The system that stores, filters and delivers an air/fuel mixture into the cylinders. Includes fuel tank, fuel lines, fuel filter, fuel pump, and either a carburetor or a fuel injection system.

fuel tank   A metal compartment in which fuel is stored. Usually located at the rear of the vehicle. Also called the gas tank.

fuse   An electrical device used to protect a circuit against an overload or malfunction. When an overload or malfunction occurs, a metal strip inside the fuse melts and breaks the circuit.

galleries   Small drilled or cast passages through which oil flows for lubrication purposes.

gap   The space between spark plug electrodes, or the space between breaker points.

gas cap   A removable cap that seals the gas tank. It contains pressure and vacuum valves to help prevent the escape of fuel vapors into the atmosphere.

gasket   A formed or shaped sealing device made of cork, rubber, paper, or metal. Used to form a seal between parts and prevent leakage.

gear   A circular wheel with teeth in its outer edge that mesh with teeth in another gear. Gears of different sizes are used to increase and decrease speeds and the amount of torque transferred between rotating parts.

gear ratio   A set of numbers that indicates the relative number of turns and rotating speeds of a driving and a driven gear.

generator   A belt-driven device that produces electricity to charge the battery.

grease   A thick, oily lubricant usually made from a petroleum base. Most often used to lubricate gears and sealed bearings.

grease fitting   A valve device that seals grease in and allows the addition of more grease to cushion and lubricate moving parts.

grease gun   A tool equipped with extenders and adapters for adding lubricant to grease fittings.

ground   A common connection through which electricity can flow to complete a circuit.

group number   A coded designation that indicates a specific length, width, and height of automotive battery. *See also* cold cranking amps; reserve capacity.

half shaft   Either of the two drive shafts that connect the transaxle to the drive wheels in front-wheel drive cars. It is also used to connect differential to the drive wheels in many four wheel and all wheel drive vehicles.

halogen   A relatively inert gas used in headlight bulbs to prevent the deterioration of metal filaments.

head gasket   The seal between the engine block and cylinder head. *See also* gasket.

headlight   The lights used to illuminate the road surface ahead of the vehicle.

heater core   A small, radiator-like unit through which hot coolant is circulated. Air passes over the fins of the unit and into the passenger compartment for heating purposes.

heater hose   A hose that connects the heater core to the engine cooling system.

horsepower (Hp)   A measure of the rate at which an engine can produce torque. Originally based on the amount of work one horse could do in a specific amount of time.

hydraulic   Operated by fluid under pressure.

hydrocarbon (HC)   A substance made up of hydrogen and carbon atoms, such as petroleum products. The presence of unburned hydrocarbons in exhaust emissions creates one type of air pollution.

hydrometer A device used to determine the specific gravity of a liquid. Specific gravity is a comparison between the density of a liquid and the density of pure water. Battery hydrometers and coolant hydrometers are used to check electrolyte and coolant concentrations, respectively.

idle The speed of the engine under a no-load condition (out of gear, no throttle applied).

idler arm Part of a steering linkage system that helps to maintain the stability of the linkage; connects between the intermediate link and the car's chassis.

ignition coil A part of the ignition system that transforms low battery voltage into high voltage for the spark plugs. *See also* coil.

ignition system The system that produces and delivers high-voltage electrical current for the spark used to ignite the air/fuel mixture in the combustion chamber. Includes the battery, ignition switch, ignition coil, distributor, spark plugs, and connecting cables and wiring.

ignition timing The moment at which the spark plugs ignite the air/fuel mixture; stated in relation to the number of degrees of crankshaft rotation before or after a piston reaches top dead center (TDC).

independent suspension A suspension system in which the left and right wheels are free to move independently; may be used on either front or rear suspension systems.

in-line engine An engine design in which the cylinders are lined up in a single row.

intake manifold A cast structure through which the air/fuel mixture passes into the cylinder head intake ports.

intake stroke The first of the four strokes of the four-stroke engine. An air and fuel mixture is drawn into the cylinder as the piston moves downward.

intake valve A metal valve that opens to allow an air/fuel mixture to enter the cylinder on the intake stroke.

intercooler a device used on some turbocharged engines to reduce the temperature of the compressed air before delivery to the intake manifold.

internal combustion engine An engine that operates on power generated by intense burning of an air/fuel mixture inside a cylinder.

The resulting expansion of the heated mixture forces a piston downward, turning a crankshaft.

jack   A device used to temporarily raise part of a car off the ground.

jack stand   A safety device to prevent a raised car from falling to the ground. Generally used in pairs.

journal   An area of the crankshaft around which a bearing allows a connecting part to turn. Connecting rods are fitted to offset areas known as rod journals; the crankshaft is supported by bearings in the crankcase area around the main journals.

jumper cables   A set of two cables, with connecting clamps at both ends, used to make connections between a charged battery and a weak battery in order to start an engine. Also known as booster cables.

knock   A form of detonation that produces a characteristic sound, heard when an engine is accelerating or is loaded heavily.

leaf spring   A springing device that consists of one or more special flat steel plates that flex to resist the motion of the wheel as it travels over bumps and into depressions.

leak-down test   A test of a cylinder's ability to hold compression pressure. Too great a percentage of leakage can indicate sealing problems in the cylinders/piston rings, intake or exhaust valves, and head gasket.

lean   Having proportionately more air and less gasoline than the optimum 14.7:1 air/fuel ratio. *See also* rich; stoichiometric.

load   An externally applied force, or resistance, against which a mechanism or process must labor; examples are electrical loads and mechanical loads, which require energy to overcome.

longitudinal   In line or parallel with the centerline of the vehicle.

lubricant   A substance added to reduce friction and wear between two surfaces.

lubrication system   The engine system that stores, cleans, and circulates oil through the engine to lubricate its moving parts. Includes the oil pump, oil pan, oil filter, and galleries.

lugs (nuts or bolts)   Nuts or bolts that hold the wheel on the wheel hub.

lug wrench   A wrench used to remove or attach lugs in changing tires.

MacPherson strut   A suspension design in which a coil spring and shock absorber unit are attached to the body of a car to locate the wheel; a single lower control arm is used to form the second attachment point to hold the wheel firmly in position. As the wheel moves up and down, the strut is compressed and expanded; a ball joint allows the wheel to pivot for steering. *See also* double A-arm suspension.

main bearing   A bearing that supports the crankshaft in the crankcase and allows it to turn freely.

malfunction indicator lamp   A lamp located on the instrument panel used to inform the driver when a fault occurs that affects the vehicle's emission levels.

manifold   *See* exhaust manifold, intake manifold.

manual transmission   A transmission system in which the driver selects and changes gears through a hand-operated shifting linkage and a foot-operated clutch. Also called a standard transmission or "stick shift." *See also* transmission.

master cylinder   A device that contains a reservoir for storing brake fluid and a pump to force hydraulic fluid through brake lines to individual wheel calipers and cylinders, which operate the brakes.

MIL   *See* malfunction indicator lamp.

misfiring   A failure of one or more cylinders to properly burn the air/fuel mixture in the combustion chamber. This failure may be intermittent or continuous. *See also* miss.

miss   A lack of power in one or more cylinders. *See also* misfiring.

mixture control solenoid   An electrically operated device in a feedback carburetor system that regulates the air/fuel mixture by positioning a tapered needle within a carburetor jet.

motor mount   A rubber pad placed between the engine and chassis, on which the engine rests to absorb vibrations.

muffler   A device that reduces the noise of the exhaust gases before they leave the exhaust system.

multiribbed belt   *See* serpentine belt.

multiviscosity oil   An oil that has been chemically modified for use at both cold and hot temperatures.

negative terminal   The battery terminal that is connected, through a cable, to the chassis of the vehicle. *See also* positive terminal.

nut   An hexagonal metal part with an internal screw thread that can be tightened onto a bolt or screw to connect two objects.

OBD II   *See* On-Board Diagnostics II.

octane rating   A number that indicates the ability of a fuel to resist detonation under the heat of compression. Higher numbers indicate fuels with more resistance to detonation.

odometer   A dashboard instrument that indicates how far the car has traveled.

OEM parts   Parts made by the original manufacturer.

oil   A petroleum-based or synthetic light liquid lubricant.

oil drain plug   A threaded plug in the bottom of the oil pan that is removed to drain the oil.

oil filler   An opening at the top of the engine, usually in the valve cover, through which new oil is added.

oil filter   A canister containing a filtering device that cleans the oil before it circulates through the engine's galleries.

oil pan   A metal reservoir, attached to the bottom of the crankcase, that stores engine oil.

oil pressure gauge   A monitoring device that indicates the pressure under which the oil is pumped through the engine.

oil pump   A small pump, located in the crankcase, that circulates oil from the oil pan through the oil filter and to the engine's moving parts.

oil pumping   A process in which lubricating oil is drawn up past worn piston rings and cylinders on the intake stroke and burned in the combustion chamber.

oil seal   A barrier formed by a device around a rotating or moving shaft that is used to prevent oil leakage.

On-Board Diagnostics II   The diagnostics system inside the computer to detect when engine or component wear or failure could cause exhaust emissions to increase more than 50%.

open loop   An operating condition for a computer-controlled process in which input signals are ignored by a computer. The computer provides a default, or preset, output signal to operate output devices at predetermined levels. *See also* closed loop; computer; electronic control unit.

output device   A device that is operated by output signals from a computer.

overflow tube   A tube protruding from the radiator filler neck that allows excess liquid to escape from the cooling system under conditions of excessive pressure. On newer cars, the tube leads to a coolant recovery system. On older cars, the tube directs excess coolant onto the ground.

overhead camshaft (OHC)   A camshaft located in the cylinder head above the combustion chambers, rather than in the cylinder block. *See also* dual overhead camshaft.

overhead valve (OHV)   An engine design in which the valves are located in the cylinder head.

oxides of nitrogen (NO_X)   An air pollutant formed by high combustion temperatures. *See also* exhaust gas recirculation (EGR).

oxidizing catalytic converter   A catalytic converter in which oxygen is added to harmful exhaust gases—unburned hydrocarbons and oxides of nitrogen—to yield water, carbon dioxide, and free nitrogen, all of which are found in the natural environment.

oxygen sensor   A sensor that is mounted in the exhaust system of a car to detect the oxygen levels in the exhaust gasses; provides input signals to the computer that modifies the air/fuel mixture for the most complete combustion.

pad   The flat frictional part of a disc brake unit that is forced against the disc to slow and stop the wheel. *See also* disc brake.

parking brake   Formerly called the emergency brake. A secondary safety system that uses a separate mechanical linkage to operate half of the brake system, typically, the rear brakes. The parking brake is used to lock the wheels when the car is parked.

PCM   *See* powertrain control module.

performance   Ability of a car to do work on demand.

ping   Preignition, a form of detonation. *See also* detonation; knock.

piston   A solid cylindrical metal part that is sealed in—and can move within—a hollow cylinder.

piston ring   A metal ring that forms a seal between the piston and cylinder wall; fitted into a groove in the piston's outside circumference.

pitman arm   An arm connected to the sector shaft of a recirculating-ball steering mechanism; transfers force from the steering gearbox to the steering linkage.

planetary gearset   Common in automatic transmissions, a group of gears named after the solar system because of their arrangement and action. All the gears mesh constantly.

port   An opening in the cylinder head through which gases can pass; sealed and opened by the action of valves.

positive crankcase ventilation (PCV)   An emissions control system that routes crankcase blowby and oil fumes into the intake manifold for burning in the cylinders.

positive terminal   The battery terminal that is connected to the starter motor and other electrical system connections. *See also* negative terminal.

power   The rate at which work can be done. *See also* energy; force; work.

power brake   An hydraulic braking system that includes a power assist unit; uses engine intake manifold vacuum and atmospheric pressure to provide force to reduce braking effort.

power steering   A system that uses an engine-driven pump to provide hydraulic pressure that moves steering linkage parts to help the driver turn the wheel.

power stroke   The third stroke of the four-stroke cycle, in which the compressed air and fuel mixture ignites, burns, heats and expands in the combustion chamber. This expansion of the burning air and fuel produces the power in the engine.

powertrain   The engine and drivetrain units considered together.

powertrain control module (PCM)   An electronic device that accepts input data in the form of electrical signals and produces output signals in response to a program, or set of instructions used to primarily control fuel injection, ignition, and other engine functions. *See also* computer; electronic control unit.

pressure   Force exerted on a unit of area, as by a compressed gas within a closed chamber.

pressure cap   A radiator cap that allows a buildup of pressure in the cooling system to raise the boiling point of the coolant liquid. A calibrated, spring-loaded valve in the cap allows excess pressure to escape. Another valve prevents the creation of a vacuum in the radiator as the system cools off.

pressure plate A spring-loaded plate that can force the clutch disc into contact with the engine flywheel. *See also* clutch; flywheel.

pressurized coolant expansion tank An addition to a cooling system that allows for coolant expansion when heated. This tank is a pressurized component of the cooling system and serves to help purge air from the cooling system.

preventive maintenance Routinely scheduled procedures designed to prevent vehicle failures.

program A set of instructions that directs the actions of a computer to accept input signals and produce planned outputs.

pushrods Metal rods that transfer the motion of the valvetrain camshaft from a lifter to a rocker arm.

R-134a The refrigerant used in the air conditioning system. It is replacing R-12 because it does not deplete the earth's ozone layer.

radiator A cooling system component with inlet and outlet tanks and a central core of finned tubes. Heated coolant flows through the tubes and transfers heat to them. Heat is transferred from the finned tubes into the air that passes around them.

radiator cap *See* pressure cap.

radiator core The portion of the radiator that consists of brass or aluminum tubes, fitted with air fins of the same material. *See also* core; radiator.

ratchet A reversible-action handle for turning sockets.

ratio A numerical comparison between two quantities, such as 2:1 or 14.7:1.

rear-wheel drive (RWD) A drivetrain arrangement in which torque is directed to turn the rear wheels of a vehicle.

refrigerant A chemical compound used in automotive air-conditioning systems. *See also* R-134a.

relay A remote-control magnetic switch that operates to complete a circuit.

release bearing The part of the clutch that pushes against and operates the pressure plate to force the clutch plate into contact with the flywheel or to release pressure. *See also* clutch.

reserve capacity An indication in minutes of how long a battery can continue to create a 25-ampere current at an electrolyte temperature of 80 degrees Fahrenheit.

residual value An agreed on amount that a vehicle will be worth at the end of the lease term.

resonator   A small auxiliary muffler on some larger cars that helps reduce exhaust noise and produce a more pleasing exhaust sound.

revolutions per minute (rpm)   A way of measuring engine running speed in terms of how many times the crankshaft revolves in one minute.

rich   Having proportionately more fuel and less air than the optimum 14.7:1 air/fuel ratio. *See also* lean; stoichiometric.

ring   *See* piston ring.

rocker arm   A pivoting metal device that changes the upward motion of the push rod into a downward push against the valve and causes the valve to open.

rotor   A device within the distributor that rotates and aligns with metal terminals inside the distributor cap. A spark jumps the gap between the tip of the rotor and the metal terminals within the cap, then travels through the spark plug cables to the spark plug gap. The same term is applied to a part of a disc braking system (*see* disc brake).

RTV   A room temperature vulcanizing gasket sealant. It is used to help seal gaskets or, in some applications, can be formed in place of a gasket.

scan tool   A portable computer used to interface with the vehicle's on-board computer for the purpose of diagnosis, troubleshooting, and malfunction code retrieval.

sealed-beam unit   A headlight that contains the bulb or bulbs, reflectors, and lenses in a unit sealed to keep out moisture and dirt.

service brakes   The braking system normally used to stop the vehicle that excludes the parking or emergency brakes.

serpentine belt   A multiribbed belt used to drive water pumps, alternators, air conditioning compressor, power steering pumps, and other belt-driven devices.

servo   A device that produces mechanical movement.

shock absorbers   An hydraulic device at each wheel that dampens and controls the bouncing action of the springing devices on an automotive suspension.

short circuit   A condition in which electricity accidentally bypasses the load of the intended circuit; the electrical energy of the battery is uncontrolled and can cause the conductors to overheat severely. *See also* accidental ground.

short- and long-arm suspension (SLA) A suspension system using an upper and lower control arm. The upper arm is shorter than the lower arm. *See also* wishbone suspension.

sludge A thick tarry substance composed of contaminated oil, acid, and water that can block oil passages and galleries and damage an engine.

socket wrench An interchangeable wrench that encircles a bolt or nut head for turning purposes; used with a ratchet handle.

soft plug A thin metal plug that seals the passages left in the block and cylinder head during the casting process. Also called a core plug or freeze plug.

solenoid A device that provides a pushing, pulling, or holding action through magnetic force when a coil is energized electrically.

spark plug The device that delivers electrical spark to the combustion chamber to ignite the air/fuel mixture.

spark plug gap *See* gap.

spindle A short shaft, or stub axle, on which a wheel rotates.

stabilizer bar A steel torsion bar connected between the left- and right-side suspension components to limit swaying and body roll on turns.

standard transmission *See* manual transmission.

starter A powerful electric motor that turns the engine's crankshaft for starting purposes.

starter solenoid A magnetically operated switch that controls the electrical circuit between the battery and the starter.

starting system The system that delivers electrical current from the battery to a motor that cranks the engine to start it running. Consists of an ignition switch, starter motor, solenoid, and battery.

steering knuckle A mounting point for a front wheel located at the end of a tie rod on the steering linkage. Typically, the front suspension system connects to the steering knuckle through a MacPherson strut or ball joint.

steering linkage The system of parts that connects the steering wheel and the front wheels, allowing the driver to change direction by turning the steering wheel.

stoichiometric A chemically perfect mixture of air and fuel that yields the best combustion. It is neither too rich or too lean. It is about 14.7 parts air to one part of gasoline.

stroke The distance a piston in a cylinder will travel in a single direction.

supercharger   A belt-driven device that forces air through the fuel delivery system under pressure for greater torque and horsepower output from an engine.

suspension system   The system of springs and linkages through which the wheels are mounted to the chassis.

switch   A device used to turn a circuit on and off.

tachometer   A device that indicates engine speed in revolutions per minute (rpm).

tail pipe   The final link in the exhaust system; conducts exhaust gases from the muffler or resonator to the atmosphere.

test light   A tool with a built-in light that illuminates when there is voltage present in the electrical system.

thermostat   A device that controls the amount of engine coolant that can pass from the cylinder head into the radiator for cooling purposes.

throttle body   An assembly on the intake manifold that houses the throttle valve.

throttle plate   The fuel-delivery system component that controls the volume of airflow through a carburetor or fuel-injection intake system. Consists of a butterfly valve near the bottom of the barrel, with a mechanical linkage to the accelerator pedal.

throwout bearing   The part of the clutch that pushes against and operates the pressure plate to force the clutch plate into contact with the flywheel or to release pressure. *See also* clutch.

tie rod   The rod at the end of the steering linkage that is connected to the steering knuckle at the wheel.

timing chain/gear/belt   The chain, gear assembly, or belt that coordinates crankshaft and camshaft rotation so that valves open and close at the proper time in relation to piston movement.

toe-in   A condition of the front suspension in which the leading edges of the wheels point inward *See also* wheel alignment.

toe-out   A condition of the front suspension in which the leading edges of the wheels point outward. *See also* wheel alignment.

top dead center (TDC)   The point at which the piston reaches the top of its stroke within a cylinder. BTDC and ATDC, meaning before and after top dead center, respectively, refer to positions specified for ignition timing.

torque   Twisting or turning force.

torque converter   A fluid-filled coupling device that transfers engine torque into an automatic transmission.

torque wrench   A tool that measures the amount of force being applied in tightening a bolt or other object.

torsion bar   A steel rod that twists to absorb the up-and-down motion of a wheel. Used in some suspension systems instead of springs.

traction   A tire's ability to hold the road surface due to friction.

transaxle   A drivetrain mechanism used on front-wheel drive cars that includes a transmission and a differential.

transfer case   A mechanism that connects a transmission or transaxle to an auxiliary drive shaft and differential on four-wheel drive vehicles.

transmission   A mechanism that can provide different gear ratios to regulate the torque applied to the driving wheels and to control the speed of the vehicle. *See also* automatic transmission, manual transmission.

transverse   Perpendicular or at a right angle to the centerline of the vehicle.

tune up   The process of replacing spark plugs to deliver the optimum spark to the cylinders.

turbocharger   A mechanism that uses the force of exhaust gases to turn a fanlike device that forces air into the fuel delivery system under pressure. This process increases the torque and horsepower output of an engine.

turning angle   The difference between the angle of the inside and the outside heels during a turn.

U-joint   Abbreviation for universal joint.

unibody   A stressed body acts as the foundation with all major components attached to mounting points on the body *See also* body-over-frame.

universal joint   A flexible driveline coupling that can transfer torque at an angle. Used to connect the drive shaft to the transmission output shaft at the front and to the differential drive pinion at the rear.

vacuum advance   A device on the side of the distributor that advances the ignition spark in response to the vacuum produced by engine operation.

valve   A metal device that can close or open a port in the cylinder head. When it is pushed open by a rocker arm or cam follower, the valve permits gases to move around it, either in or out of a cylinder.

valve body   The mechanism within an automatic transmission that controls the operation of clutches, bands, and gearsets through hydraulic pressures.

valve cover   A metal housing attached to the cylinder head to cover an overhead valve or overhead camshaft valve train.

valve guide   The hollow tube that guides and supports the valve stem in the cylinder head.

valve lifter   The metal part that rides on the camshaft and transfers motion to a pushrod to open a valve. Also called a cam follower or tappet. May be mechanical or operated by hydraulic (oil) pressure.

valve seat   The machined, angled hole in the cylinder head that the valve head rests against when the valve is closed.

valve spring   A coil of metal used to close a valve causing it to seal the combustion chamber.

valve stem   The long thin body of the valve that fits into the valve guide in the cylinder head.

valvetrain   The system of parts that opens and closes the valves at the proper times during the four-stroke cycle. Can include the crankshaft gear and camshaft gear (which may be positioned so they mesh, or connected by a chain or a cog belt), camshaft, cam lobes, valve lifters, push rods, rocker arms, and valves. In an overhead-cam engine, the push rods and valve lifters are eliminated.

vapor control system   An emissions control system that stores fuel vapors and allows them to be drawn into the intake manifold and burned when the engine runs.

vaporization   The changing of a liquid into a gas.

VECI   *See* vehicle emission control information.

vehicle emission control information (VECI)   A decal found under the hood on all vehicles that includes important information concerning the emission systems found on the vehicle, such as vacuum hose routing and engine displacement.

venturi   A narrowed portion in the air passage that causes the air speed to increase, thereby creating a low pressure area.

viscosity   The ability of a liquid to flow, determined by its thickness or thinness. Engine oil must be able to flow freely in cold temperatures while retaining enough thickness for safe lubrication in high temperatures.

voltage regulator   The device that controls, or regulates, the amount of electrical current generated by the alternator. Usually built in the alternator.

voltmeter   A tool used to measure the voltage in the electrical system.

V-type engine   An engine configuration in which the cylinders are lined up in two rows, or banks, set at an angle to each other. Most common are V-6 and V-8 engines, although V-4s and V-2s also are in use.

water jacket   Hollow passages in the engine block and cylinder head through which liquid coolant circulates to absorb heat from hot engine parts.

water pump   *See* coolant pump.

wheel alignment   The process of adjusting the position of the front wheels for optimum handling and tire wear. *See also* camber; caster; toe-in; toe-out.

wheel balancing   The process of ensuring even distribution of weight around the circumference of a wheel to provide smooth performance and even tire wear.

wheelbase   The distance between the center of the front wheel and the center of the rear wheel.

wheel cylinder   A small cylinder containing a piston that operates under hydraulic pressure to activate a brake. Brake fluid is pumped into the cylinder from the master cylinder. The piston forces the shoe against the drum on a drum brake, or the pads against the rotor on a disc brake.

wishbone suspension   A suspension system where the wheel is guided by two triangulated lateral control arms and a tie rod. It is similar to the short- and long-arm suspension. *See also* short- and long-arm suspension.

work   The action of applying force to an object and causing it to move. *See also* energy; force; power.

zerk fitting   *See* grease fitting.